COMPUTATIONAL FLUID–STRUCTURE INTERACTION

WILEY SERIES IN COMPUTATIONAL MECHANICS

Series Advisors:

René de Borst
Perumal Nithiarasu
Tayfun E. Tezduyar
Genki Yagawa
Tarek Zohdi

Computational Fluid–Structure Interaction: Methods and Applications	Bazilevs, Takizawa and Tezduyar	January 2013
Introduction to Finite Strain Theory for Continuum Elasto-Plasticity	Hashiguchi and Yamakawa	November 2012
Nonlinear Finite Element Analysis of Solids and Structures: Second edition	De Borst, Crisfield, Remmers and Verhoosel	August 2012
An Introduction to Mathematical Modeling: A Course in Mechanics	Oden	November 2011
Computational Mechanics of Discontinua	Munjiza, Knight and Rougier	November 2011
Introduction to Finite Element Analysis: Formulation, Verification and Validation	Szabó and Babuška	March 2011

COMPUTATIONAL FLUID–STRUCTURE INTERACTION
METHODS AND APPLICATIONS

Yuri Bazilevs
Department of Structural Engineering
University of California, San Diego, USA

Kenji Takizawa
Department of Modern Mechanical Engineering and
Waseda Institute for Advanced Study
Waseda University, Japan

Tayfun E. Tezduyar
Department of Mechanical Engineering and Materials Science
Rice University, USA

A John Wiley & Sons, Ltd., Publication

This edition first published 2013
© 2013, John Wiley & Sons Ltd

Registered office
John Wiley & Sons Ltd, The Atrium, Southern Gate, Chichester, West Sussex, PO19 8SQ, United Kingdom

For details of our global editorial offices, for customer services and for information about how to apply for permission to reuse the copyright material in this book please see our website at www.wiley.com.

The right of the author to be identified as the author of this work has been asserted in accordance with the Copyright, Designs and Patents Act 1988.

All rights reserved. No part of this publication may be reproduced, stored in a retrieval system, or transmitted, in any form or by any means, electronic, mechanical, photocopying, recording or otherwise, except as permitted by the UK Copyright, Designs and Patents Act 1988, without the prior permission of the publisher.

Wiley also publishes its books in a variety of electronic formats. Some content that appears in print may not be available in electronic books.

Designations used by companies to distinguish their products are often claimed as trademarks. All brand names and product names used in this book are trade names, service marks, trademarks or registered trademarks of their respective owners. The publisher is not associated with any product or vendor mentioned in this book. This publication is designed to provide accurate and authoritative information in regard to the subject matter covered. It is sold on the understanding that the publisher is not engaged in rendering professional services. If professional advice or other expert assistance is required, the services of a competent professional should be sought.

Library of Congress Cataloging-in-Publication Data

Bazilevs, Yuri.
  Computational fluid-structure interaction : methods and applications / Yuri Bazilevs, Kenji Takizawa, Tayfun E. Tezduyar.
       pages cm
       Includes bibliographical references and index.
   ISBN 978-0-470-97877-1 (hardback)
  1. Fluid-structure interaction–Data processing.   2. Fluid-structure interaction–Mathematical models.
 I. Takizawa, Kenji.   II. Tezduyar, T. E. (Tayfun E.)   III. Title.
   TA357.5.F58B39 2013
   624.1'71–dc23
                                                                                            2012030898

A catalogue record for this book is available from the British Library.

Print ISBN: 9780470978771

Typeset in Times 10/12pt size by Laserwords, India

Contents

Series Preface		xi
Preface		xiii
Acknowledgements		xix

1	**Governing Equations of Fluid and Structural Mechanics**	**1**
1.1	Governing Equations of Fluid Mechanics	1
	1.1.1 Strong Form of the Navier–Stokes Equations of Incompressible Flows	1
	1.1.2 Model Differential Equations	5
	1.1.3 Nondimensional Equations and Numbers	6
	1.1.4 Some Specific Boundary Conditions	7
	1.1.5 Weak Form of the Navier–Stokes Equations	10
1.2	Governing Equations of Structural Mechanics	12
	1.2.1 Kinematics	12
	1.2.2 Principle of Virtual Work and Variational Formulation of Structural Mechanics	14
	1.2.3 Conservation of Mass	15
	1.2.4 Structural Mechanics Formulation in the Current Configuration	15
	1.2.5 Structural Mechanics Formulation in the Reference Configuration	17
	1.2.6 Additional Boundary Conditions of Practical Interest	18
	1.2.7 Some Constitutive Models	19
	1.2.8 Linearization of the Structural Mechanics Equations: Tangent Stiffness and Equations of Linear Elasticity	22
	1.2.9 Thin Structures: Shell, Membrane, and Cable Models	25
1.3	Governing Equations of Fluid Mechanics in Moving Domains	31
	1.3.1 Kinematics of ALE and Space–Time Descriptions	31
	1.3.2 ALE Formulation of Fluid Mechanics	33

2	**Basics of the Finite Element Method for Nonmoving-Domain Problems**	**37**
2.1	An Abstract Variational Formulation for Steady Problems	37
2.2	FEM Applied to Steady Problems	38
2.3	Construction of Finite Element Basis Functions	42
	2.3.1 Construction of Element Shape Functions	43
	2.3.2 Finite Elements Based on Lagrange Interpolation Functions	46
	2.3.3 Construction of Global Basis Functions	49

		2.3.4	Element Matrices and Vectors and their Assembly into the Global Equation System	51
2.4	Finite Element Interpolation and Numerical Integration			53
	2.4.1		Interpolation by Finite Elements	53
	2.4.2		Numerical Integration	55
2.5	Examples of Finite Element Formulations			58
	2.5.1		Galerkin Formulation of the Advection–Diffusion Equation	58
	2.5.2		Stabilized Formulation of the Advection–Diffusion Equation	59
	2.5.3		Galerkin Formulation of Linear Elastodynamics	62
2.6	Finite Element Formulation of the Navier–Stokes Equations			65
	2.6.1		Standard Essential Boundary Conditions	65
	2.6.2		Weakly Enforced Essential Boundary Conditions	70

3 Basics of the Isogeometric Analysis   73
3.1 B-Splines in 1D   74
3.2 NURBS Basis Functions, Curves, Surfaces, and Solids   75
3.3 h-, p-, and k-Refinement of NURBS Meshes   77
3.4 NURBS Analysis Framework   78

4 ALE and Space–Time Methods for Moving Boundaries and Interfaces   83
4.1 Interface-Tracking (Moving-Mesh) and Interface-Capturing (Nonmoving-Mesh) Techniques   83
4.2 Mixed Interface-Tracking/Interface-Capturing Technique (MITICT)   84
4.3 ALE Methods   84
4.4 Space–Time Methods   86
4.5 Advection–Diffusion Equation   89
    4.5.1 ALE Formulation   89
    4.5.2 Space–Time Formulation   91
4.6 Navier–Stokes Equations   92
    4.6.1 ALE Formulation   92
    4.6.2 Generalized-α Time Integration of the ALE Equations   95
    4.6.3 Space–Time Formulation   98
4.7 Mesh Moving Methods   106

5 ALE and Space–Time Methods for FSI   111
5.1 FSI Formulation at the Continuous Level   111
5.2 ALE Formulation of FSI   114
    5.2.1 Spatially-Discretized ALE FSI Formulation with Matching Fluid and Structure Discretizations   114
    5.2.2 Generalized-α Time Integration of the ALE FSI Equations   118
    5.2.3 Predictor–Multicorrector Algorithm and Linearization of the ALE FSI Equations   120
5.3 Space–Time Formulation of FSI   123
    5.3.1 Core Formulation   123
    5.3.2 Interface Projection Techniques for Nonmatching Fluid and Structure Interface Discretizations   127

5.4	Advanced Mesh Update Techniques		129
	5.4.1 Solid-Extension Mesh Moving Technique (SEMMT)		129
	5.4.2 Move-Reconnect-Renode Mesh Update Method (MRRMUM)		132
	5.4.3 Pressure Clipping		134
5.5	FSI Geometric Smoothing Technique (FSI-GST)		136

6 Advanced FSI and Space–Time Techniques — 139

6.1	Solution of the Fully-Discretized Coupled FSI Equations	139
	6.1.1 Block-Iterative Coupling	140
	6.1.2 Quasi-Direct Coupling	141
	6.1.3 Direct Coupling	142
6.2	Segregated Equation Solvers and Preconditioners	144
	6.2.1 Segregated Equation Solver for Nonlinear Systems (SESNS)	144
	6.2.2 Segregated Equation Solver for Linear Systems (SESLS)	145
	6.2.3 Segregated Equation Solver for Fluid–Structure Interactions (SESFSI)	146
6.3	New-Generation Space–Time Formulations	149
	6.3.1 Mesh Representation	150
	6.3.2 Momentum Equation	150
	6.3.3 Incompressibility Constraint	151
6.4	Time Representation	151
	6.4.1 Time Marching Problem	151
	6.4.2 Design of Temporal NURBS Basis Functions	153
	6.4.3 Approximation in Time	154
	6.4.4 An Example: Circular-Arc Motion	154
6.5	Simple-Shape Deformation Model (SSDM)	157
6.6	Mesh Update Techniques in the Space–Time Framework	158
	6.6.1 Mesh Computation and Representation	158
	6.6.2 Remeshing Technique	158
6.7	Fluid Mechanics Computation with Temporal NURBS Mesh	159
	6.7.1 No-Slip Condition on a Prescribed Boundary	159
	6.7.2 Starting Condition	160
6.8	The Surface-Edge-Node Contact Tracking (SENCT-FC) Technique	163
	6.8.1 Contact Detection and Node Sets	164
	6.8.2 Contact Force and Reaction Force	165
	6.8.3 Solving for the Contact Force	167

7 General Applications and Examples of FSI Modeling — 171

7.1	2D Flow Past an Elastic Beam Attached to a Fixed, Rigid Block	171
7.2	2D Flow Past an Airfoil Attached to a Torsion Spring	174
7.3	Inflation of a Balloon	175
7.4	Flow Through and Around a Windsock	177
7.5	Aerodynamics of Flapping Wings	181
	7.5.1 Surface and Volume Meshes	181
	7.5.2 Flapping-Motion Representation	185

		7.5.3	Mesh Motion	186

		7.5.4	Fluid Mechanics Computation	187
8	**Cardiovascular FSI**			**191**
8.1	Special Techniques			194
		8.1.1	Mapping Technique for Inflow Boundaries	194
		8.1.2	Preconditioning Technique	195
		8.1.3	Calculation of Wall Shear Stress	195
		8.1.4	Calculation of Oscillatory Shear Index	196
		8.1.5	Boundary Condition Techniques for Inclined Inflow and Outflow Planes	197
8.2	Blood Vessel Geometry, Variable Wall Thickness, Mesh Generation, and Estimated Zero-Pressure (EZP) Geometry			198
		8.2.1	Arterial-Surface Extraction from Medical Images	198
		8.2.2	Mesh Generation and EZP Arterial Geometry	199
		8.2.3	Blood Vessel Wall Thickness Reconstruction	201
8.3	Blood Vessel Tissue Prestress			203
		8.3.1	Tissue Prestress Formulation	203
		8.3.2	Linearized Elasticity Operator	204
8.4	Fluid and Structure Properties and Boundary Conditions			205
		8.4.1	Fluid and Structure Properties	205
		8.4.2	Boundary Conditions	205
8.5	Simulation Sequence			209
8.6	Sequentially-Coupled Arterial FSI (SCAFSI) Technique			210
8.7	Multiscale Versions of the SCAFSI Technique			213
8.8	Computations with the SSTFSI Technique			215
		8.8.1	Performance Tests for Structural Mechanics Meshes	215
		8.8.2	Multiscale SCAFSI Computations	218
		8.8.3	WSS Calculations with Refined Meshes	222
		8.8.4	Computations with New Surface Extraction, Mesh Generation, and Boundary Condition Techniques	225
		8.8.5	Computations with the New Techniques for the EZP Geometry, Wall Thickness, and Boundary-Layer Element Thickness	230
8.9	Computations with the ALE FSI Technique			233
		8.9.1	Cerebral Aneurysms: Tissue Prestress	236
		8.9.2	Total Cavopulmonary Connection	240
		8.9.3	Left Ventricular Assist Device	250
9	**Parachute FSI**			**259**
9.1	Parachute Specific FSI-DGST			261
9.2	Homogenized Modeling of Geometric Porosity (HMGP)			262
		9.2.1	HMGP in its Original Form	265
		9.2.2	HMGP-FG	266
		9.2.3	Periodic n-Gore Model	267
9.3	Line Drag			269
9.4	Starting Point for the FSI Computation			271

9.5	"Symmetric FSI" Technique		274
9.6	Multiscale SCFSI M2C Computations		275
	9.6.1	Structural Mechanics Solution for the Reefed Stage	275
	9.6.2	Fabric Stress Computations	278
9.7	Single-Parachute Computations		280
	9.7.1	Various Canopy Configurations	280
	9.7.2	Various Suspension Line Length Ratios	288
9.8	Cluster Computations		293
	9.8.1	Starting Conditions	294
	9.8.2	Computational Conditions	295
	9.8.3	Results	297
9.9	Techniques for Dynamical Analysis and Model-Parameter Extraction		299
	9.9.1	Contributors to Parachute Descent Speed	299
	9.9.2	Added Mass	311
10	**Wind-Turbine Aerodynamics and FSI**		**315**
10.1	Aerodynamics Simulations of a 5MW Wind-Turbine Rotor		317
	10.1.1	5MW Wind-Turbine Rotor Geometry Definition	317
	10.1.2	ALE-VMS Simulations Using NURBS-based IGA	322
	10.1.3	Computations with the DSD/SST Formulation Using Finite Elements	325
10.2	NREL Phase VI Wind-Turbine Rotor: Validation and the Role of Weakly-Enforced Essential Boundary Conditions		328
10.3	Structural Mechanics of Wind-Turbine Blades		334
	10.3.1	The Bending-Strip Method	334
	10.3.2	Time Integration of the Structural Mechanics Equations	340
10.4	FSI Coupling and Aerodynamics Mesh Update		342
10.5	FSI Simulations of a 5MW Wind-Turbine Rotor		343
10.6	Pre-Bending of the Wind-Turbine Blades		344
	10.6.1	Problem Statement and the Pre-Bending Algorithm	346
	10.6.2	Pre-Bending Results for the NREL 5MW Wind-Turbine Blade	349
References			**353**
Index			**373**

Series Preface

The series on *Computational Mechanics* is a conveniently identifiable set of books covering interrelated subjects that have been receiving much attention in recent years and need to have a place in senior undergraduate and graduate school curricula, and in engineering practice. The subjects will cover applications and methods categories. They will range from biomechanics to fluid–structure interactions to multiscale mechanics and from computational geometry to meshfree techniques to parallel and iterative computing methods. Application areas will be across the board in a wide range of industries, including civil, mechanical, aerospace, automotive, environmental and biomedical engineering. Practicing engineers, researchers and software developers at universities, industry and government laboratories, and graduate students will find this book series to be an indispensible source for new engineering approaches, interdisciplinary research, and a comprehensive learning experience in computational mechanics.

Computational Fluid–Structure Interaction has seen rapid developments over the past decade. The present book, written by a team of prominent researchers, gives a comprehensive treatment of modern developments in the field and fills a gap in the literature. Starting from basic concepts in computational fluid and structural mechanics, it explains the computational methods used in the analysis of fluid–structure interaction problems in a clear and accessible manner. The book stands out in that not only standard finite element methods are used for the spatial discretization, but that space–time and isogeometric finite element methods are also covered, thus emphasizing the state-of-the-art character of the book. The later chapters provide a plethora of examples and applications – ranging from cardiovascular problems, the aerodynamics of flapping wings, wind turbine blades to the simulation of parachutes – all taken from the vast research experience of the authors in the field.

Preface

Significance of FSI

Fluid–structure interaction (FSI) is a class of problems with mutual dependence between the fluid and structural mechanics parts. The flow behavior depends on the shape of the structure and its motion, and the motion and deformation of the structure depend on the fluid mechanics forces acting on the structure. We see FSI almost everywhere in engineering, sciences, and medicine, and also in our daily lives. The FSI effects become more significant and noticeable when the dependence between the influence and response becomes stronger. The fluttering of aircraft wings, flapping of an airport windsock, deflection of wind-turbine blades, falling of a leaf, inflation of automobile airbags, dynamics of spacecraft parachutes, rocking motion of ships, pumping of blood by the ventricles of the human heart, accompanied by the opening and closing of the heart valves, and blood flow and arterial dynamics in cerebral aneurysms, are all FSI examples. In engineering applications, FSI plays an important role and influences the decisions that go into the design of systems of contemporary interest. Therefore, truly predictive FSI methods, which help address these problems of interest, are in high demand in industry, research laboratories, medical fields, space exploration, and many other contexts.

Role of Computational FSI

The inherently nonlinear and time-dependent nature of FSI makes it very difficult to use analytical methods in this class of problems. Only a handful of cases have been studied analytically, where simplifying assumptions have been invoked to arrive at closed-form solutions of the underlying partial differential equations. While we see some use of analytical methods in solution of fluid-only or structure-only problems, there are very few such developments in solution of FSI problems. In contrast, there have been significant advances in computational FSI research, especially in recent decades, in both core FSI methods forming a general framework and special FSI methods targeting specific classes of problems (see, for example, Tezduyar, 1992; Tezduyar et al., 1992a,c; Morand and Ohayon, 1995; Tezduyar, 2003a; Michler et al., 2003, 2004; van Brummelen and de Borst, 2005; Lohner et al., 2006; Dettmer and Peric, 2006; Tezduyar et al., 2006a; Tezduyar and Sathe, 2007; Bazilevs et al., 2007a, 2008; Dettmer and Peric, 2008; Idelsohn et al., 2008a,b; Cottrell et al., 2009; Takizawa and Tezduyar, 2011, 2012a; Takizawa et al., 2012a; Takizawa and Tezduyar, 2012b; Bazilevs et al., 2012b). Computational methods, which are robust, efficient, and capable of accurately modeling in 3D FSI with geometrically complex configurations at full spatial scales, have been the focal point of these advances.

Computational FSI Challenges

The challenges involved in computational FSI can be categorized into three areas: *problem formulation*, *numerical discretization*, and *fluid–structure coupling*. These challenges are summarized here.

The problem formulation takes place at the continuous level, before the discretization. However, one must keep in mind that the modeling choices made at the continuous level have implications for the numerical discretizations that are most suitable for the case at hand. In a typical single-field mechanics problem, such as a fluid-only or structure-only problem, one begins with a set of governing differential equations in the problem domain and a set of boundary conditions on the domain boundary. The domain may or may not be in motion. The situation is more complicated in an FSI problem. The sets of differential equations and boundary conditions associated with the fluid and structure domains must be satisfied simultaneously. The domains do not overlap, and the two systems are coupled at the fluid–structure interface, which requires a set of physically meaningful interface conditions. These coupling conditions are the compatibility of the kinematics and tractions at the fluid–structure interface. The structure domain is in motion and, in most cases, its motion follows the material particles, or points, which constitute the structure. This is known as the *Lagrangian description* of the structural motion. As the structure moves through space, the shape of the fluid subdomain changes to conform to the motion of the structure. The motion of the fluid mechanics domain needs to be accounted for in the differential equations and boundary conditions. There are two major classes of methods for this, which are known in the discrete setting as the *nonmoving-grid* and *moving-grid* approaches. Furthermore, the motion of the fluid domain is not known a priori. It is a function of the unknown structural displacement. This makes FSI a three-field problem, where the third unknown is the motion of the fluid domain.

All the issues related to the numerical discretization of a single-field problem, such as the accuracy, stability, robustness, speed of execution, and the ability to handle complex geometries, are likewise present in an FSI problem. The additional challenges in FSI come from the discretization at the fluid–structure interface. The most flexible option is, of course, to have separate fluid and structure discretizations for the individual subproblems, which results in nonmatching meshes at the interface. In this case, one needs to ensure that, despite the nonmatching interface meshes, the fluid and structure have the correct coupling of the kinematics and tractions. A simpler option is to have matching discretizations at the fluid–structure interface. In this case, the satisfaction of the FSI coupling conditions is much less challenging. However, this choice leads to a lack of flexibility in the discretization choices and mesh refinement levels for the fluid and structure subproblems. That flexibility becomes increasingly important as the complexity of the fluid–structure interface geometry increases. On the other hand, there are situations where having matching discretizations at the interface is the most effective approach. Another computational challenge in some FSI applications is the need to accommodate very large structural motions. In this case, one needs a robust mesh moving technique and the option to periodically regenerate the fluid mechanics mesh (i.e., remesh) to preserve the mesh quality and consequently the accuracy of the FSI computations. The remeshing procedure requires the interpolation of the solution from the old mesh to the new one. Remeshing and data interpolation are also necessary for fluid-only computations over domains with known motion. The difference between that and FSI is that the remeshing can be precomputed in such fluid-only simulations, while in the case of FSI the fluid mechanics mesh quality depends on the unknown structural displacements, and the decision to remesh is made "on the fly."

There are two major classes of FSI coupling techniques: loosely-coupled and strongly-coupled, which are also referred to as staggered and monolithic, respectively. Monolithic coupling often refers to strong coupling with matching interface discretizations. In loosely-coupled approaches, the equations of fluid mechanics, structural mechanics, and mesh moving are solved sequentially. For a given time step, a typical loosely-coupled algorithm involves the solution of the fluid mechanics equations with the velocity boundary conditions coming from the extrapolated structure displacement rate at the interface, followed by the solution of the structural mechanics equations with the updated fluid mechanics interface traction, and followed by the solution of the mesh moving equations with the updated structural displacement at the interface. This enables the use of existing fluid and structure solvers, a significant motivation for adopting this approach. In addition, for several problems the staggered approach works well and is very efficient. However, convergence difficulties are encountered sometimes, most-commonly when the structure is light and the fluid is heavy, and when an incompressible fluid is fully enclosed by the structure. In strongly-coupled approaches, the equations of fluid, structure, and mesh moving are solved simultaneously, in a fully-coupled fashion. The main advantage is that strongly-coupled solvers are more robust. Many of the problems encountered with the staggered approaches are avoided. However, strongly-coupled approaches necessitate writing a fully-integrated FSI solver, virtually precluding the use of existing fluid and structure solvers. There are three categories of coupling techniques in strongly-coupled FSI methods: block-iterative, quasi-direct, and direct coupling. The methods are ranked according to the level of coupling between the blocks of the left-hand-side matrix. In all three cases, iterations are performed within a time step to simultaneously converge the solutions of all the equations involved.

Organization of the Chapters

The three categories of FSI computational challenges outlined above constitute the bulk of the book's content.

In Chapter 1, the boundary value problems associated with the fluid and structural mechanics are stated. The fluid mechanics modeling is restricted to incompressible flows. This is mainly due to the principal research interests of the authors. The presentation of the structural mechanics covers 3D solids and thin structures. The latter includes shells, membranes, and cables. The equations of fluid mechanics in a moving spatial domain are also presented, and the fundamental concepts of space–time and Arbitrary Lagrangian–Eulerian (ALE) formulations are introduced. Both the conservative and convective forms of the Navier–Stokes equations of incompressible flows, in the ALE frame, are derived, and the implications for the conservation properties of the corresponding discrete FSI formulations are discussed.

Chapter 2 is on the basics of the finite element method (FEM). The presentation is confined to nonmoving spatial domains. Examples of time-dependent advection–diffusion, linear-elasticity, and Navier–Stokes equations discretized with the FEM are presented. In this book the FEM and Isogeometric Analysis (IGA) are employed for the discretization of the fluid and structural mechanics equations. IGA is a newly developed computational method that is based on the basis function technology of Computer-Aided Design (CAD) and Computer Graphics (CG), which was developed to provide a tighter integration between engineering design and analysis. In this chapter, the basics of stabilized and multiscale methods for fluid mechanics are also introduced. These methods possess superior stability properties compared to their Galerkin counterparts, while retaining the full order of accuracy. The chapter concludes with the presentation of weakly enforced essential boundary conditions, which enhance the accuracy of the stabilized and multiscale formulations in the presence of unresolved thin boundary layers, which are seen near solid boundaries.

The basic concepts of IGA are presented in Chapter 3. The presentation is focused on IGA based on non-Uniform Rational B-Splines (NURBS). For more in-depth understanding of this new computational technology, the reader is referred to Cottrell *et al.* (2009).

Chapter 4 is focused on the FEM for fluid mechanics problems with moving boundaries and interfaces. ALE and space–time FEM are covered in detail. The fully discrete stabilized and multiscale formulations of the fluid mechanics problem are presented for both approaches. In the case of the ALE approach, the FEM semi-discretization is followed by a presentation of the generalized-α time integration algorithm. In the case of space–time FEM, because both space and time behavior is approximated with finite element functions, the method leads to a fully discrete system of nonlinear equations at every time step. The standard mesh update strategy, which is based on the equations of linear elastostatics driven by the time-dependent displacement of the fluid mechanics domain boundary, is discussed at the end of the chapter.

In Chapter 5, the FSI problem is formulated at the continuous level and in the weak form. It is shown that the weak form, with the appropriate kinematic constraints on the trial and test function sets, gives the FSI formulation with the correct coupling conditions at the fluid–structure interface. The semi-discrete ALE FSI formulation, with the assumption of matching fluid–structure discretizations at the interface, is presented. The generalized-α method and the corresponding predictor–multicorrector algorithm are extended to the FSI problem. The linearization of the discrete FSI equation system is also discussed in the context of the ALE formulation. The space–time FSI formulation is presented next. The requirement of the matching interface discretizations is removed and several treatment options for nonmatching interface discretizations are discussed. Advanced mesh moving and remeshing techniques are outlined at the end of the chapter.

Chapter 6 presents the advanced FSI and space–time techniques the authors introduced, which include FSI coupling techniques, matrix-free computation techniques, segregated linear equation solvers, and preconditioning strategies. The chapter includes methods introduced for using temporal NURBS basis functions, in the context of the space–time formulation, in representation of the motion and deformation of the moving surfaces and volume meshes, and in remeshing. The chapter concludes with an FSI contact algorithm.

Chapter 7 presents a selection of representative FSI computations, which are used to explain some of the computational challenges faced in real-world applications. The test cases presented, and the computational methods used to solve them, address the challenges of FSI in fully-enclosed domains, as well as in the presence of structures that undergo large deformations and topological changes at the fluid–structure interface. Computational aerodynamics of flapping-wings with video-captured wing motion and deformation patterns of an actual locust is also presented. This illustrates the challenges of moving-domain fluid mechanics simulations in the presence of multiple surfaces undergoing large relative motions.

Chapters 8–10 present the applications of the FSI methods developed by the authors and their research groups to cardiovascular biomechanics, parachutes, and wind-turbine rotors. In all cases, the modeling and simulations are performed in 3D and at full spatial and temporal scales involved in these physical systems. In all three classes of FSI problems, the *core FSI technologies* are those presented in the earlier chapters. However, successful FSI computations require also the development of special FSI techniques targeting these specific classes of problems. The special techniques for each of the three classes of FSI problems are presented together with the computational results.

Acknowledgements

We are grateful for the privilege of being associated with Thomas J.R. Hughes. He taught us and inspired us. We would not be where we are today in FSI research if it were not for what we learned from him and how we were inspired by him.

Many of our collaborators, associates, and students contributed to the computations presented in the book, by computational-technology development, by computation, or by providing data. We thank them for that. Computational-technology development and computation: Sunil Sathe and Ming-Chen Hsu. Computational-technology development: James Liou, Sanjay Mittal, Vinay Kalro, Yasuo Osawa, Timothy Cragin, Dave Benson, Ido Akkerman, and Josef Kiendl. Computation: Keith Stein, Bryan Nanna, Jason Pausewang, Matthew Schwaab, Jason Christopher, Samuel Wright, Creighton Moorman, Bradley Henicke, Timothy Spielman, Tyler Brummer, Anthony Puntel, and Darren Montes. Data: Ryo Torii, Jessica Zhang, and Alison Marsden.

We would not have had the chapter on parachute FSI if it were not for the encouragement, parachute data, and guidance we received from Ricardo Machin and Jay LeBeau at NASA Johnson Space Center. We thank them for that.

1

Governing Equations of Fluid and Structural Mechanics

In this chapter, we introduce the partial differential equations that govern the fluid and structural mechanics parts of the fluid–structure interaction (FSI) problem. The fluid and structural mechanics equations are complemented by the applicable boundary conditions and constitutive models. For the structural mechanics part of the problem, we adopt mostly a 3D solid description, but include a discussion of thin structures such as shells and membranes. Both strong and weak (or variational) forms of the fluid and structural mechanics equations are presented. We conclude the section with an Arbitrary Lagrangian–Eulerian (ALE) description of the fluid mechanics equations suitable for moving-domain computations.

1.1 Governing Equations of Fluid Mechanics

The fluid mechanics part of the FSI problem is governed by the Navier–Stokes equations of incompressible flows. In what follows, we present the strong and weak forms of these equations and discuss the applicable boundary conditions.

1.1.1 Strong Form of the Navier–Stokes Equations of Incompressible Flows

Let $\Omega_t \in \mathbb{R}^{n_{\mathrm{sd}}}$, $n_{\mathrm{sd}} = 2, 3$, be the spatial fluid mechanics domain with boundary Γ_t at time $t \in (0, T)$ (see Figure 1.1 for an illustration). The subscript t indicates that the fluid mechanics spatial domain is time-dependent. The Navier–Stokes equations of incompressible flows[1] may be written on Ω_t and $\forall\, t \in (0, T)$ as

$$\frac{\partial (\rho \mathbf{u})}{\partial t} + \nabla \cdot (\rho \mathbf{u} \otimes \mathbf{u} - \boldsymbol{\sigma}) - \rho \mathbf{f} = \mathbf{0}, \tag{1.1}$$

$$\nabla \cdot \mathbf{u} = 0, \tag{1.2}$$

---

[1] Although the term "incompressible Navier–Stokes equations" is often employed instead of "Navier–Stokes equations of incompressible flows," we prefer the latter because it is the flows and not the equations that are incompressible.

Computational Fluid–Structure Interaction: Methods and Applications, First Edition.
Yuri Bazilevs, Kenji Takizawa and Tayfun E. Tezduyar.
© 2013 John Wiley & Sons, Ltd. Published 2013 by John Wiley & Sons, Ltd.

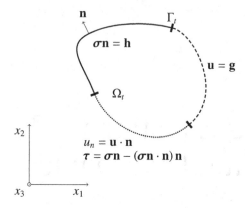

Figure 1.1 Fluid mechanics spatial domain and its boundary

where ρ, $\mathbf{u}$, and $\mathbf{f}$ are the density, velocity, and the external force (per unit mass), respectively, and the stress tensor $\boldsymbol{\sigma}$ is defined as

$$\boldsymbol{\sigma}(\mathbf{u}, p) = -p\mathbf{I} + 2\mu\boldsymbol{\varepsilon}(\mathbf{u}). \tag{1.3}$$

Here p is the pressure, $\mathbf{I}$ is the identity tensor, μ is the dynamic viscosity, and $\boldsymbol{\varepsilon}(\mathbf{u})$ is the strain-rate tensor given by

$$\boldsymbol{\varepsilon}(\mathbf{u}) = \frac{1}{2}\left(\boldsymbol{\nabla}\mathbf{u} + \boldsymbol{\nabla}\mathbf{u}^T\right). \tag{1.4}$$

Equations (1.1) and (1.2) represent the local balance of linear momentum and mass, respectively, and the momentum balance equation is written in the so-called *conservative form*. The local mass balance for incompressible flows states that the velocity field must be divergence-free at every point in space and time, which is also known as the incompressibility constraint (see Equation (1.2)).

REMARK 1.1 *In Equation (1.1), and everywhere in this book, we denote by $\boldsymbol{\nabla}$ the gradient with respect to the spatial coordinates $\mathbf{x}$. We also denote by $\frac{\partial}{\partial t}$ the time derivative that is taken holding $\mathbf{x}$ fixed. If coordinates other than $\mathbf{x}$ are used, the gradient operator will be assigned the appropriate subscript.*

For incompressible flows, we can write the momentum equation also as

$$\rho\left(\frac{\partial \mathbf{u}}{\partial t} + \boldsymbol{\nabla}\cdot(\mathbf{u}\otimes\mathbf{u}) - \mathbf{f}\right) - \boldsymbol{\nabla}\cdot\boldsymbol{\sigma} = \mathbf{0}, \tag{1.5}$$

For constant density, Equation (1.5) represents the conservative form of the momentum equation. Starting from Equation (1.1) and using the conservation of mass, or starting from Equation (1.5) and using $\boldsymbol{\nabla}\cdot\mathbf{u} = 0$, we can obtain

$$\rho\left(\frac{\partial \mathbf{u}}{\partial t} + \mathbf{u}\cdot\boldsymbol{\nabla}\mathbf{u} - \mathbf{f}\right) - \boldsymbol{\nabla}\cdot\boldsymbol{\sigma} = \mathbf{0}, \tag{1.6}$$

$$\boldsymbol{\nabla}\cdot\mathbf{u} = 0. \tag{1.7}$$

Governing Equations of Fluid and Structural Mechanics

In this case, the momentum balance equation is written in the so-called *convective form*. We use the convective form of the Navier–Stokes equations in the rest of the section. The convective form of the Navier–Stokes equations can further be simplified by using the incompressibility constraint in the viscous part of the stress tensor. In this case, assuming constant viscosity:

$$\nabla \cdot \sigma(\mathbf{u}, p) = -\nabla p + \mu \Delta \mathbf{u}, \tag{1.8}$$

which leads to

$$\rho \left(\frac{\partial \mathbf{u}}{\partial t} + \mathbf{u} \cdot \nabla \mathbf{u} - \mathbf{f} \right) + \nabla p - \mu \Delta \mathbf{u} = \mathbf{0}. \tag{1.9}$$

Although this form of the linear-momentum balance equation is often used in the computations reported in the literature, we do not favor this choice because it leads to a non-objective definition of the Cauchy stress. For the importance of objectivity in fluid mechanics simulations, see, e.g., Limache *et al.* (2008) and references therein.

Assuming a fixed Cartesian basis on $\mathbb{R}^{n_{sd}}$, we let indices i and j take on the values $1, \ldots, n_{sd}$. We focus on the case of the spatial dimension $n_{sd} = 3$. We let u_i denote the ith Cartesian component of $\mathbf{u}$, and let x_i denote the ith component of $\mathbf{x}$. We denote differentiation by a comma (e.g., $u_{i,j} = u_{i,x_j} = \partial u_i / \partial x_j$). We will also use the summation convention, in which repeated indices imply summation; e.g., in $\mathbb{R}^3$,

$$u_{i,jj} = u_{i,11} + u_{i,22} + u_{i,33} = \frac{\partial^2 u_i}{\partial x_1^2} + \frac{\partial^2 u_i}{\partial x_2^2} + \frac{\partial^2 u_i}{\partial x_3^2}. \tag{1.10}$$

Using index notation, the Navier–Stokes equations of incompressible flows, as given by Equations (1.6) and (1.7), can be rewritten as:

$$\rho \left(u_{i,t} + u_j u_{i,j} - f_i \right) - \sigma_{ij,j} = 0, \tag{1.11}$$

$$u_{i,i} = 0, \tag{1.12}$$

where

$$\sigma_{ij} = -p \delta_{ij} + 2\mu \varepsilon_{ij}, \tag{1.13}$$

$$\varepsilon_{ij} = \frac{1}{2} \left(u_{i,j} + u_{j,i} \right), \tag{1.14}$$

and δ_{ij} is the Kronecker delta (i.e., $\delta_{ij} = 1$ if $i = j$, and $\delta_{ij} = 0$ if $i \neq j$).

REMARK 1.2 *Note the use of indices on the Cauchy stress σ_{ij} in Equation (1.11). Here, and in what follows, we adopt the convention where the first index, in this case i, indicates the direction in which the stress is acting, while the second index, j, indicates that the stress is acting on a plane normal to the x_j-axis. This is the opposite of the convention used in the papers written by the second and third authors of the book. However, we also note that as a consequence of the local moment equilibrium, the Cauchy stress is symmetric, which implies $\sigma_{ij} = \sigma_{ji}$. This, in turn, implies that the roles of indices in the Cauchy stress may be interchanged without any effect on the governing equations.*

REMARK 1.3 *We also declare the notations we have adopted for the representations related to the gradient and divergence operations on vectors and tensors. The gradient of a vector field is represented as*

$$\nabla \mathbf{a} \equiv \partial \mathbf{a}/\partial \mathbf{x} = \frac{\partial a_i}{\partial x_j} \mathbf{e}_i \otimes \mathbf{e}_j, \quad (1.15)$$

where $\frac{\partial a_i}{\partial x_j}$ are the Cartesian components of $\nabla \mathbf{a}$. We note that the first index corresponds to the vector component, while the second index corresponds to the spatial derivative. The divergence of a tensor is represented as

$$\nabla \cdot \mathbf{A} = \frac{\partial A_{ij}}{\partial x_j} \mathbf{e}_i, \quad (1.16)$$

where the contraction occurs on the second index. In the special case $\mathbf{A} = \mathbf{a} \otimes \mathbf{b}$, that representation becomes

$$\nabla \cdot (\mathbf{a} \otimes \mathbf{b}) = \frac{\partial (a_i b_j)}{\partial x_j} \mathbf{e}_i. \quad (1.17)$$

The advective term, however, is represented as

$$\mathbf{b} \cdot \nabla \mathbf{a} = b_j \frac{\partial a_i}{\partial x_j} \mathbf{e}_i. \quad (1.18)$$

To complete the statement of the fluid mechanics problem, we need to specify the boundary conditions. In general, on a given part of the spatial boundary, either kinematic or traction boundary conditions are prescribed. Kinematic boundary conditions are also referred to as essential or Dirichlet, while traction boundary conditions are also called natural or Neumann. Because the unknown velocity is a vector, one needs to generalize the boundary conditions to the vector case. The essential and natural boundary conditions for Equation (1.11) are

$$u_i = g_i \quad \text{on} \quad (\Gamma_t)_{g_i}, \quad (1.19)$$

$$\sigma_{ij} n_j = h_i \quad \text{on} \quad (\Gamma_t)_{h_i}, \quad (1.20)$$

where, for every velocity component i, $(\Gamma_t)_{g_i}$ and $(\Gamma_t)_{h_i}$ are the complementary subsets of the domain boundary Γ_t, n_i's are components of the unit outward normal vector $\mathbf{n}$, and g_i and h_i are given functions.

REMARK 1.4 *Equations (1.19) and (1.20) pertain to boundary condition specification for individual Cartesian components of the velocity and traction vectors. This is sufficient for many cases of interest. However, more general boundary conditions are possible and will be discussed in the later sections of this book.*

In the case where the fluid velocity vector is specified on the entire boundary of the fluid domain, the pressure is determined up to an arbitrary constant (that is, if p satisfies Equations (1.6) and (1.7), then so does $p + C$, where C is an arbitrary constant over Ω_t) and

$$\int_{\Gamma_t} \mathbf{g} \cdot \mathbf{n} \, d\Gamma = \int_{\Gamma_t} g_i n_i \, d\Gamma = 0, \quad (1.21)$$

which is a consequence of the incompressibility constraint. Equation (1.21) is often the source of difficulty for FSI coupling in the case of flows in enclosed domains.

1.1.2 Model Differential Equations

The Navier–Stokes equations of incompressible flows describe a wide range of behavior in viscous incompressible flows. Several simplifications of these equations are considered in the literature (and in practice) to better understand and model the physical phenomena involved. The two important special cases of the Navier–Stokes equations are the Stokes and Euler equations of incompressible flows.

Stokes equations. The Stokes equations are obtained by neglecting the convective terms in Equation (1.6), that is,

$$\rho\left(\frac{\partial \mathbf{u}}{\partial t} - \mathbf{f}\right) - \nabla \cdot \boldsymbol{\sigma} = \mathbf{0}, \tag{1.22}$$

$$\nabla \cdot \mathbf{u} = 0. \tag{1.23}$$

The above model is used for describing very slow (e.g., "creeping") flows. Note that the Stokes equations are linear with respect to both velocity and pressure, while the Navier–Stokes equations are not.

Euler equations. The other special case corresponds to inviscid flows described by the Euler equations of incompressible flows, namely

$$\rho\left(\frac{\partial \mathbf{u}}{\partial t} + \mathbf{u} \cdot \nabla \mathbf{u} - \mathbf{f}\right) - \nabla p = \mathbf{0}, \tag{1.24}$$

$$\nabla \cdot \mathbf{u} = 0. \tag{1.25}$$

The Euler equations retain the quadratic nonlinearity of the convective term.

Advection–diffusion equation. The *linear* advection–diffusion equation can be seen as an equation obtained by relaxing the incompressibility constraint (and, as a result, neglecting the pressure, which is also the Lagrange multiplier that enforces the incompressibility constraint) and "freezing" the advective velocity $\mathbf{u}$, namely

$$\rho\left(\frac{\partial \boldsymbol{\phi}}{\partial t} + \mathbf{u} \cdot \nabla \boldsymbol{\phi} - \mathbf{f}\right) - \mu \Delta \boldsymbol{\phi} = \mathbf{0}, \tag{1.26}$$

where we replaced $\mathbf{u}$ with a vector $\boldsymbol{\phi}$. Dividing Equation (1.26) by the density and realizing that the vector components of $\boldsymbol{\phi}$ are decoupled, we obtain a classical form of the scalar, time-dependent advection–diffusion equation:

$$\frac{\partial \phi}{\partial t} + \mathbf{u} \cdot \nabla \phi - \nu \Delta \phi - f = 0, \tag{1.27}$$

where $\nu = \mu/\rho$ is the kinematic viscosity. The advection–diffusion equation (Equation (1.27)) generally models transport of species, with concentration denoted by ϕ, in the presence of molecular diffusion. In this case ν is replaced with molecular diffusivity κ. Equation (1.27) is often employed as a starting point for the development of numerical formulations in fluid mechanics.

1.1.3 Nondimensional Equations and Numbers

To nondimensionalize the equations of fluid mechanics, we select a characteristic flow speed U and length scale L. We define

$$\mathbf{u} = \mathbf{u}^* U, \quad (1.28)$$

$$\boldsymbol{\nabla} = \boldsymbol{\nabla}^* \frac{1}{L}, \quad (1.29)$$

where $\mathbf{u}^*$ is the nondimensional flow velocity and $\boldsymbol{\nabla}^*$ is the nondimensional gradient operator. Equation (1.29) implies

$$\Delta = \boldsymbol{\nabla} \cdot \boldsymbol{\nabla} = \frac{1}{L^2}(\boldsymbol{\nabla}^*) \cdot (\boldsymbol{\nabla}^*) = \frac{1}{L^2}\Delta^*, \quad (1.30)$$

where Δ^* is the nondimensional Laplace operator. Starting from the steady version of the advection–diffusion equation (Equation (1.27)), assuming zero forcing, and introducing the definitions given by Equations (1.28)–(1.30), we obtain

$$\frac{U}{L}(\mathbf{u}^* \cdot \boldsymbol{\nabla}^*)\phi - \frac{\nu}{L^2}(\Delta^*)\phi = 0. \quad (1.31)$$

Rearranging the terms in the above equation gives

$$(\mathbf{u}^* \cdot \boldsymbol{\nabla}^*)\phi - \frac{1}{\mathrm{Pe}}(\Delta^*)\phi = 0, \quad (1.32)$$

where

$$\mathrm{Pe} = \frac{UL}{\nu} \quad (1.33)$$

is the *Peclet number* that represents the significance of advection relative to diffusion. For large Pe advection dominates, while for small Pe diffusion dominates. The case Pe $= \infty$ corresponds to pure advection. In the case of pure advection ϕ may only be set on the inflow part of the boundary Γ_t^-, which is defined as

$$\Gamma_t^- = \{\mathbf{x} \mid \mathbf{u} \cdot \mathbf{n} \leq 0 \; \forall \mathbf{x} \subset \Gamma_t\}. \quad (1.34)$$

The advection-dominant case gives rise to interior and boundary layers in ϕ, which causes difficulties in the numerical approximation of advection–diffusion equations. Diffusion dominance precludes the formation of thin layers in the solution and presents very little challenge to the numerical approximation. This situation is illustrated in Figure 1.2.

Performing the same analysis for the Navier–Stokes equations, we obtain the analog of the Peclet number, the *Reynolds number* Re, given by

$$\mathrm{Re} = \frac{UL}{\nu} = \frac{\rho UL}{\mu}. \quad (1.35)$$

The limit Re $\to 0$ yields the Stokes flow, while large Re values give rise to turbulent solutions to the Navier–Stokes equations. Turbulence may be characterized as a continuous spectrum of spatial and temporal scales in the velocity and pressure fields, and also presents a significant challenge for accurate approximation of the solutions to the Navier–Stokes equations in this regime.

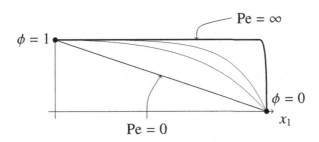

Figure 1.2 Illustration of the solution behavior for the advection–diffusion equation at Pe number ranging from zero to nearly infinity. In the figure, 1D setup is assumed, advective velocity is constant and points from left to right, and ϕ is set to unity on the left side of the interval and zero on the right side. For Pe = 0 the analytical solution is a straight line connecting the prescribed boundary values. As the Pe number increases, the solution forms a thin boundary layer on the right side of the domain. Such boundary layers present a source of difficulty for numerical approximation of the advection–diffusion equation

1.1.4 Some Specific Boundary Conditions

In this section we give a detailed account of the boundary conditions that are most often used in fluid mechanics simulations.

Solid surface. At a solid surface it is convenient to split the velocity vector into its normal and tangential components. For this, along with the normal vector $\mathbf{n}$, we define in 3D two orthonormal tangential vectors $\mathbf{t}_1$ and $\mathbf{t}_2$ (see Figure 1.3). In this new basis the velocity vector components become

$$u_n = \mathbf{u} \cdot \mathbf{n}, \tag{1.36}$$

$$u_{t_1} = \mathbf{u} \cdot \mathbf{t}_1, \tag{1.37}$$

$$u_{t_2} = \mathbf{u} \cdot \mathbf{t}_2. \tag{1.38}$$

Independent of whether the flow is viscous or inviscid, the no-penetration boundary condition becomes

$$u_n = g_n, \tag{1.39}$$

where g_n is the normal velocity of the solid surface. In the case of viscous flows, the remaining velocity components are set to

$$u_{t_1} = g_{t_1}, \tag{1.40}$$

$$u_{t_2} = g_{t_2}, \tag{1.41}$$

where g_t's are the tangential velocities of the solid surface. This results in the so-called no-slip boundary condition. However, in the case of turbulent boundary layers, or in the presence of "rough" surfaces, the tangential traction boundary conditions are adopted in place of no-slip boundary conditions, leading to so-called wall function formulations (see, e.g., Wilcox, 1998).

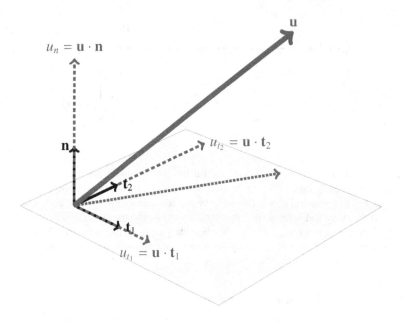

Figure 1.3 The normal vector **n** and two orthonormal vectors $\mathbf{t}_1$ and $\mathbf{t}_2$

Free surface. Another case of interest are the boundary conditions at a fluid surface that is free to deform, or the free surface (see, e.g., Tezduyar, 1992; Tezduyar *et al.*, 1993, 1996; Johnson and Tezduyar, 1994; Guler *et al.*, 1999; Akin *et al.*, 2007; Takizawa *et al.*, 2007a,b; Akkerman *et al.*, 2011). In this case, the following traction boundary condition holds:

$$\sigma\mathbf{n} = -p_{\text{atm}}\mathbf{n}, \qquad (1.42)$$

where p_{atm} is the atmospheric pressure. The pressure may be scaled such that $p_{\text{atm}} = 0$, in which case a homogeneous traction boundary condition is enforced.

External boundaries. This situation presents the most commonly encountered setup in computational fluid mechanics. One is interested in computing the flow over an object placed in a free stream. A truncated problem domain is created, which encloses the object and contains external boundaries. Typically, the external boundaries contain the inflow, outflow, and lateral (or side) boundaries (see Figure 1.4). No-slip condition is applied on the object. The external boundaries are placed sufficiently far from the object to approximate the free-stream conditions. The free-stream conditions are

$$\mathbf{u} = \mathbf{u}_\infty, \qquad (1.43)$$

or

$$\sigma\mathbf{n} = \sigma_\infty\mathbf{n}, \qquad (1.44)$$

Governing Equations of Fluid and Structural Mechanics

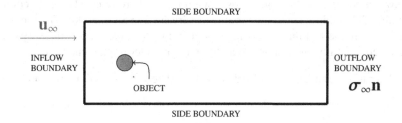

Figure 1.4 External boundaries (inflow, outflow, and side)

where the subscript ∞ is used to indicate quantities far from the object. Using index notation, the free-stream boundary conditions may be expressed as

$$u_i = (u_i)_\infty \quad \text{on} \quad (\Gamma_t)_{gi}, \tag{1.45}$$

$$\sigma_{ij} n_j = \left(\sigma_{ij}\right)_\infty n_j \quad \text{on} \quad (\Gamma_t)_{hi}. \tag{1.46}$$

In most cases,

$$\mathbf{u}_\infty = \begin{Bmatrix} U \\ 0 \\ 0 \end{Bmatrix}, \tag{1.47}$$

which leads to

$$\boldsymbol{\sigma}_\infty = -p_\infty \mathbf{I}, \tag{1.48}$$

and, consequently, to

$$\boldsymbol{\sigma}_\infty \mathbf{n} = -p_\infty \mathbf{n}. \tag{1.49}$$

It is often possible to scale the pressure so that $p_\infty = 0$. With this, the recommended boundary conditions at the external boundaries become:

- At the inflow boundary the entire velocity vector is prescribed:

$$\begin{Bmatrix} u_1 \\ u_2 \\ u_3 \end{Bmatrix} = \begin{Bmatrix} U \\ 0 \\ 0 \end{Bmatrix}. \tag{1.50}$$

- At the outflow boundary, free-stream traction boundary conditions are prescribed:

$$\boldsymbol{\sigma} \mathbf{n} = \mathbf{0}. \tag{1.51}$$

Assuming the normal vector is $\mathbf{n} = (1, 0, 0)^T$, we obtain, in component form,

$$\begin{Bmatrix} \sigma_{11} \\ \sigma_{21} \\ \sigma_{31} \end{Bmatrix} = \begin{Bmatrix} 0 \\ 0 \\ 0 \end{Bmatrix}. \tag{1.52}$$

Outflow boundary conditions given by Equation (1.51) are referred to as traction-free or "do-nothing" boundary conditions (see, e.g., Gresho and Sani, 2000). The second phrase originates from the fact that in the finite element method zero-stress boundary conditions are satisfied naturally and require no additional computer implementation.

- At the lateral boundaries zero normal velocity and zero tangential traction are prescribed, namely

$$\mathbf{u} \cdot \mathbf{n} = 0, \tag{1.53}$$
$$\mathbf{t}_1 \cdot \sigma \mathbf{n} = 0, \tag{1.54}$$
$$\mathbf{t}_2 \cdot \sigma \mathbf{n} = 0. \tag{1.55}$$

For the top and bottom lateral boundaries, where the normal vector is $\mathbf{n} = (0, \pm 1, 0)^T$, and the tangential vectors are $\mathbf{t}_1 = (1, 0, 0)^T$ and $\mathbf{t}_2 = (0, 0, 1)^T$, we obtain

$$\begin{Bmatrix} \sigma_{12} \\ u_2 \\ \sigma_{32} \end{Bmatrix} = \begin{Bmatrix} 0 \\ 0 \\ 0 \end{Bmatrix}. \tag{1.56}$$

For the span-wise lateral boundaries, where the normal vector is $\mathbf{n} = (0, 0, \pm 1)^T$, and the tangential vectors are $\mathbf{t}_1 = (1, 0, 0)^T$ and $\mathbf{t}_2 = (0, 1, 0)^T$, we obtain

$$\begin{Bmatrix} \sigma_{13} \\ \sigma_{23} \\ u_3 \end{Bmatrix} = \begin{Bmatrix} 0 \\ 0 \\ 0 \end{Bmatrix}. \tag{1.57}$$

We will consider other, more specialized, boundary conditions in the chapters on specific applications of FSI.

1.1.5 Weak Form of the Navier–Stokes Equations

We denote by $\mathcal{S}_u$ and $\mathcal{S}_p$ the sets of infinite-dimensional trial functions for the velocity and pressure. The function sets $\mathcal{S}_u$ and $\mathcal{S}_p$ are defined as

$$\mathcal{S}_u = \left\{ \mathbf{u} \mid \mathbf{u}(\cdot, t) \in (H^1(\Omega_t))^{n_{sd}},\ u_i = g_i \text{ on } (\Gamma_t)_{g_i} \right\} \tag{1.58}$$

and

$$\mathcal{S}_p = \left\{ p \mid p(\cdot) \in L^2(\Omega_t),\ \int_{\Omega_t} p\, d\Omega = 0 \text{ if } \Gamma_t = (\Gamma_t)_g \right\}. \tag{1.59}$$

Here $L^2(\Omega_t)$ denotes the space of scalar-valued functions that are square-integrable on Ω_t, and $(H^1(\Omega_t))^{n_{sd}}$ denotes the space of vector-valued functions with square-integrable derivatives on Ω_t. The functions in $\mathcal{S}_u$ satisfy the essential boundary conditions of the fluid mechanics problem. In the case when the essential boundary conditions are set on all of Γ_t, we require that the average of the pressure field over Ω_t is zero, which is built into the definition of $\mathcal{S}_p$.

In conjunction with $\mathcal{S}_u$ and $\mathcal{S}_p$, we define the sets of test functions (also called "weighting functions" in this book) for the linear-momentum and continuity equations, denoted by $\mathcal{V}_u$ and $\mathcal{V}_p$ as

$$\mathcal{V}_u = \left\{ \mathbf{w} \mid \mathbf{w}(\cdot) \in (H^1(\Omega_t))^{n_{\mathrm{sd}}}, \; w_i = 0 \text{ on } (\Gamma_t)_{gi} \right\} \tag{1.60}$$

and

$$\mathcal{V}_p = \mathcal{S}_p. \tag{1.61}$$

Note that the function sets $\mathcal{S}_u$ and $\mathcal{V}_u$ only differ in the definition of boundary conditions, that is, the test functions for the linear-momentum balance equations vanish on the parts of the boundary where the fluid velocity is prescribed. The sets of pressure trial functions and the continuity equation test functions are coincident.

To derive the weak form of the fluid mechanics equations, following the standard approach, we multiply Equations (1.6) and (1.7) by the linear-momentum and continuity equation test functions respectively, integrate over Ω_t, and add the equations to obtain

$$\int_{\Omega_t} \mathbf{w} \cdot \rho \left(\frac{\partial \mathbf{u}}{\partial t} + \mathbf{u} \cdot \nabla \mathbf{u} - \mathbf{f} \right) d\Omega - \int_{\Omega_t} \mathbf{w} \cdot (\nabla \cdot \boldsymbol{\sigma}(\mathbf{u}, p)) \, d\Omega + \int_{\Omega_t} q \nabla \cdot \mathbf{u} \, d\Omega = 0. \tag{1.62}$$

We integrate by parts the Cauchy stress terms in Equation (1.62) and apply the homogeneous form of the essential boundary conditions on $\mathbf{w}$ to get

$$-\int_{\Omega_t} \mathbf{w} \cdot (\nabla \cdot \boldsymbol{\sigma}(\mathbf{u}, p)) \, d\Omega = \int_{\Omega_t} \boldsymbol{\varepsilon}(\mathbf{w}) : \boldsymbol{\sigma}(\mathbf{u}, p) \, d\Omega - \int_{(\Gamma_t)_h} \mathbf{w} \cdot \boldsymbol{\sigma}(\mathbf{u}, p) \mathbf{n} \, d\Gamma, \tag{1.63}$$

where $(\Gamma_t)_h$ is the abstract representation of the natural boundary of the fluid mechanics domain. We replace the traction vector $\boldsymbol{\sigma}(\mathbf{u}, p) \mathbf{n}$ with its prescribed value $\mathbf{h}$ on $(\Gamma_t)_h$ in the last term on the right-hand-side of Equation (1.63) to obtain

$$-\int_{\Omega_t} \mathbf{w} \cdot (\nabla \cdot \boldsymbol{\sigma}(\mathbf{u}, p)) \, d\Omega = \int_{\Omega_t} \boldsymbol{\varepsilon}(\mathbf{w}) : \boldsymbol{\sigma}(\mathbf{u}, p) \, d\Omega - \int_{(\Gamma_t)_h} \mathbf{w} \cdot \mathbf{h} \, d\Gamma. \tag{1.64}$$

Combining Equations (1.62) and (1.64) results in the weak form of the Navier–Stokes equations: find $\mathbf{u} \in \mathcal{S}_u$ and $p \in \mathcal{S}_p$, such that $\forall \, \mathbf{w} \in \mathcal{V}_u$ and $q \in \mathcal{V}_p$:

$$\int_{\Omega_t} \mathbf{w} \cdot \rho \left(\frac{\partial \mathbf{u}}{\partial t} + \mathbf{u} \cdot \nabla \mathbf{u} - \mathbf{f} \right) d\Omega + \int_{\Omega_t} \boldsymbol{\varepsilon}(\mathbf{w}) : \boldsymbol{\sigma}(\mathbf{u}, p) \, d\Omega$$
$$- \int_{(\Gamma_t)_h} \mathbf{w} \cdot \mathbf{h} \, d\Gamma + \int_{\Omega_t} q \nabla \cdot \mathbf{u} \, d\Omega = 0. \tag{1.65}$$

The weak formulation given by Equation (1.65) is the point of departure for the finite element formulations of the fluid mechanics problem. The details of the finite element method will be presented in later chapters. Note that while the velocity boundary conditions for the weak formulation of the fluid mechanics equations are built into the corresponding function sets, the traction boundary conditions are imposed weakly as a consequence of the integration-by-parts procedure described by Equations (1.63) and (1.64).

1.2 Governing Equations of Structural Mechanics

In this section we present the governing equations of structural mechanics. The equations are derived on the basis of 3D continuum modeling. We conclude this section with a presentation of the governing equations for shell, membrane, and cable structures.

1.2.1 Kinematics

Let $\Omega_0 \in \mathbb{R}^{n_{sd}}$ be the material domain of a structure in the reference configuration, and let Γ_0 be its boundary. Let $\Omega_t \in \mathbb{R}^{n_{sd}}$, $t \in (0, T)$, be the material domain of a structure in the current configuration, and let Γ_t be its boundary. For the upcoming developments we assume that the reference configuration coincides with the initial configuration or the configuration of the structure taken at $t = 0$. Let $\mathbf{X}$ be the coordinates of the initial or reference configuration, and let $\mathbf{y}$ be the displacement with respect to the initial configuration. We think of $\mathbf{y} = \mathbf{y}(\mathbf{X}, t)$ as a time-varying vector field over Ω_0 and define a mapping

$$\mathbf{x}(\mathbf{X}, t) = \mathbf{X} + \mathbf{y}(\mathbf{X}, t), \tag{1.66}$$

which maps the coordinates of material points in the reference configuration to their counterparts in the current configuration. We also denote by $\mathbf{x}$ the coordinates of the current configuration. Because this "abuse of notation" is standard practice in continuum mechanics, we adopt it here as well. The setup is illustrated in Figure 1.5.

The velocity $\mathbf{u}$ and acceleration $\mathbf{a}$ of the structure are obtained by differentiating the displacement $\mathbf{y}$ with respect to time holding the material coordinate $\mathbf{X}$ fixed, namely

$$\mathbf{u} = \frac{d\mathbf{y}}{dt} \tag{1.67}$$

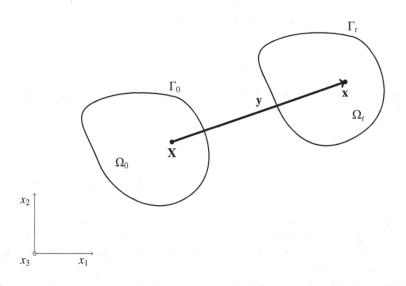

Figure 1.5 Reference and current configurations

and

$$\mathbf{a} = \frac{d^2 \mathbf{y}}{dt^2}. \tag{1.68}$$

REMARK 1.5 *Here, and everywhere in the book, $\frac{d}{dt}$ will denote the total time derivative, or the time derivative taken holding the material coordinate $\mathbf{X}$ fixed.*

The deformation gradient $\mathbf{F}$ is given by

$$\mathbf{F} = \frac{\partial \mathbf{x}}{\partial \mathbf{X}} = \mathbf{I} + \frac{\partial \mathbf{y}}{\partial \mathbf{X}}, \tag{1.69}$$

which we use to define the Cauchy–Green deformation tensor $\mathbf{C}$ as

$$\mathbf{C} = \mathbf{F}^T \mathbf{F}, \tag{1.70}$$

and the Green–Lagrange strain tensor $\mathbf{E}$ as

$$\mathbf{E} = \frac{1}{2}(\mathbf{C} - \mathbf{I}). \tag{1.71}$$

The determinant of the deformation gradient J is given by

$$J = \det \mathbf{F}. \tag{1.72}$$

We now introduce the index notation for the structural mechanics. Due to the presence of the reference and current configurations, the quantities referring to the reference configuration are typically subscripted with upper-case indices (e.g., I, J, K), and those referring to the current configuration with lower-case indices (e.g., i, j, k). However, despite the use of different index types, we assume that the vector and tensor components in the reference and current configurations are referred to a fixed Cartesian basis. The summation convention applies to the upper- and lower-case indices *separately*, and all indices take on the values $1, \ldots, n_{\mathrm{sd}}$.

The Cartesian components of the deformation gradient become

$$F_{iI} = \frac{\partial x_i}{\partial X_I} = \delta_{iI} + \frac{\partial y_i}{\partial X_I}, \tag{1.73}$$

which means that one "leg" of the deformation gradient tensor is in the reference and another is in the current configuration. The components of the Cauchy–Green deformation tensor and the Green–Lagrange strain tensor are given by

$$C_{IJ} = F_{iI} F_{iJ} \tag{1.74}$$

and

$$E_{IJ} = \frac{1}{2}(C_{IJ} - \delta_{IJ}), \tag{1.75}$$

respectively. Note that the tensors $\mathbf{C}$ and $\mathbf{E}$ are completely defined in the reference or undeformed configuration.

1.2.2 Principle of Virtual Work and Variational Formulation of Structural Mechanics

The starting point for the structural mechanics formulations is the *principle of virtual work* (see, e.g., Belytschko *et al.*, 2000):

$$\delta W = \delta W_{\text{int}} + \delta W_{\text{ext}} = 0, \tag{1.76}$$

where W, W_{int}, and W_{ext} are the total, internal, and external work, respectively, and δ denotes their variation with respect to the virtual displacement $\mathbf{w}$. Given the structural displacement $\mathbf{y}$, δW is computed by taking the directional derivative of W as

$$\delta W = \left. \frac{\mathrm{d}}{\mathrm{d}\epsilon} W(\mathbf{y} + \epsilon \mathbf{w}) \right|_{\epsilon=0}. \tag{1.77}$$

Here δW_{ext} includes the virtual work done by the inertial and body forces and surface tractions, and is given by

$$\delta W_{\text{ext}} = \int_{\Omega_t} \mathbf{w} \cdot \rho (\mathbf{f} - \mathbf{a}) \, \mathrm{d}\Omega + \int_{(\Gamma_t)_h} \mathbf{w} \cdot \mathbf{h} \, \mathrm{d}\Gamma, \tag{1.78}$$

where ρ is the mass density of the structure in the current configuration, $\mathbf{f}$ is the body force per unit mass, and $\mathbf{h}$ is the external traction vector applied on the subset $(\Gamma_t)_h$ of the total boundary Γ_t.

The virtual work done by the internal stresses, δW_{int}, may be computed as

$$\delta W_{\text{int}} = - \int_{\Omega_0} \delta \mathbf{E} : \mathbf{S} \, \mathrm{d}\Omega. \tag{1.79}$$

Here $\mathbf{S}$ is the second Piola–Kirchhoff stress tensor, which is symmetric and work-conjugate to $\mathbf{E}$. Although $\mathbf{S}$ is not possible to measure experimentally, it plays a prominent role in constitutive modeling of materials.

In Equation (1.79), $\delta \mathbf{E}$ is the variation of the Green–Lagrange strain tensor, which is also referred to as the virtual strain. Putting Equations (1.76)–(1.79) together, and recognizing that $\mathbf{w}$ is arbitrary, we arrive at the variational formulation of the structural mechanics problem: find the structural displacement $\mathbf{y} \in \mathcal{S}_y$, such that $\forall\, \mathbf{w} \in \mathcal{V}_y$:

$$\int_{\Omega_t} \mathbf{w} \cdot \rho \mathbf{a} \, \mathrm{d}\Omega + \int_{\Omega_0} \delta \mathbf{E} : \mathbf{S} \, \mathrm{d}\Omega - \int_{\Omega_t} \mathbf{w} \cdot \rho \mathbf{f} \, \mathrm{d}\Omega - \int_{(\Gamma_t)_h} \mathbf{w} \cdot \mathbf{h} \, \mathrm{d}\Gamma = 0. \tag{1.80}$$

Here $\mathcal{S}_y$ and $\mathcal{V}_y$ are the sets of trial and test functions for the structural mechanics problem, defined as

$$\mathcal{S}_y = \left\{ \mathbf{y} \mid \mathbf{y}(\cdot, t) \in (H^1(\Omega_t))^{n_{\text{sd}}}, \; y_i = g_i \text{ on } (\Gamma_t)_{gi} \right\}, \tag{1.81}$$

and

$$\mathcal{V}_y = \left\{ \mathbf{w} \mid \mathbf{w}(\cdot) \in (H^1(\Omega_t))^{n_{\text{sd}}}, \; w_i = 0 \text{ on } (\Gamma_t)_{gi} \right\}. \tag{1.82}$$

Here, for each i, $(\Gamma_t)_{gi}$ and $(\Gamma_t)_{hi}$ are the complementary subsets of the domain boundary Γ_t, and g_i is a given function. Note that the essential boundary conditions are built into the definition of the function sets of the structural mechanics problem. The variational formulation given by Equation (1.80) is the point of departure for the structural mechanics finite element formulations.

1.2.3 Conservation of Mass

In Equation (1.80) the structural density ρ in the current configuration is not known a priori. To derive the dependence of the structural density on the structural displacement, we first define the structural mass m as

$$m = \int_{\Omega_t} \rho \, d\Omega_t. \tag{1.83}$$

We assume that the structural mass is conserved at all times, which may be expressed as

$$\frac{dm}{dt} = 0. \tag{1.84}$$

Introducing Equation (1.83) into Equation (1.84), changing variables to the reference configuration, and taking the time derivative inside the integral, we obtain

$$\frac{dm}{dt} = \frac{d}{dt} \int_{\Omega_t} \rho \, d\Omega_t = \int_{\Omega_0} \frac{d\rho J}{dt} \, d\Omega_0 = 0. \tag{1.85}$$

Because Ω_0 is arbitrary, we can localize the results to any material point in the structure as

$$\frac{d\rho J}{dt} = 0. \tag{1.86}$$

This, in turn, means that the product ρJ is only a function of the material point, namely $\rho J = \rho J(\mathbf{X})$. At $t = 0$ the structure is undeformed, meaning $J = 1$. Defining $\rho_0 = \rho_0(\mathbf{X})$ to be the structural mass density in the undeformed configuration, we obtain the following point-wise statement of the conservation of mass:

$$\rho_0 = \rho J. \tag{1.87}$$

Because ρ_0 is considered to be known, given the structural displacement field, Equation (1.87) may be used to obtain the density at a material point in the current configuration using this simple algebraic expression. The relationship given by Equation (1.87) is known as the Lagrangian description of mass conservation.

1.2.4 Structural Mechanics Formulation in the Current Configuration

In the variational formulation given by Equation (1.80), the stress terms are written with respect to the reference configuration, while the remaining terms are expressed in the current configuration. In order to have a formulation that is written purely in the current configuration, we proceed as follows. We first make explicit the dependence of the virtual strain $\delta \mathbf{E}$ on the virtual displacement $\mathbf{w}$. Starting with the definition of $\mathbf{E}$ in Equation (1.71) and taking the variation as in Equation (1.77) gives

$$\delta \mathbf{E} = \frac{1}{2} \left(\mathbf{F}^T \nabla_X \mathbf{w} + \nabla_X \mathbf{w}^T \mathbf{F} \right), \tag{1.88}$$

where ∇_X denotes the gradient taken with respect to the spatial coordinates of the reference configuration. Due to the symmetry of $\mathbf{S}$, the scalar product $\delta \mathbf{E} : \mathbf{S}$ simplifies to

$$\delta \mathbf{E} : \mathbf{S} = \nabla_X \mathbf{w} : \mathbf{P}, \tag{1.89}$$

where

$$P = FS \tag{1.90}$$

is the first Piola–Kirchhoff stress tensor, which is nonsymmetric. With these definitions the variational formulation of the structural mechanics problem becomes: find $\mathbf{y} \in \mathcal{S}_y$, such that $\forall\, \mathbf{w} \in \mathcal{V}_y$:

$$\int_{\Omega_t} \mathbf{w} \cdot \rho \mathbf{a}\, d\Omega + \int_{\Omega_0} \nabla_X \mathbf{w} : \mathbf{P}\, d\Omega - \int_{\Omega_t} \mathbf{w} \cdot \rho \mathbf{f}\, d\Omega - \int_{(\Gamma_t)_h} \mathbf{w} \cdot \mathbf{h}\, d\Gamma = 0. \tag{1.91}$$

We now change variables in the stress terms in Equation (1.91) to obtain

$$\int_{\Omega_0} \nabla_X \mathbf{w} : \mathbf{P}\, d\Omega = \int_{\Omega_0} \nabla_X \mathbf{w} : (\mathbf{FS})\, d\Omega = \int_{\Omega_0} \frac{\partial w_i}{\partial X_J} \frac{\partial x_i}{\partial X_I} S_{IJ}\, d\Omega \tag{1.92}$$

$$= \int_{\Omega_t} \frac{\partial w_i}{\partial x_j} \left(\frac{\partial x_i}{\partial X_I} S_{IJ} \frac{\partial x_j}{\partial X_J} J^{-1} \right) d\Omega \tag{1.93}$$

$$= \int_{\Omega_t} \nabla \mathbf{w} : \left(J^{-1} \mathbf{F S F}^T \right) d\Omega, \tag{1.94}$$

where in the last term we recognize the Cauchy stress tensor $\boldsymbol{\sigma}$:

$$\boldsymbol{\sigma} = J^{-1} \mathbf{F S F}^T. \tag{1.95}$$

Using index notation, the component form of Equation (1.95) may be written as

$$\sigma_{ij} = J^{-1} F_{iI} S_{IJ} F_{jJ}. \tag{1.96}$$

The Cauchy stress, unlike the second Piola–Kirchhoff stress, may be measured experimentally. Due to the the symmetry of the Cauchy stress tensor, we can write

$$\int_{\Omega_t} \nabla \mathbf{w} : \boldsymbol{\sigma}\, d\Omega = \int_{\Omega_t} \boldsymbol{\varepsilon}(\mathbf{w}) : \boldsymbol{\sigma}\, d\Omega, \tag{1.97}$$

where, as before,

$$\boldsymbol{\varepsilon}(\mathbf{w}) = \frac{1}{2} \left(\nabla \mathbf{w} + \nabla \mathbf{w}^T \right). \tag{1.98}$$

Combining Equations (1.91), (1.92), (1.95), and (1.97), we obtain the structural mechanics variational formulation in the current configuration: find $\mathbf{y} \in \mathcal{S}_y$, such that $\forall\, \mathbf{w} \in \mathcal{V}_y$:

$$\int_{\Omega_t} \mathbf{w} \cdot \rho \mathbf{a}\, d\Omega + \int_{\Omega_t} \boldsymbol{\varepsilon}(\mathbf{w}) : \boldsymbol{\sigma}\, d\Omega - \int_{\Omega_t} \mathbf{w} \cdot \rho \mathbf{f}\, d\Omega - \int_{(\Gamma_t)_h} \mathbf{w} \cdot \mathbf{h}\, d\Gamma = 0. \tag{1.99}$$

An equivalent variational formulation of the structural mechanics problem may be developed in the reference configuration. We present that in the next section.

To infer the strong formulation of the structural problem from Equation (1.99), we integrate by parts the stress terms, apply the homogeneous form of the essential boundary conditions on $\mathbf{w}$, and group the interior and boundary integral terms to obtain

$$\int_{\Omega_t} \mathbf{w} \cdot (\rho(\mathbf{a} - \mathbf{f}) - \nabla \cdot \boldsymbol{\sigma})\, d\Omega + \int_{(\Gamma_t)_h} \mathbf{w} \cdot (\boldsymbol{\sigma} \mathbf{n} - \mathbf{h})\, d\Gamma = 0. \tag{1.100}$$

Because Equation (1.100) holds for all admissible **w**, we conclude that

$$\rho(\mathbf{a} - \mathbf{f}) - \nabla \cdot \boldsymbol{\sigma} = \mathbf{0} \tag{1.101}$$

at every point inside Ω_t and

$$\boldsymbol{\sigma}\mathbf{n} - \mathbf{h} = \mathbf{0} \tag{1.102}$$

at every point on the traction boundary $(\Gamma_t)_h$. Equations (1.101) and (1.102) constitute the point-wise balance of linear momentum and traction boundary condition, respectively, in the current configuration.

Using index notation, the strong form of the structural mechanics equations may be written as

$$\rho(a_i - f_i) - \sigma_{ij,j} = 0 \text{ in } \Omega_t, \tag{1.103}$$

$$y_i = g_i \text{ on } (\Gamma_t)_{gi}, \tag{1.104}$$

$$\sigma_{ij}n_j = h_i \text{ on } (\Gamma_t)_{hi}. \tag{1.105}$$

Here, n_i's are components of the unit outward normal vector **n** in the current configuration, and h_i's are given functions. Note that the boundary conditions for the displacement lead to the boundary conditions for the velocity and acceleration, namely

$$u_i = \frac{dg_i}{dt} \text{ on } (\Gamma_t)_{gi}, \tag{1.106}$$

and

$$a_i = \frac{d^2 g_i}{dt^2} \text{ on } (\Gamma_t)_{gi}. \tag{1.107}$$

1.2.5 Structural Mechanics Formulation in the Reference Configuration

To infer the weak form of the structural mechanics equations in the reference configuration Ω_0, we again start with the variational formulation given by Equation (1.80), and change variables in the inertial and body force terms as

$$\int_{\Omega_t} \mathbf{w} \cdot \rho(\mathbf{a} - \mathbf{f}) \, d\Omega = \int_{\Omega_0} \mathbf{w} \cdot \rho_0 (\mathbf{a} - \mathbf{f}) \, d\Omega. \tag{1.108}$$

Mass conservation given by Equation (1.87) is also employed to arrive at the above result. Combining Equations (1.80), (1.88), (1.89), and (1.108), we obtain the following variational formulation of the structural mechanics problem posed in Ω_0: find $\mathbf{y} \in \mathcal{S}_y$, such that $\forall \, \mathbf{w} \in \mathcal{V}_y$:

$$\int_{\Omega_0} \mathbf{w} \cdot \rho_0 \mathbf{a} \, d\Omega + \int_{\Omega_0} \nabla_X \mathbf{w} : \mathbf{P} \, d\Omega - \int_{\Omega_0} \mathbf{w} \cdot \rho_0 \mathbf{f} \, d\Omega - \int_{(\Gamma_0)_h} \mathbf{w} \cdot \hat{\mathbf{h}} \, d\Gamma = 0, \tag{1.109}$$

where $\hat{\mathbf{h}}$, is the traction vector acting in the reference configuration. Integrating by parts the stress terms in Equation (1.109), applying the homogeneous form of the essential boundary conditions on **w**, and grouping the interior and boundary integrals, we obtain

$$\int_{\Omega_0} \mathbf{w} \cdot (\rho_0 (\mathbf{a} - \mathbf{f}) - \nabla_X \cdot \mathbf{P}) \, d\Omega + \int_{(\Gamma_0)_h} \mathbf{w} \cdot \left(\mathbf{P}\hat{\mathbf{n}} - \hat{\mathbf{h}} \right) d\Gamma = 0, \tag{1.110}$$

which holds for all admissible **w**, and where $\hat{\mathbf{n}}$ is the unit normal in the reference configuration. From Equation (1.110) we infer the point-wise balance of linear momentum in Ω_0:

$$\rho_0 (\mathbf{a} - \mathbf{f}) - \nabla_X \cdot \mathbf{P} = \mathbf{0}, \quad (1.111)$$

and the traction boundary condition on $(\Gamma_0)_h$:

$$\mathbf{P}\hat{\mathbf{n}} - \hat{\mathbf{h}} = \mathbf{0}. \quad (1.112)$$

Using index notation, the strong form of the structural mechanics boundary value problem may be written as

$$\rho_0 (a_i - f_i) - P_{iI,I} = 0 \text{ in } \Omega_0, \quad (1.113)$$
$$y_i = g_i \text{ on } (\Gamma_0)_{gi}, \quad (1.114)$$
$$P_{iI}\hat{n}_I = \hat{h}_i \text{ on } (\Gamma_0)_{hi}. \quad (1.115)$$

REMARK 1.6 *The variational statement given by Equation (1.99) is sometimes called the updated Lagrangian formulation of structural mechanics, while Equation (1.109) corresponds to the so-called total Lagrangian formulation of structural mechanics (see, e.g., Belytschko et al., 2000). Both formulations are equivalent in that they produce identical solutions for the same input. The choice of the formulation is often dictated by constitutive modeling, boundary conditions, ease of computer implementation, and other factors.*

1.2.6 Additional Boundary Conditions of Practical Interest

In this section we briefly describe two cases of structural mechanics boundary conditions that are often employed in practice. These are the follower pressure load and elastic-foundation boundary conditions.

Follower pressure load. This case presents a situation where the structural deformation is driven by the external applied pressure load on $(\Gamma_t)_h$. In this case, the stress vector **h** becomes

$$\mathbf{h} = -p\mathbf{n}, \quad (1.116)$$

where p is the magnitude of the applied pressure. Because the pressure is applied to the part of the domain boundary that is in motion, this boundary condition leads to a nonlinearity that needs to be handled in the computation. Changing variables to the reference configuration and using Nanson's formula (see. e.g., Holzapfel, 2000)

$$\mathbf{n} \, d\Gamma_t = J\mathbf{F}^{-T}\hat{\mathbf{n}} \, d\Gamma_0, \quad (1.117)$$

where $d\Gamma_t$ and $d\Gamma_0$ are the differential surface area elements in the current and reference configurations, respectively, we obtain

$$\int_{(\Gamma_t)_h} \mathbf{w} \cdot \mathbf{h} \, d\Gamma_t = -\int_{(\Gamma_t)_h} \mathbf{w} \cdot p\mathbf{n} \, d\Gamma_t = -\int_{(\Gamma_0)_h} \mathbf{w} \cdot pJ\mathbf{F}^{-T}\hat{\mathbf{n}} \, d\Gamma_0. \quad (1.118)$$

From Equation (1.118) we conclude that the resultant reference configuration traction vector $\hat{\mathbf{h}}$ for the follower pressure load is

$$\hat{\mathbf{h}} = -pJ\mathbf{F}^{-T}\hat{\mathbf{n}}, \tag{1.119}$$

and the corresponding boundary condition becomes

$$\mathbf{P}\hat{\mathbf{n}} + pJ\mathbf{F}^{-T}\hat{\mathbf{n}} = \mathbf{0}, \tag{1.120}$$

or, using index notation,

$$P_{il}\hat{n}_I + pJF_{Ii}^{-1}\hat{n}_I = 0. \tag{1.121}$$

Elastic foundation. In this case, the structure is assumed to be supported by an elastic-foundation, which is modeled using the spring analogy. The traction vector is made proportional to the structural displacement as

$$\mathbf{h} = -k\mathbf{y}, \tag{1.122}$$

where $k > 0$ is the spring constant. In this case, the traction term in the variational equations of structural mechanics is replaced with

$$\int_{(\Gamma_t)_h} \mathbf{w} \cdot \mathbf{h} \, d\Gamma = -\int_{(\Gamma_t)_h} \mathbf{w} \cdot k\mathbf{y} \, d\Gamma. \tag{1.123}$$

Note that the limit $k \to \infty$ represents a rigid foundation and the limit $k \to 0$ gives a zero-traction boundary condition. In some cases it is desirable to only employ the spring analogy in the direction normal to the boundary. In this case, the boundary condition (1.122) is replaced by

$$\mathbf{h} = -k(\mathbf{y} \cdot \mathbf{n})\,\mathbf{n}, \tag{1.124}$$

which yields the following traction term:

$$\int_{(\Gamma_t)_h} \mathbf{w} \cdot \mathbf{h} \, d\Gamma = -\int_{(\Gamma_t)_h} (\mathbf{w} \cdot \mathbf{n}) k(\mathbf{y} \cdot \mathbf{n}) \, d\Gamma. \tag{1.125}$$

1.2.7 Some Constitutive Models

To present the constitutive models in structural mechanics, we restrict the presentation to the class of hyperelastic materials. More complicated cases, such as inelastic materials, may be found in Simo and Hughes (1998). The theory of hyperelasticity assumes the existence of a stored elastic-energy density per unit volume of the undeformed configuration, φ, expressed as a function of the strain as

$$\varphi = \varphi(\mathbf{E}). \tag{1.126}$$

The second Piola–Kirchhoff stress $\mathbf{S}$ is obtained by differentiating φ with respect to $\mathbf{E}$ as

$$\mathbf{S}(\mathbf{E}) = \frac{\partial \varphi(\mathbf{E})}{\partial \mathbf{E}}. \tag{1.127}$$

Given the second Piola–Kirchhoff stress, the Cauchy stress is computed according to Equation (1.95) or (1.96). The tensor of elastic moduli, which plays an important role in the linearization of the structural mechanics equations, is defined as the second derivative of φ with respect to $\mathbf{E}$, namely,

$$\mathbb{C}(\mathbf{E}) = \frac{\partial^2 \varphi(\mathbf{E})}{\partial \mathbf{E} \partial \mathbf{E}}. \tag{1.128}$$

Different forms of $\varphi(\mathbf{E})$ in Equation (1.126) lead to different constitutive relationships between stress and strain. Here we present a few important cases.

St. Venant–Kirchhoff model. This model is characterized by $\varphi(\mathbf{E})$ defined as

$$\varphi(\mathbf{E}) = \frac{1}{2} \mathbf{E} : \mathbb{C}\mathbf{E}, \tag{1.129}$$

where $\mathbb{C}$ is a fourth-rank tensor of elastic moduli that is *independent of the state of deformation*. Using Equation (1.127), it is easy to see that this choice of the elastic-energy density leads to a linear relationship between the Green–Lagrange strain and the second Piola–Kirchhoff stress:

$$\mathbf{S}(\mathbf{E}) = \mathbb{C}\mathbf{E}. \tag{1.130}$$

Because $\mathbf{E}$ respects the principle of objectivity (see e.g., Holzapfel, 2000), this constitutive relationship is applicable to modeling structures in the regime of large displacements in that all rigid-body motions (i.e., translations and large rotations) produce no strains. However, the model is only physically valid in the small-strain regime as many materials deviate from a linear stress-strain relationship, even for very modest strain levels.

The constitutive tensor $\mathbb{C}$ may be designed to represent various types of material anisotropy (e.g., composite materials). The simplest case is the isotropic material for which the constitutive tensor, in the component form, becomes

$$\mathbb{C}_{IJKL} = \left(\kappa - \frac{2}{3}\mu\right)\delta_{IJ}\delta_{KL} + \mu(\delta_{IK}\delta_{JL} + \delta_{IL}\delta_{JK}), \tag{1.131}$$

where κ and μ are the material bulk and shear moduli, respectively. These are related to the material Young's modulus, E, and Poisson's ratio, ν, as

$$\kappa = \lambda + \frac{2}{3}\mu, \tag{1.132}$$

$$\mu = \frac{E}{2(1+\nu)}, \tag{1.133}$$

$$\lambda = \frac{\nu E}{(1+\nu)(1-2\nu)}, \tag{1.134}$$

where μ and λ are the well-known Lamé constants of the linear-elasticity model.

Neo-Hookean model with dilatational penalty. It is well known that the St. Venant–Kirchhoff model is unstable in the regime of strong compression (see Holzapfel, 2000).

A constitutive model that addresses this drawback was proposed in Simo and Hughes (1998). The model is isotropic, and the elastic-energy density takes the form

$$\varphi(\mathbf{C}, J) = \frac{1}{2}\mu\left(J^{-2/3}\text{tr}\mathbf{C} - 3\right) + \frac{1}{2}\kappa\left(\frac{1}{2}\left(J^2 - 1\right) - \ln J\right). \quad (1.135)$$

Note that the elastic-energy density is written with respect to the Cauchy–Green stress tensor $\mathbf{C}$ and the determinant of the Jacobian of the deformation gradient J. The $J^2 - 1$ term penalizes the deviation of J from unity and the $\ln J$ term stabilizes the formulation for the regime of strong compression. With this definition of the elastic-energy density, the second Piola–Kirchhoff stress tensor $\mathbf{S}$ and the tensor of elastic moduli $\mathbb{C}$ may be explicitly computed, and are given by

$$\mathbf{S} = \mu J^{-2/3}\left(\mathbf{I} - \frac{1}{3}\text{tr}\mathbf{C}\,\mathbf{C}^{-1}\right) + \frac{1}{2}\kappa\left(J^2 - 1\right)\mathbf{C}^{-1} \quad (1.136)$$

and

$$\mathbb{C} = \left(\frac{2}{9}\mu J^{-2/3}\text{tr}\mathbf{C} + \kappa J^2\right)\mathbf{C}^{-1} \otimes \mathbf{C}^{-1}$$
$$+ \left(\frac{2}{3}\mu J^{-2/3}\text{tr}\mathbf{C} - \kappa\left(J^2 - 1\right)\right)\mathbf{C}^{-1} \odot \mathbf{C}^{-1}$$
$$- \frac{2}{3}\mu J^{-2/3}\left(\mathbf{I} \otimes \mathbf{C}^{-1} + \mathbf{C}^{-1} \otimes \mathbf{I}\right). \quad (1.137)$$

In (1.137), the symbols $\otimes$ and $\odot$ are defined as

$$(\mathbf{A} \otimes \mathbf{B})_{IJKL} = (\mathbf{A})_{IJ}(\mathbf{B})_{KL}, \quad (1.138)$$

$$\left(\mathbf{C}^{-1} \odot \mathbf{C}^{-1}\right)_{IJKL} = \frac{\left(\mathbf{C}^{-1}\right)_{IK}\left(\mathbf{C}^{-1}\right)_{JL} + \left(\mathbf{C}^{-1}\right)_{IL}\left(\mathbf{C}^{-1}\right)_{JK}}{2}. \quad (1.139)$$

In this model, κ and μ are also interpreted as the material bulk and shear moduli, respectively. One can see this by evaluating the material constitutive tensor $\mathbb{C}$ from Equation (1.137) for the case when the reference and current configurations coincide. In this case, $\mathbf{x} = \mathbf{X}$, $\mathbf{F} = \mathbf{C} = \mathbf{I}$, and the material modulus from Equation (1.137) reduces to the form given by Equation (1.131).

Mooney–Rivlin model. For the compressible Mooney–Rivlin material, the expression for $\mathbf{S}$, in component form, is given as

$$S_{IJ} = 2\left(C_1 + C_2 C_{KK}\right)\delta_{IJ} - 2C_2 C_{IJ} + \left(K_{\text{PEN}} \ln J - 2\left(C_1 + 2C_2\right)\right)C_{IJ}^{-1}, \quad (1.140)$$

where C_1 and C_2 are the Mooney–Rivlin material constants. The near-incompressibility is enforced with the penalty term $K_{\text{PEN}} \ln J$ (see Betsch et al., 1996), where K_{PEN} is a penalty parameter determined based on the expression given in Stuparu (2002) for the bulk modulus:

$$K_{\text{PEN}} = \frac{2\left(C_1 + C_2\right)}{\left(1 - 2\nu_{\text{PEN}}\right)}. \quad (1.141)$$

Here ν_{PEN} (with a value close to 0.50) is the "penalty" Poisson's ratio used in the expression in place of the actual Poisson's ratio.

Fung model. For the Fung material, the expression for **S** is given as

$$S_{IJ} = 2D_1 D_2 (e^{D_2(C_{KK}-3)} \delta_{IJ} - C_{IJ}^{-1}) + K_{\text{PEN}} \ln J\, C_{IJ}^{-1}, \tag{1.142}$$

where D_1 and D_2 are the Fung material constants, and K_{PEN} is defined as

$$K_{\text{PEN}} = \frac{2D_1 D_2}{(1 - 2\nu_{\text{PEN}})}. \tag{1.143}$$

1.2.8 Linearization of the Structural Mechanics Equations: Tangent Stiffness and Equations of Linear Elasticity

In this section we linearize the structural mechanics equations. Linearization gives a tangent stiffness operator that is used in the implementation of the Newton–Raphson method to solve the nonlinear structural equations. Linearization also gives rise to the equations of linear elastodynamics, which are often used in structural modeling.

To arrive at the linearized structural mechanics problem, we "perturb" the structure around its deformed state and only keep the terms that are linear in the displacement perturbation. Namely, we set

$$\delta W(\mathbf{w}, \bar{\mathbf{y}}) + \frac{d}{d\epsilon} \delta W(\mathbf{w}, \bar{\mathbf{y}} + \epsilon \mathbf{y}) \bigg|_{\epsilon=0} = 0, \tag{1.144}$$

where $\bar{\mathbf{y}}$ is the structural displacement that defines its deformed state, $\mathbf{y}$ now plays the role of a small displacement perturbation, and

$$\delta W(\mathbf{w}, \bar{\mathbf{y}}) = \int_{\Omega_0} \mathbf{w} \cdot \rho_0 \bar{\mathbf{a}}\, d\Omega + \int_{\Omega_0} \boldsymbol{\nabla}_X \mathbf{w} : \bar{\mathbf{P}}\, d\Omega - \int_{\Omega_0} \mathbf{w} \cdot \rho_0 \bar{\mathbf{f}}\, d\Omega - \int_{(\Gamma_0)_h} \mathbf{w} \cdot \bar{\mathbf{h}}\, d\Gamma \tag{1.145}$$

are the structural mechanics variational equations evaluated at $\bar{\mathbf{y}}$. The superimposed bar denotes quantities evaluated at the deformed state.

Linearization of the internal virtual-work terms gives

$$\frac{d}{d\epsilon} \delta W_{\text{int}}(\mathbf{w}, \bar{\mathbf{y}} + \epsilon \mathbf{y}) \bigg|_{\epsilon=0} = \delta \int_{\Omega_0} \mathbf{F}^T \boldsymbol{\nabla}_X \mathbf{w} : \mathbf{S}\, d\Omega \tag{1.146}$$

$$= \int_{\Omega_0} \left(\delta \mathbf{F}^T \boldsymbol{\nabla}_X \mathbf{w} : \bar{\mathbf{S}} + \bar{\mathbf{F}}^T \boldsymbol{\nabla}_X \mathbf{w} : \delta \mathbf{S} \right) d\Omega \tag{1.147}$$

$$= \int_{\Omega_0} \left(\boldsymbol{\nabla}_X \mathbf{w} : \boldsymbol{\nabla}_X \mathbf{y} \bar{\mathbf{S}} + \bar{\mathbf{F}}^T \boldsymbol{\nabla}_X \mathbf{w} : \left(\frac{\partial \mathbf{S}}{\partial \mathbf{E}} \right) \frac{1}{2} \left(\bar{\mathbf{F}}^T \boldsymbol{\nabla}_X \mathbf{y} + \boldsymbol{\nabla}_X \mathbf{y}^T \bar{\mathbf{F}} \right) \right) d\Omega. \tag{1.148}$$

Recognizing that

$$\frac{\partial \mathbf{S}}{\partial \mathbf{E}} = \frac{\partial^2 \varphi}{\partial \mathbf{E} \partial \mathbf{E}} = \mathbb{C}, \tag{1.149}$$

and using the minor symmetry of $\mathbb{C}$, Equation (1.148) may be written as

$$\int_{\Omega_0} \left(\bar{\mathbf{F}}^T \nabla_X \mathbf{w} : \bar{\mathbb{C}} \bar{\mathbf{F}}^T \nabla_X \mathbf{y} + \nabla_X \mathbf{w} : \nabla_X \mathbf{y} \bar{\mathbf{S}} \right) d\Omega. \tag{1.150}$$

Using index notation, Equation (1.150) may be expressed as

$$\int_{\Omega_0} w_{i,J} \bar{D}_{iJkL} y_{k,L} \, d\Omega, \tag{1.151}$$

where $\bar{D}_{iJkL}$'s are the components of the *tangent stiffness* tensor given by

$$\bar{D}_{iJkL} = \bar{F}_{iI} \bar{\mathbb{C}}_{IJKL} \bar{F}_{kK} + \delta_{ik} \bar{S}_{JL}. \tag{1.152}$$

The first term on the right-hand-side of Equation (1.152) is the *material stiffness*, while the second term is the *geometric stiffness* contribution to the tangent stiffness tensor.

Linearization of the external virtual-work gives

$$\frac{d}{d\epsilon} \delta W_{\text{ext}}(\mathbf{w}, \bar{\mathbf{y}} + \epsilon \mathbf{y}) \bigg|_{\epsilon=0} = \int_{\Omega_0} \mathbf{w} \cdot \rho_0 \mathbf{a} \, d\Omega - \int_{\Omega_0} \mathbf{w} \cdot \rho_0 \mathbf{f} \, d\Omega - \int_{(\Gamma_0)_h} \mathbf{w} \cdot \hat{\mathbf{h}} \, d\Gamma, \tag{1.153}$$

where $\mathbf{a}$, $\mathbf{f}$, and $\hat{\mathbf{h}}$ are now the increments of acceleration, body force, and surface traction, respectively.

Using Equation (1.144) and the above derivations, we arrive at the variational statement of a complete linearized problem: given the structural displacement state $\bar{\mathbf{y}}$, find the displacement perturbation $\mathbf{y} \in \mathcal{S}_y$, such that $\forall \, \mathbf{w} \in \mathcal{V}_y$:

$$\int_{\Omega_0} \mathbf{w} \cdot \rho_0 \bar{\mathbf{a}} \, d\Omega + \int_{\Omega_0} \nabla_X \mathbf{w} : \bar{\mathbf{P}} \, d\Omega - \int_{\Omega_0} \mathbf{w} \cdot \rho_0 \bar{\mathbf{f}} \, d\Omega - \int_{(\Gamma_0)_h} \mathbf{w} \cdot \bar{\mathbf{h}} \, d\Gamma$$
$$+ \int_{\Omega_0} \mathbf{w} \cdot \rho_0 \mathbf{a} \, d\Omega + \int_{\Omega_0} \left(\bar{\mathbf{F}}^T \nabla_X \mathbf{w} : \bar{\mathbb{C}} \bar{\mathbf{F}}^T \nabla_X \mathbf{y} + \nabla_X \mathbf{w} : \nabla_X \mathbf{y} \bar{\mathbf{S}} \right) d\Omega$$
$$- \int_{\Omega_0} \mathbf{w} \cdot \rho_0 \mathbf{f} \, d\Omega - \int_{(\Gamma_0)_h} \mathbf{w} \cdot \hat{\mathbf{h}} \, d\Gamma = 0. \tag{1.154}$$

If the structural mechanics equations are linearized around an equilibrium configuration (i.e., the displacement state $\bar{\mathbf{y}}$ satisfies the variational equations), then the linearized problem given by Equation (1.154) reduces to: given $\bar{\mathbf{y}}$, find $\mathbf{y} \in \mathcal{S}_y$, such that $\forall \, \mathbf{w} \in \mathcal{V}_y$:

$$\int_{\Omega_0} \mathbf{w} \cdot \rho_0 \mathbf{a} \, d\Omega + \int_{\Omega_0} \left(\bar{\mathbf{F}}^T \nabla_X \mathbf{w} : \bar{\mathbb{C}} \bar{\mathbf{F}}^T \nabla_X \mathbf{y} + \nabla_X \mathbf{w} : \nabla_X \mathbf{y} \bar{\mathbf{S}} \right) d\Omega$$
$$- \int_{\Omega_0} \mathbf{w} \cdot \rho_0 \mathbf{f} \, d\Omega - \int_{(\Gamma_0)_h} \mathbf{w} \cdot \hat{\mathbf{h}} \, d\Gamma = 0. \tag{1.155}$$

An important special case of the formulation given by Equation (1.155) is obtained when the virtual-work equations are linearized around the stress-free undisplaced configuration. In this case, $\bar{\mathbf{y}} = \mathbf{0}$, $\bar{\mathbf{F}} = \mathbf{I}$, $\bar{\mathbf{E}} = \mathbf{0}$, $\bar{\mathbf{S}} = \mathbf{0}$, and we obtain

$$\int_{\Omega} \mathbf{w} \cdot \rho \mathbf{a} \, d\Omega + \int_{\Omega} \boldsymbol{\varepsilon}(\mathbf{w}) : \mathbb{C} \boldsymbol{\varepsilon}(\mathbf{y}) \, d\Omega - \int_{\Omega} \mathbf{w} \cdot \rho \mathbf{f} \, d\Omega - \int_{\Gamma_h} \mathbf{w} \cdot \mathbf{h} \, d\Gamma = 0, \tag{1.156}$$

where $\boldsymbol{\varepsilon}(\mathbf{y})$ is the linear, or infinitesimal, strain, which is a linearization of the Green–Lagrange strain about an undeformed configuration. Infinitesimal strain vanishes for the rigid-body translation, however, it does not vanish for the rigid-body rotation. As a result, the formulation given by Equation (1.156) is not suitable for structural mechanics problems where large deformations are expected (see Figure 1.6). Equation (1.156) represents the well-known variational formulation of linear elastodynamics. The equations of linear elastostatics are obtained by omitting the inertial term in Equation (1.156).

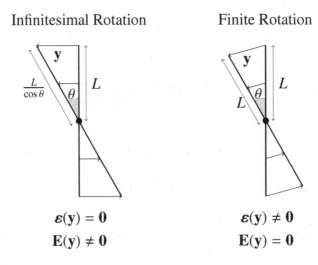

Figure 1.6 Elastic bar undergoing infinitesimal and finite rotations. In the case of infinitesimal rotation, linear-elastic analysis produces zero strain and, as a result, zero stress, which is nonphysical since the bar elongates. In the case of finite rotation, linear-elastic analysis produces a nonzero strain and stress, which is also nonphysical. In general, structural analysis with large displacements requires an objective strain measure to produce physically correct results

REMARK 1.7 *Note that because the reference and current configurations coincide, we no longer distinguish between* $\mathbf{x}$ *and* $\mathbf{X}$ *in Equation (1.156). As a result, we remove the subscript from* ∇_X, *and set* $\Omega = \Omega_0 = \Omega_t$ *and* $\Gamma = \Gamma_0 = \Gamma_t$.

We conclude this section with the linearization of the follower pressure load. For convenience, we employ index notation. We begin with the expression for the follower-pressure-load boundary condition in the reference configuration and first compute

$$\delta\left[JF_{Ii}^{-1}\right] = \delta J F_{Ii}^{-1} + J\delta F_{Ii}^{-1} = JF_{Jj}^{-1}y_{j,J}F_{Ii}^{-1} - JF_{Ij}^{-1}y_{j,J}F_{Ji}^{-1}. \tag{1.157}$$

Introducing the above variation into Equation (1.118), we obtain

$$-\int_{(\Gamma_0)_h} w_i p \delta\left[JF_{Ii}\right]\hat{n}_I \, d\Gamma = -\int_{(\Gamma_0)_h} w_i p J \left(F_{Jj}^{-1}F_{Ii}^{-1} - F_{Ij}^{-1}F_{Ji}^{-1}\right)\hat{n}_I y_{j,J} \, d\Gamma. \tag{1.158}$$

Changing variables in the above expression to the current configuration, we obtain

$$-\int_{(\Gamma_t)_h} \left(w_i p n_i y_{j,j} - w_i p n_j y_{j,i}\right) d\Gamma, \tag{1.159}$$

which is somewhat simpler than Equation (1.158). The above linearization is employed in the implementation of consistent tangent stiffness matrices in nonlinear structural analysis. An alternative linearization of the follower pressure load, which uses the parametric coordinates of the boundary surface, may be found in Wriggers (2008).

1.2.9 Thin Structures: Shell, Membrane, and Cable Models

1.2.9.1 Kirchhoff–Love Shell Model

In this section we follow the developments of Kiendl et al. (2009, 2010) and Bazilevs et al. (2011c) that present the governing equations of the Kirchhoff–Love shell theory. The theory is appropriate for thin-shell structures and, when discretized using smooth basis functions, requires no rotational degrees of freedom.

In the case of shells, the 3D continuum description is reduced to that of the shell midsurface, and the transverse normal stress is neglected. Furthermore, the Kirchhoff–Love theory assumes that the shell director remains normal to its middle surface during the deformation, which implies that the transverse shear strains are zero. As a result, only in-plane stress and strain tensors are considered, and the indices $\alpha = 1, 2$ and $\beta = 1, 2$ are employed to denote their components. We denote by Γ_0^s and Γ_t^s the shell midsurface in the reference and deformed configurations, respectively. Furthermore, h_{th} is the (variable) shell thickness, and $\xi_3 \in [-h_{\text{th}}/2, h_{\text{th}}/2]$ is the through-thickness coordinate.

We introduce the following standard shell kinematic quantities and relationships (see Bischoff et al., 2004; Kiendl et al., 2009 for more details):

$$E_{\alpha\beta} = \varepsilon_{\alpha\beta} + \xi_3 \kappa_{\alpha\beta}, \tag{1.160}$$

$$\varepsilon_{\alpha\beta} = \frac{1}{2}\left(\mathbf{g}_\alpha \cdot \mathbf{g}_\beta - \mathbf{G}_\alpha \cdot \mathbf{G}_\beta\right), \tag{1.161}$$

$$\kappa_{\alpha\beta} = -\frac{\partial \mathbf{g}_\alpha}{\partial \xi_\beta} \cdot \mathbf{g}_3 - \left(-\frac{\partial \mathbf{G}_\alpha}{\partial \xi_\beta} \cdot \mathbf{G}_3\right), \tag{1.162}$$

$$\mathbf{g}_\alpha = \frac{\partial \mathbf{x}}{\partial \xi_\alpha}, \tag{1.163}$$

$$\mathbf{G}_\alpha = \frac{\partial \mathbf{X}}{\partial \xi_\alpha}, \tag{1.164}$$

$$\mathbf{g}_3 = \frac{\mathbf{g}_1 \times \mathbf{g}_2}{\|\mathbf{g}_1 \times \mathbf{g}_2\|}, \tag{1.165}$$

$$\mathbf{G}_3 = \frac{\mathbf{G}_1 \times \mathbf{G}_2}{\|\mathbf{G}_1 \times \mathbf{G}_2\|}, \tag{1.166}$$

$$\mathbf{G}^\alpha = (\mathbf{G}_\alpha \cdot \mathbf{G}_\beta)^{-1} \mathbf{G}_\beta. \tag{1.167}$$

Here, $E_{\alpha\beta}$, $\varepsilon_{\alpha\beta}$, and $\kappa_{\alpha\beta}$ are the contravariant components of the in-plane Green–Lagrange strain, membrane strain, and curvature tensors, respectively. The spatial coordinates of the *shell midsurface* in the current and reference configurations are $\mathbf{x} = \mathbf{x}(\xi_1, \xi_2)$ and $\mathbf{X} = \mathbf{X}(\xi_1, \xi_2)$, parameterized by ξ_1 and ξ_2. The covariant surface basis vectors in the current and reference configurations are $\mathbf{g}_\alpha$ and $\mathbf{G}_\alpha$. The unit outward normal vectors to the shell midsurface in the current and reference configurations are $\mathbf{g}_3$ and $\mathbf{G}_3$. The contravariant surface basis vectors in the reference configuration are denoted by $\mathbf{G}^\alpha$ (see Figure 1.7).

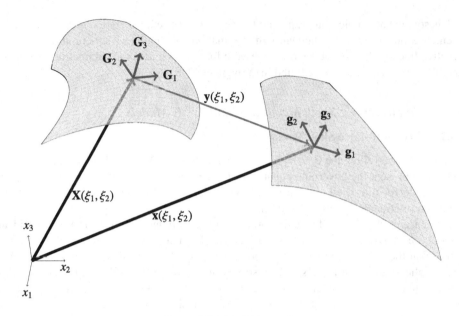

Figure 1.7 Shell kinematics

We select the *local Cartesian basis* vectors as follows:

$$\bar{\mathbf{e}}_1 = \frac{\mathbf{G}_1}{\|\mathbf{G}_1\|}, \tag{1.168}$$

$$\bar{\mathbf{e}}_2 = \frac{\mathbf{G}_2 - (\mathbf{G}_2 \cdot \bar{\mathbf{e}}_1)\bar{\mathbf{e}}_1}{\|\mathbf{G}_2 - (\mathbf{G}_2 \cdot \bar{\mathbf{e}}_1)\bar{\mathbf{e}}_1\|}, \tag{1.169}$$

that is, the first local basis vector is the normalized first covariant basis vector in the reference configuration. The local Cartesian basis vectors $\bar{\mathbf{e}}_\alpha$ are used in expressing a constitutive relationship for the shell. Because the local basis is orthonormal, we make no distinction between covariant and contravariant quantities, which are expressed with respect to it.

With the above definitions, we calculate the components of the Green–Lagrange strain tensor and its variation in the local coordinate system as

$$\bar{E}_{\alpha\beta} = \bar{\varepsilon}_{\alpha\beta} + \xi_3 \bar{\kappa}_{\alpha\beta}, \tag{1.170}$$

$$\delta\bar{E}_{\alpha\beta} = \delta\bar{\varepsilon}_{\alpha\beta} + \xi_3 \delta\bar{\kappa}_{\alpha\beta}, \tag{1.171}$$

$$\bar{\varepsilon}_{\alpha\beta} = \varepsilon_{\gamma\delta}(\mathbf{G}^\gamma \cdot \bar{\mathbf{e}}_\alpha)(\mathbf{G}^\delta \cdot \bar{\mathbf{e}}_\beta), \tag{1.172}$$

$$\bar{\kappa}_{\alpha\beta} = \kappa_{\gamma\delta}(\mathbf{G}^\gamma \cdot \bar{\mathbf{e}}_\alpha)(\mathbf{G}^\delta \cdot \bar{\mathbf{e}}_\beta), \tag{1.173}$$

$$\delta\bar{\varepsilon}_{\alpha\beta} = \delta\varepsilon_{\gamma\delta}(\mathbf{G}^\gamma \cdot \bar{\mathbf{e}}_\alpha)(\mathbf{G}^\delta \cdot \bar{\mathbf{e}}_\beta), \tag{1.174}$$

$$\delta\bar{\kappa}_{\alpha\beta} = \delta\kappa_{\gamma\delta}(\mathbf{G}^\gamma \cdot \bar{\mathbf{e}}_\alpha)(\mathbf{G}^\delta \cdot \bar{\mathbf{e}}_\beta). \tag{1.175}$$

The variations $\delta\varepsilon_{\gamma\delta}$ and $\delta\kappa_{\gamma\delta}$ may be computed directly by taking the variational derivatives of the expressions given by Equations (1.161) and (1.162) with respect to the displacement vector.

We define the vectors of membrane strain and curvature components in the local coordinate system as

$$\bar{\boldsymbol{\varepsilon}} = \begin{bmatrix} \bar{\varepsilon}_{11} \\ \bar{\varepsilon}_{22} \\ \bar{\varepsilon}_{12} \end{bmatrix} \quad (1.176)$$

and

$$\bar{\boldsymbol{\kappa}} = \begin{bmatrix} \bar{\kappa}_{11} \\ \bar{\kappa}_{22} \\ \bar{\kappa}_{12} \end{bmatrix}, \quad (1.177)$$

together with a Green–Lagrange strain vector

$$\bar{\mathbf{E}} = \bar{\boldsymbol{\varepsilon}} + \xi_3 \bar{\boldsymbol{\kappa}}. \quad (1.178)$$

We assume St. Venant–Kirchhoff material law and write the following stress–strain relationship in the local coordinate system:

$$\bar{\mathbf{S}} = \bar{\mathbb{C}} \, \bar{\mathbf{E}}, \quad (1.179)$$

where $\bar{\mathbf{S}}$ is a vector of components of the second Piola–Kirchhoff stress tensor in the local coordinate system, and $\bar{\mathbb{C}}$ is a constitutive material matrix, which is symmetric. Introducing Equations (1.178) and (1.179) into the expression for the internal virtual work given by Equation (1.79), we obtain

$$\delta W_{\text{int}} = -\int_{\Omega_0} \delta \bar{\mathbf{E}} \cdot \bar{\mathbf{S}} \, d\Omega \quad (1.180)$$

$$= -\int_{\Gamma_0^s} \left(\int_{h_{\text{th}}} \delta \bar{\mathbf{E}} \cdot \bar{\mathbb{C}} \, \bar{\mathbf{E}} \, d\xi_3 \right) d\Gamma \quad (1.181)$$

$$= -\int_{\Gamma_0^s} \delta \bar{\boldsymbol{\varepsilon}} \cdot \left(\left(\int_{h_{\text{th}}} \bar{\mathbb{C}} \, d\xi_3 \right) \bar{\boldsymbol{\varepsilon}} + \left(\int_{h_{\text{th}}} \xi_3 \bar{\mathbb{C}} \, d\xi_3 \right) \bar{\boldsymbol{\kappa}} \right) d\Gamma$$

$$- \int_{\Gamma_0^s} \delta \bar{\boldsymbol{\kappa}} \cdot \left(\left(\int_{h_{\text{th}}} \xi_3 \bar{\mathbb{C}} \, d\xi_3 \right) \bar{\boldsymbol{\varepsilon}} + \left(\int_{h_{\text{th}}} \xi_3^2 \bar{\mathbb{C}} \, d\xi_3 \right) \bar{\boldsymbol{\kappa}} \right) d\Gamma. \quad (1.182)$$

For a general orthotropic material,

$$\bar{\mathbb{C}}_{\text{ort}} = \begin{bmatrix} \dfrac{E_1}{(1-\nu_{12}\nu_{21})} & \dfrac{\nu_{21}E_1}{(1-\nu_{12}\nu_{21})} & 0 \\ \dfrac{\nu_{12}E_2}{(1-\nu_{12}\nu_{21})} & \dfrac{E_2}{(1-\nu_{12}\nu_{21})} & 0 \\ 0 & 0 & G_{12} \end{bmatrix}. \quad (1.183)$$

In Equation (1.183), E_1 and E_2 are the Young's moduli in the directions defined by the local basis vectors, ν_{12} and ν_{21} are the Poisson's ratios, G_{12} is the shear modulus, and $\nu_{21}E_1 = \nu_{12}E_2$

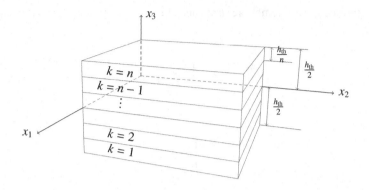

Figure 1.8 Schematic of a composite laminate

to ensure the symmetry of the constitutive material matrix $\overline{\mathbb{C}}_{\text{ort}}$. In the case of an isotropic material, $E_1 = E_2 = E$, $\nu_{21} = \nu_{12} = \nu$, and $G_{12} = E/(2(1 + \nu))$.

In the case of composite materials, we assume that the structure is composed of a set of plies, each modeled as an orthotropic material. We use the classical laminated-plate theory Reddy, 2004, and homogenize the material through-thickness constitutive behavior for a given composite ply layout. Let k denotes the kth ply (or lamina) and let n be the total number of plies (see Figure 1.8). We assume each ply has the same thickness h_{th}/n. Pre-integrating through the shell thickness in Equation (1.182), the extensional stiffness $\mathbf{K}_{\text{exte}}$, coupling stiffness $\mathbf{K}_{\text{coup}}$, and bending stiffness $\mathbf{K}_{\text{bend}}$ are given by

$$\mathbf{K}_{\text{exte}} = \int_{h_{\text{th}}} \overline{\mathbb{C}} \, d\xi_3 = \frac{h_{\text{th}}}{n} \sum_{k=1}^{n} \overline{\mathbb{C}}_k, \tag{1.184}$$

$$\mathbf{K}_{\text{coup}} = \int_{h_{\text{th}}} \xi_3 \overline{\mathbb{C}} \, d\xi_3 = \frac{h_{\text{th}}^2}{n^2} \sum_{k=1}^{n} \overline{\mathbb{C}}_k \left(k - \frac{n}{2} - \frac{1}{2} \right), \tag{1.185}$$

$$\mathbf{K}_{\text{bend}} = \int_{h_{\text{th}}} \xi_3^2 \overline{\mathbb{C}} \, d\xi_3 = \frac{h_{\text{th}}^3}{n^3} \sum_{k=1}^{n} \overline{\mathbb{C}}_k \left(\left(k - \frac{n}{2} - \frac{1}{2} \right)^2 + \frac{1}{12} \right), \tag{1.186}$$

where

$$\overline{\mathbb{C}}_k = \mathbf{T}^T(\phi_k) \overline{\mathbb{C}}_{\text{ort}} \mathbf{T}(\phi_k), \tag{1.187}$$

$$\mathbf{T}(\phi) = \begin{bmatrix} \cos^2 \phi & \sin^2 \phi & \sin \phi \cos \phi \\ \sin^2 \phi & \cos^2 \phi & -\sin \phi \cos \phi \\ -2 \sin \phi \cos \phi & 2 \sin \phi \cos \phi & \cos^2 \phi - \sin^2 \phi \end{bmatrix}. \tag{1.188}$$

In the above equations, ϕ is the fiber orientation angle in each ply, Equation (1.187) transforms $\overline{\mathbb{C}}_{\text{ort}}$ from the principal material coordinates to the laminate coordinates (defined by the local Cartesian basis) for each ply, and $\overline{\mathbb{C}}_k$ is constant within each ply. Note that, setting $n = 1$ and $\overline{\mathbb{C}}_k = \overline{\mathbb{C}}_{\text{ort}}$ in Equations (1.184)–(1.186), we get $\mathbf{K}_{\text{coup}} = \mathbf{0}$ and

$$\mathbf{K}_{\text{exte}} = h_{\text{th}} \overline{\mathbb{C}}_{\text{ort}}, \tag{1.189}$$

Governing Equations of Fluid and Structural Mechanics

$$\mathbf{K}_{\text{bend}} = \frac{h_{\text{th}}^3}{12} \overline{\mathbf{C}}_{\text{ort}}, \qquad (1.190)$$

which are the classical membrane and bending stiffnesses for an orthotropic shell.

With the above definitions, the expression for the internal virtual work for a composite shell may now be compactly written as

$$\delta W_{\text{int}} = -\int_{\Gamma_0^s} \delta \overline{\boldsymbol{\varepsilon}} \cdot \left(\mathbf{K}_{\text{exte}} \overline{\boldsymbol{\varepsilon}} + \mathbf{K}_{\text{coup}} \overline{\boldsymbol{\kappa}}\right) d\Gamma - \int_{\Gamma_0^s} \delta \overline{\boldsymbol{\kappa}} \cdot \left(\mathbf{K}_{\text{coup}} \overline{\boldsymbol{\varepsilon}} + \mathbf{K}_{\text{bend}} \overline{\boldsymbol{\kappa}}\right) d\Gamma. \qquad (1.191)$$

The complete variational formulation of the Kirchhoff–Love shell is given by: find the displacement of the shell midsurface $\mathbf{y} \in \mathcal{S}_y$, such that $\forall \, \mathbf{w} \in \mathcal{V}_y$:

$$\int_{\Gamma_0^s} \mathbf{w} \cdot h_{\text{th}} \overline{\rho}_0 (\mathbf{a} - \mathbf{f}) \, d\Gamma$$

$$+ \int_{\Gamma_0^s} \delta \overline{\boldsymbol{\varepsilon}} \cdot \left(\mathbf{K}_{\text{exte}} \overline{\boldsymbol{\varepsilon}} + \mathbf{K}_{\text{coup}} \overline{\boldsymbol{\kappa}}\right) d\Gamma$$

$$+ \int_{\Gamma_0^s} \delta \overline{\boldsymbol{\kappa}} \cdot \left(\mathbf{K}_{\text{coup}} \overline{\boldsymbol{\varepsilon}} + \mathbf{K}_{\text{bend}} \overline{\boldsymbol{\kappa}}\right) d\Gamma - \int_{(\Gamma_t^s)_h} \mathbf{w} \cdot \mathbf{h} \, d\Gamma = 0, \qquad (1.192)$$

where $(\Gamma_t^s)_h$ is the shell subdomain with a prescribed traction boundary condition, and $\overline{\rho}_0$ is the through-thickness-averaged shell density given by

$$\overline{\rho}_0 = \frac{1}{h_{\text{th}}} \int_{h_{\text{th}}} \rho_0 \, d\xi_3. \qquad (1.193)$$

Note that, for simplicity of the exposition, in Equation (1.192), we omitted the terms corresponding to the prescribed traction on the edges of the shell. Although not presented here, such terms are implemented in our structural analysis programs.

1.2.9.2 Membrane Model

The membrane formulation is obtained by neglecting the curvature tensor in the definition of the in-plane Green–Lagrange strain (see Equation (1.160)) in the equations of the Kirchhoff–Love shell. This results in a simplified structural model, in which the bending effects are neglected. The variational formulation for membrane structures may be stated as follows: find $\mathbf{y} \in \mathcal{S}_y$, such that $\forall \, \mathbf{w} \in \mathcal{V}_y$:

$$\int_{\Gamma_0^s} \mathbf{w} \cdot h_{\text{th}} \overline{\rho}_0 (\mathbf{a} - \mathbf{f}) \, d\Gamma + \int_{\Gamma_0^s} \delta \overline{\boldsymbol{\varepsilon}} \cdot \mathbf{K}_{\text{exte}} \overline{\boldsymbol{\varepsilon}} \, d\Gamma - \int_{(\Gamma_t^s)_h} \mathbf{w} \cdot \mathbf{h} \, d\Gamma = 0. \qquad (1.194)$$

Since only the first derivatives with respect to the parametric domain coordinates are employed in this variational formulation, C^0-continuous basis functions may be used to discretize the membrane equations.

The membrane formulation, in the case of an isotropic material, may be written without using the local coordinate system. In this case, the variational formulation of the membrane model becomes: find $\mathbf{y} \in \mathcal{S}_y$, such that $\forall \, \mathbf{w} \in \mathcal{V}_y$:

$$\int_{\Gamma_0^s} \mathbf{w} \cdot h_{\text{th}} \overline{\rho}_0 (\mathbf{a} - \mathbf{f}) \, d\Gamma + \int_{\Gamma_0^s} \delta \varepsilon_{\alpha\beta} h_{\text{th}} S^{\alpha\beta} \, d\Gamma - \int_{(\Gamma_t^s)_h} \mathbf{w} \cdot \mathbf{h} \, d\Gamma = 0, \qquad (1.195)$$

where

$$S^{\alpha\beta} = \left(\bar{\lambda}G^{\alpha\beta}G^{\gamma\delta} + \mu\left(G^{\alpha\gamma}G^{\beta\delta} + G^{\alpha\delta}G^{\beta\gamma}\right)\right)\varepsilon_{\gamma\delta}, \quad (1.196)$$

$\varepsilon_{\gamma\delta}$ are the in-plane components of the Green–Lagrange strain tensor given by Equation (1.161), $\bar{\lambda} = 2\lambda\mu/(\lambda + 2\mu)$, and $G^{\alpha\beta}$ are the contravariant metric tensor components in the undeformed configuration. Just as in the case of the shell model, for simplicity of the exposition, we omitted the edge-traction terms from Equations (1.194) and (1.195).

1.2.9.3 Cable Model

For cables, under the assumption of uniaxial tension, the parametric domain reduces to a line (see Figure 1.9), the indices $\alpha = \beta = 1$, and the bending effects are also neglected in the developments in Section 1.2.9.1. Furthermore, the Poisson's effect is also neglected, which leaves the Young's modulus E_c as the only material parameter in the cable constitutive model. The variational formulation for cable structures may be stated as follows: find $\mathbf{y} \in \mathcal{S}_y$, such that $\forall \, \mathbf{w} \in \mathcal{V}_y$:

$$\int_{S_0} \mathbf{w} \cdot A_c\bar{\rho}_0 (\mathbf{a} - \mathbf{f}) \, dS + \int_{S_0} \delta\bar{\varepsilon}_{11} A_c E_c \bar{\varepsilon}_{11} \, dS - \int_{(S_t)_h} \mathbf{w} \cdot \mathbf{h} \, dS = 0, \quad (1.197)$$

where S_0 and S_t are the curves that define the cable axis in the reference and deformed configuration, respectively, $(S_t)_h$ is the part of S_t with a prescribed traction boundary condition, and, in this case, $\mathbf{h}$ has the dimensions of force per unit length.

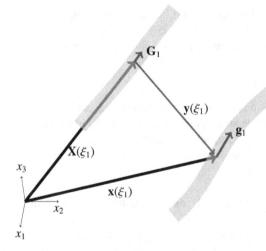

Figure 1.9 Cable kinematics

The cable formulation may be stated without using the local coordinate system as follows: find $\mathbf{y} \in \mathcal{S}_y$, such that $\forall \, \mathbf{w} \in \mathcal{V}_y$:

$$\int_{S_0} \mathbf{w} \cdot A_c\bar{\rho}_0 (\mathbf{a} - \mathbf{f}) \, dS + \int_{S_0} \delta\varepsilon_{11} A_c E_c G^{11} G^{11} \varepsilon_{11} \, dS - \int_{(S_t)_h} \mathbf{w} \cdot \mathbf{h} \, dS = 0. \quad (1.198)$$

Governing Equations of Fluid and Structural Mechanics

Although the terms representing the traction boundary conditions at the end points are not explicitly included in Equation (1.198), they are implemented in our analysis programs.

1.3 Governing Equations of Fluid Mechanics in Moving Domains

In this section we revisit the fluid mechanics governing equations and recast them in the ALE framework. The ALE form of the fluid mechanics equations is often used to simulate flows on moving domains, including FSI.

1.3.1 Kinematics of ALE and Space–Time Descriptions

The ALE description also makes use of a reference domain. However, the major difference from the Lagrangian approach, which is typically adopted for structural mechanics, is that the motion of the fluid mechanics problem reference domain does not follow the motion of the fluid itself. For this reason, we denote this reference domain by $\hat{\Omega} \in \mathbb{R}^{n_{sd}}$ and the coordinates in this reference domain by $\hat{\mathbf{x}}$. See Figure 1.10 for an illustration. The fluid spatial domain Ω_t is given by

$$\Omega_t = \left\{ \mathbf{x} \mid \mathbf{x} = \boldsymbol{\phi}(\hat{\mathbf{x}}, t) \ \forall \hat{\mathbf{x}} \in \hat{\Omega}, t \in (0, T) \right\}. \tag{1.199}$$

The mapping given by Equation (1.199) takes the form

$$\boldsymbol{\phi}(\hat{\mathbf{x}}, t) = \hat{\mathbf{x}} + \hat{\mathbf{y}}(\hat{\mathbf{x}}, t), \tag{1.200}$$

where $\hat{\mathbf{y}}$ is the time-dependent displacement of the reference fluid domain. With this definition of the ALE map, the fluid domain velocity is given by

$$\hat{\mathbf{u}} = \left. \frac{\partial \hat{\mathbf{y}}}{\partial t} \right|_{\hat{\mathbf{x}}}, \tag{1.201}$$

where $\left.\right|_{\hat{\mathbf{x}}}$ denotes the time derivative taken holding $\hat{\mathbf{x}}$ fixed, the deformation gradient is defined as

$$\hat{\mathbf{F}} = \frac{\partial \mathbf{x}}{\partial \hat{\mathbf{x}}} = \mathbf{I} + \frac{\partial \hat{\mathbf{y}}}{\partial \hat{\mathbf{x}}}, \tag{1.202}$$

and $\hat{J} = \det \hat{\mathbf{F}}$ is the Jacobian of the deformation gradient.

We recall the Piola transformation, a classical result in continuum mechanics (see, e.g., Wriggers, 2008). Given an arbitrary vector field $\boldsymbol{\gamma}$ defined on the spatial domain, we define a vector field on the reference domain as

$$\hat{\boldsymbol{\gamma}} = \hat{J} \hat{\mathbf{F}}^{-1} \boldsymbol{\gamma}. \tag{1.203}$$

In this case, the following equality holds:

$$\int_{\Omega_t} \nabla \cdot \boldsymbol{\gamma} \, d\Omega = \int_{\hat{\Omega}} \nabla_{\hat{\mathbf{x}}} \cdot \hat{\boldsymbol{\gamma}} \, d\hat{\Omega}. \tag{1.204}$$

Relationship (1.203) is the Piola transformation, which preserves the divergence, or conservation, structure of the vector field in the reference configuration (see Equation (1.204)). The Piola transformation also applies to tensor-valued quantities.

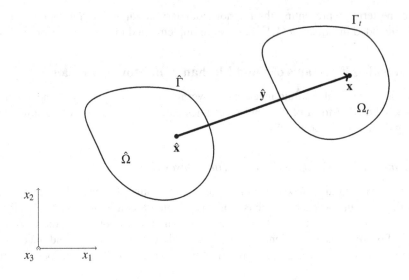

Figure 1.10 Reference and fluid spatial domains

To derive the ALE equations of fluid mechanics, we introduce the notion of a space–time domain, which we will also use in the later sections when discussing the space–time method for moving-domain problems. We begin with the fluid domain reference configuration $\hat{\Omega}$ and define its space–time counterpart $\hat{Q}$ by extruding $\hat{\Omega}$ along the time axis as

$$\hat{Q} = \hat{\Omega} \times (0, T) = \left\{ (\hat{\mathbf{x}}, t) \mid \forall \hat{\mathbf{x}} \in \hat{\Omega}, t \in (0, T) \right\}. \tag{1.205}$$

The space–time domain in the current configuration Q_t is defined by

$$Q_t = \left\{ (\mathbf{x}, t) \,\middle|\, \begin{Bmatrix} t \\ \mathbf{x} \end{Bmatrix} = \begin{Bmatrix} t \\ \boldsymbol{\phi}(\hat{\mathbf{x}}, t) \end{Bmatrix} \, \forall (\hat{\mathbf{x}}, t) \in \hat{Q} \right\}. \tag{1.206}$$

Note that time is synchronized in both configurations. The deformation gradient from $\hat{Q}$ to Q_t is given directly by

$$\begin{Bmatrix} \frac{\partial t}{\partial t} & \frac{\partial t}{\partial \hat{\mathbf{x}}} \\ \frac{\partial \mathbf{x}}{\partial t} & \frac{\partial \mathbf{x}}{\partial \hat{\mathbf{x}}} \end{Bmatrix} = \begin{Bmatrix} 1 & \mathbf{0}^T \\ \hat{\mathbf{u}} & \hat{\mathbf{F}} \end{Bmatrix}, \tag{1.207}$$

and its determinant is coincident with that of the spatial mapping

$$\det \begin{Bmatrix} 1 & \mathbf{0}^T \\ \hat{\mathbf{u}} & \hat{\mathbf{F}} \end{Bmatrix} = \hat{J}. \tag{1.208}$$

Just as in the spatial case (see Equation (1.203)), there is an analog of the Piola transformation for space–time domains (see Bazilevs *et al.*, 2008). Given a vector field, $(\gamma_0, \boldsymbol{\gamma})^T : Q_t \to \mathbb{R}^{n_{sd}+1}$, we define a vector field, $(\hat{\gamma}_0, \hat{\boldsymbol{\gamma}})^T : \hat{Q} \to \mathbb{R}^{n_{sd}+1}$ as

$$\begin{Bmatrix} \hat{\gamma}_0 \\ \hat{\boldsymbol{\gamma}} \end{Bmatrix} = \hat{J} \begin{Bmatrix} 1 & \mathbf{0}^T \\ \hat{\mathbf{u}} & \hat{\mathbf{F}} \end{Bmatrix}^{-1} \begin{Bmatrix} \gamma_0 \\ \boldsymbol{\gamma} \end{Bmatrix} = \hat{J} \begin{Bmatrix} 1 & \mathbf{0}^T \\ -\hat{\mathbf{F}}^{-1}\hat{\mathbf{u}} & \hat{\mathbf{F}}^{-1} \end{Bmatrix} \begin{Bmatrix} \gamma_0 \\ \boldsymbol{\gamma} \end{Bmatrix} = \begin{Bmatrix} \hat{J}\gamma_0 \\ \hat{J}\hat{\mathbf{F}}^{-1}(\boldsymbol{\gamma} - \gamma_0 \hat{\mathbf{u}}) \end{Bmatrix}. \tag{1.209}$$

Governing Equations of Fluid and Structural Mechanics

In this case, the following space–time integral relationship holds:

$$\int_{Q_t} \left(\frac{\partial \gamma_0}{\partial t} + \nabla \cdot \boldsymbol{\gamma} \right) dQ = \int_{\hat{Q}} \left(\left. \frac{\partial \hat{\gamma}_0}{\partial t} \right|_{\hat{x}} + \nabla_{\hat{x}} \cdot \hat{\boldsymbol{\gamma}} \right) d\hat{Q}, \tag{1.210}$$

which shows that the space–time Piola transformation preserves the conservation structure of a vector field in space–time.

Using index notation, the space–time Piola transformation given by Equation (1.209) becomes

$$\begin{Bmatrix} \hat{\gamma}_0 \\ \hat{\gamma}_I \end{Bmatrix} = \begin{Bmatrix} \hat{J} \gamma_0 \\ \hat{J} \hat{F}_{Ii}^{-1} (\gamma_i - \gamma_0 \hat{u}_i) \end{Bmatrix}, \tag{1.211}$$

where the upper- and lower-case indices refer to the referential and spatial quantities, respectively, and the integral equality (1.210) may be written as

$$\int_{Q_t} (\gamma_{0,t} + \gamma_{i,i}) \, dQ = \int_{\hat{Q}} \left(\left. \hat{\gamma}_{0,t} \right|_{\hat{x}} + \hat{\gamma}_{I,I} \right) d\hat{Q}. \tag{1.212}$$

1.3.2 ALE Formulation of Fluid Mechanics

We begin with the conservative form of the linear-momentum equation written on the spatial domain Ω_t, $t \in (0, T)$ (see Equation (1.1)):

$$\frac{\partial (\rho \mathbf{u})}{\partial t} + \nabla \cdot (\rho \mathbf{u} \otimes \mathbf{u} - \boldsymbol{\sigma}) - \rho \mathbf{f} = \mathbf{0}. \tag{1.213}$$

This equation is a suitable starting point for space–time finite element discretizations, which approximate both the space and time behavior using basis functions. However, if one wants to use a more standard semi-discrete approach, in which the space part is handled with finite elements and the time part is handled with a finite-difference-like time integration method, Equation (1.213) is not a convenient starting point. To arrive at a form of the differential equations that is suitable for a semi-discrete approach, we first integrate Equation (1.213) over Q_t:

$$\int_{Q_t} \left(\frac{\partial (\rho \mathbf{u})}{\partial t} + \nabla \cdot (\rho \mathbf{u} \otimes \mathbf{u} - \boldsymbol{\sigma}) - \rho \mathbf{f} \right) dQ = \mathbf{0}, \tag{1.214}$$

and then rewrite the result using index notation as

$$\int_{Q_t} \left((\rho u_i)_{,t} + (\rho u_i u_j - \sigma_{ij})_{,j} - \rho f_i \right) dQ = 0. \tag{1.215}$$

Changing variables $Q_t \to \hat{Q}$ in Equation (1.215) and applying the space–time Piola transformation given by Equation (1.209) (with $\gamma_0 = \rho u_i$ and $\gamma_j = \rho u_i u_j - \sigma_{ij}$ for each component i), we obtain

$$\int_{\hat{Q}} \left(\left. (\hat{J} \rho u_i)_{,t} \right|_{\hat{x}} + \left(\hat{J} (\rho u_i (u_j - \hat{u}_j) - \sigma_{ij}) \hat{F}_{Jj}^{-1} \right)_{,J} - \hat{J} \rho f_i \right) d\hat{Q} = 0. \tag{1.216}$$

Because $\int_{\hat{Q}} = \int_0^T \int_{\hat{\Omega}}$, and $\hat{\mathbf{x}}$ and t are independent variables, we can interchange the order of space and time integrations, namely $\int_{\hat{Q}} = \int_{\hat{\Omega}} \int_0^T$. We note that this cannot be done for $\int_{Q_t}$, because $\mathbf{x}$ and t are not independent. Furthermore, because the time interval is arbitrary, we can "localize" Equation (1.216) in time to obtain

$$\int_{\hat{\Omega}} \left(\left. (\hat{J}\rho u_i) \right|_{,t|\hat{x}} + \left(\hat{J} \left(\rho u_i (u_j - \hat{u}_j) - \sigma_{ij} \right) \hat{F}_{Jj}^{-1} \right)_{,J} - \hat{J}\rho f_i \right) d\hat{\Omega} = 0. \quad (1.217)$$

Changing variables $\hat{\Omega} \to \Omega_t$ in Equation (1.217) and applying the spatial Piola transformation given by Equation (1.203) gives

$$\int_{\Omega_t} \left(\frac{1}{\hat{J}} \left. (\hat{J}\rho u_i) \right|_{,t|\hat{x}} + \left(\rho u_i(u_j - \hat{u}_j) - \sigma_{ij} \right)_{,j} - \rho f_i \right) d\Omega = 0. \quad (1.218)$$

Localizing Equation (1.218) in space gives a point-wise balance of linear momentum:

$$\frac{1}{\hat{J}} \left. (\hat{J}\rho u_i) \right|_{,t|\hat{x}} + \left(\rho u_i(u_j - \hat{u}_j) - \sigma_{ij} \right)_{,j} - \rho f_i = 0, \quad (1.219)$$

which may be rewritten using vector notation as

$$\frac{1}{\hat{J}} \left. \frac{\partial \hat{J}\rho \mathbf{u}}{\partial t} \right|_{\hat{x}} + \nabla \cdot (\rho \mathbf{u} \otimes (\mathbf{u} - \hat{\mathbf{u}}) - \boldsymbol{\sigma}) - \rho \mathbf{f} = \mathbf{0}. \quad (1.220)$$

Equation (1.220) above is the *conservative form* of the linear-momentum balance equation of fluid mechanics in the ALE description. This form is often taken as the starting point of the numerical formulations of fluid mechanics on moving domains (see, e.g., Le Tallec and Mouro, 2001).

A so-called *convective form* of the ALE equations may be obtained from the conservative form as follows. We first differentiate through the time derivative and convective terms in Equation (1.220) to obtain

$$\rho \left(\frac{1}{\hat{J}} \left. \frac{\partial \hat{J}}{\partial t} \right|_{\hat{x}} \mathbf{u} + \left. \frac{\partial \mathbf{u}}{\partial t} \right|_{\hat{x}} \right) + \rho \left(\mathbf{u} \nabla \cdot (\mathbf{u} - \hat{\mathbf{u}}) + (\mathbf{u} - \hat{\mathbf{u}}) \cdot \nabla \mathbf{u} \right) - \nabla \cdot \boldsymbol{\sigma} - \rho \mathbf{f} = \mathbf{0}, \quad (1.221)$$

where we also assume that the density ρ is constant. Using a well-known identity in continuum mechanics (see, e.g., Wriggers, 2008)

$$\left. \frac{\partial \hat{J}}{\partial t} \right|_{\hat{x}} = \hat{J} \nabla \cdot \hat{\mathbf{u}} \quad (1.222)$$

and the incompressibility constraint $\nabla \cdot \mathbf{u} = 0$ in Equation (1.221), we obtain

$$\rho \left(\left. \frac{\partial \mathbf{u}}{\partial t} \right|_{\hat{x}} + (\mathbf{u} - \hat{\mathbf{u}}) \cdot \nabla \mathbf{u} - \mathbf{f} \right) - \nabla \cdot \boldsymbol{\sigma} = \mathbf{0}. \quad (1.223)$$

This is the convective form of the linear-momentum balance equation of incompressible flows in the ALE description. Note that Equation (1.223) is a substantially simplified version of

Equation (1.220). While in the fully continuous setting the conservative and convective forms of the fluid mechanics equations are equivalent, this is not always the case in the discrete setting. See the remarks below for further elaboration.

REMARK 1.8 *The discrete geometric conservation law is satisfied if the numerical formulation preserves a constant fluid velocity in space and time when there are no body forces and the stress tensor is self-equilibrating (see, e.g., Farhat et al., 2001 and references therein). If the constant fluid velocity is assumed, it is easily seen that Equation (1.223), corresponding to the convective form of the linear-momentum equations, is identically satisfied. Furthermore, assuming that the time integration method "respects" a constant solution, that is, if the velocity field is constant in time the discrete approximation to the time derivative is zero, then the formulation satisfies the geometric conservation law at the fully discrete level. Any reasonable time integration method should satisfy this condition.*

REMARK 1.9 *The satisfaction of the discrete geometric conservation law for the ALE formulation based on the conservative form of the linear-momentum balance given by Equation (1.220) depends on whether the identity given by Equation (1.222) holds at the fully discrete level. Due to the different treatment of space and time discretizations in ALE methods, Equation (1.222) may not be satisfied in the fully discrete case.*

REMARK 1.10 *The situation is reversed for the global conservation of linear momentum. The conservative form of the linear-momentum equations typically leads to fully discrete formulations that are globally momentum-conserving. Global momentum conservation for ALE formulations based on the convective form of the equations typically holds only up to the time discretization (see, e.g., Bazilevs et al., 2008).*

REMARK 1.11 *Because in space–time formulations the basis functions depend on both space and time, Equation (1.222) holds at the fully discrete level. As a result, space–time formulations naturally satisfy the discrete geometric conservation law and the global conservation of linear momentum.*

2

Basics of the Finite Element Method for Nonmoving-Domain Problems

In the previous chapter we introduced the governing equations of fluid and structural mechanics. The equations of interest are nonlinear, time-dependent, and are posed on moving spatial domains. In this chapter we provide the basics of the finite element method (FEM), focusing on nonmoving-domain problems. The moving-domain case will be discussed in the next chapter. We begin with the Galerkin finite element technique, which is a generally accepted computational methodology in structural mechanics. In the case of fluids, Galerkin's methods suffers a series of shortcomings, which are addressed by means of stabilized and multiscale finite element methods. We briefly present these methods for the linear advection–diffusion equation and the Navier–Stokes equations of incompressible flows.

We begin the discussion of the FEM method for steady problems. In this simpler setting we are able to introduce most of the basic constructions that we will employ in the remainder of this book. We then proceed to give examples that include time-dependent behavior, however, the governing differential equations are still assumed to be posed on a fixed spatial domain.

2.1 An Abstract Variational Formulation for Steady Problems

Like in the previous chapter, in the case of steady problems, we assume the existence of the spatial domain $\Omega \in \mathbb{R}^{n_{\mathrm{sd}}}$ and its boundary Γ. Note that Ω and Γ are fixed in space. The boundary Γ is subdivided into two complementary subsets Γ_g and Γ_h, the essential and natural parts of Γ. Over the domain Ω, we define $\mathcal{S}$ and $\mathcal{V}$ to be the sets of trial and test functions, respectively. The sets $\mathcal{S}$ and $\mathcal{V}$ are infinite-dimensional, and are comprised of functions that possess the necessary smoothness (or differentiability) requirements. The essential boundary conditions are built into the definitions of the function sets, that is,

$$u = g \quad \text{on} \quad \Gamma_g, \forall u \in \mathcal{S}, \tag{2.1}$$

$$w = 0 \quad \text{on} \quad \Gamma_g, \forall w \in \mathcal{V}, \tag{2.2}$$

where g is a given function. Given the domain, its boundary, and the sets of trial and weighting functions,[1] an *abstract* weak or variational formulation of the boundary value problem (BVP) reads: find $u \in \mathcal{S}$, such that $\forall\, w \in \mathcal{V}$:

$$B(w, u) - F(w) = 0. \tag{2.3}$$

In Equation (2.3), $B(\cdot, \cdot) : \mathcal{V} \times \mathcal{S} \to \mathbb{R}$ is a semilinear form (always linear in the test function argument, but, generally, not in the trial function argument), and $F(\cdot) : \mathcal{V} \to \mathbb{R}$ is a linear functional, which contains the natural boundary conditions. Equation (2.3), together with the definition of the function sets $\mathcal{S}$ and $\mathcal{V}$ and the boundary conditions (2.1) and (2.2), constitute the weak, or variational, statement of the *continuous* BVP. Here u and w may be scalar- or vector-valued functions. Examples of weak formulations of fluid and structural mechanics equations that fit into this framework were given in the previous chapter.

2.2 FEM Applied to Steady Problems

FEM is a collection of procedures to generate approximate solutions to BVPs stated in a weak form as per Equation (2.3). Although the FEM is most mature for problems in structural and solid mechanics, the developments over the last several decades have enabled the application of FEM to problems in many other fields of mechanics and engineering, and, in particular, fluid mechanics. Currently the FEM is accepted to be the most general approximation method for a large class of problems that are governed by partial differential equations posed in complex-geometry configurations.

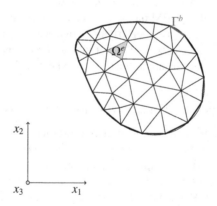

Figure 2.1 Finite elements Ω^e's and domain boundaries Γ^b's

In what follows, we present the basic ideas behind the FEM. We will attempt to preserve as much generality as we can, and supply the discussion with clarifying examples. In the

---

[1] In the literature, the test functions are also called weighting functions. Both terms will be used in this book interchangeably.

Basics of the Finite Element Method for Nonmoving-Domain Problems

FEM, as illustrated in Figure 2.1, the problem domain Ω is approximated by a collection of subdomains Ω^e's as

$$\Omega \approx \Omega^h = \bigcup_{e=1}^{n_{el}} \Omega^e, \qquad (2.4)$$

where Ω^h is the domain of the finite element problem, the subdomains Ω^e's are called finite elements, n_{el} is their total number, and the superscript h refers to a representative size (e.g., the diameter) of the finite elements. The subdomains Ω^e's take on simple shapes, which are typically triangles or quadrilaterals in 2D, and tetrahedra or hexahedra in 3D. The collection of Ω^e's is referred to as the finite element mesh. Due to the simple shapes of the elements, for complex-geometry Ω, it is often the case that Ω^h is coincident with Ω in the limit of mesh refinement.

The decomposition of the problem domain into finite elements leads to the decomposition of the problem domain boundary as

$$\Gamma \approx \Gamma^h = \bigcup_{b=1}^{n_{eb}} \Gamma^b, \qquad (2.5)$$

where Γ^b's are the edges (in 2D) or faces (in 3D) of the domain boundary. As in the case of the finite element domain, the boundary is represented approximately due to the simple element shapes involved.

Given the finite element decomposition of the domain and its boundary, the *finite dimensional* finite element function sets $\mathcal{S}^h$ and $\mathcal{V}^h$, the discrete counterparts of $\mathcal{S}$ and $\mathcal{V}$, are constructed. To define $\mathcal{S}^h$, we first construct the finite element basis functions for $\mathcal{S}^h$, $N_A(\mathbf{x})$, $A \in \eta^s$, where η^s is the index set. In most cases, each $N_A(\mathbf{x})$ is associated to its unique mesh node A, and thus η^s is the index set of those mesh nodes.[2] The functions $N_A(\mathbf{x})$ are chosen such that $\mathcal{S}^h$ has good approximation properties, and such that they have *local support*. Local support means that each of the N_A's is supported on a small subset of elements in the finite element mesh. We will postpone the presentation of examples of such basis functions until later sections.

Given the basis functions, any finite element function $u^h \in \mathcal{S}^h$ may be represented as

$$u^h = v^h + g^h, \qquad (2.6)$$

$$v^h = \sum_{A \in (\eta^s - \eta^s_g)} u_A N_A(\mathbf{x}), \qquad (2.7)$$

$$g^h = \sum_{A \in \eta^s_g} g_A N_A(\mathbf{x}), \qquad (2.8)$$

---

[2] In FEM, basis functions are not necessarily associated with mesh nodes. In the case of low-order mixed FEM, some basis functions are associated with element interiors. Other examples of FEM basis functions not associated with mesh nodes include discontinuous enrichment functions in XFEM Moes et al., 1999, and NURBS and T-spline basis functions in isogeometric analysis. The latter are associated with mesh *control points* (see Chapter 3).

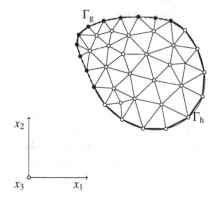

Figure 2.2 The nodes with known and unknown solution coefficients are represented by solid and empty circles, respectively

with the additional requirement that

$$v^h = 0 \text{ on } \Gamma_g^h, \tag{2.9}$$

$$g^h = \Pi^h g \text{ on } \Gamma_g^h. \tag{2.10}$$

The solution coefficients u_A's in Equation (2.7) are the unknowns of the finite element problem, referred to as the *degrees-of-freedom*. The index set η_g^s in Equation (2.8) collects the prescribed solution coefficients g_A's that are responsible for the imposition of the essential boundary conditions. The coefficients g_A's are considered known. Figure 2.2 illustrates the concept of nodes with known and unknown solution coefficients. The essential boundary conditions are set according to Equation (2.10), where $\Pi^h g$ is the interpolation, or projection, of the essential boundary condition g onto the space of finite element functions that are supported on the boundary Γ_g^h.

Analogously to $\mathcal{S}^h$, an arbitrary element w^h of $\mathcal{V}^h$, the discrete counterpart of $\mathcal{V}$, may be represented as

$$w^h = \sum_{A \in \eta^w} w_A Q_A(\mathbf{x}), \tag{2.11}$$

with the additional requirement[3] that

$$w^h = 0 \text{ on } \Gamma_g^h. \tag{2.12}$$

---

[3] Depending on the nature of the BVP, and the details of the construction of the finite element function sets, this requirement may not be necessary. Consider, for example, a pure advection problem in 1D. For simplicity, it is assumed that the problem is posed on a unit interval and the advective velocity is positive everywhere. As a result, to have a well-posed problem, the solution is prescribed only at the left end of the interval. There are many possible FEM discretizations for this case. One such discretization involves $\mathcal{S}^h$ comprised of continuous, element-wise linear functions, and $\mathcal{V}^h$ comprised of discontinuous, element-wise constant functions. In this case, the test functions do not vanish on the essential boundary. Furthermore, for this case, the basis functions in $\mathcal{S}^h$ are associated with the mesh nodes, and the basis functions in $\mathcal{V}^h$ are associated with the mesh elements.

In Equation (2.11), Q_A is the basis function for node A, and η^w is the index set for those nodes.[4] They are also chosen such that the $\mathcal{V}^h$ has favorable approximation properties and the functions have local support.

With the above definitions of the domain and the basis functions, the discrete counterpart of the variational formulation (2.3) may be written as: find $u^h \in \mathcal{S}^h$, such that $\forall\, w^h \in \mathcal{V}^h$:

$$B^h(w^h, u^h) - F^h(w^h) = 0, \tag{2.13}$$

where the semilinear form $B^h(\cdot, \cdot) : \mathcal{V}^h \times \mathcal{S}^h \to \mathbb{R}$ and the linear form $F^h(\cdot) : \mathcal{V}^h \to \mathbb{R}$ are the discrete counterparts of B and F from Equation (2.3). Introducing the sum from Equation (2.11) into Equation (2.13), using the semilinearity of B^h, the linearity of F^h, and the fact that the discrete variational statement (2.13) holds *for all* weighting functions $w^h \in \mathcal{V}^h$, we arrive at a discrete equation for each degree-of-freedom A, given by

$$B^h(Q_A, u^h) - F^h(Q_A) = 0 \quad \forall A \in \eta^w. \tag{2.14}$$

The equation system (2.14) is solved by finding the coefficients u_B^h, $B \in \eta^s - \eta_g^s$. We note that we require

$$\dim \eta^w = \dim(\eta^s - \eta_g^s). \tag{2.15}$$

In the case that B^h is a *bilinear* form (i.e., it is linear in both arguments), a substitution of Equations (2.6)–(2.8) into Equation (2.14) gives the following matrix problem:

$$\mathbf{K}\mathbf{U} = \mathbf{F}, \tag{2.16}$$

where

$$\mathbf{K} = [K_{AB}], \tag{2.17}$$

$$\mathbf{U} = [u_B], \tag{2.18}$$

$$\mathbf{F} = [F_A], \tag{2.19}$$

$$K_{AB} = B^h(Q_A, N_B), \tag{2.20}$$

$$F_A = F^h(Q_A) - \sum_{C \in \eta_g^s} K_{AC} g_C, \tag{2.21}$$

for all $A \in \eta^w$ and $B \in \eta^s - \eta_g^s$. The linear equation system (2.16) is solved for the unknown solution vector $\mathbf{U}$. The solution field u^h is obtained by taking a linear combination of u_A's and N_A's and augmenting it with the boundary condition contribution according to Equations (2.6)–(2.8).

In the case when B^h is only semilinear, a *Newton–Raphson procedure* is typically employed to obtain the solution of Equation (2.14). Given the initial guess $\mathbf{U}^0$, the following iteration is performed for $i = 0, 1, \ldots, (i_{\max} - 1)$:

$$\mathbf{K}^i \Delta \mathbf{U}^i = -\mathbf{R}^i, \tag{2.22}$$

---

[4] See Notes 2 and 3 in this chapter.

where

$$\mathbf{K}^i = [\mathrm{K}^i_{AB}], \tag{2.23}$$

$$\mathbf{R}^i = [\mathrm{r}^i_A], \tag{2.24}$$

$$\mathrm{r}^i_A = B^h\left(Q_A, \sum_{C \in \eta^s - \eta^s_g} u^i_C N_C + \sum_{C \in \eta^s_g} g_C N_C\right) - F^h(Q_A), \tag{2.25}$$

$$\mathrm{K}^i_{AB} = \left(\frac{\partial \mathrm{r}_A}{\partial u_B}\right)\bigg|_i, \tag{2.26}$$

and the $i + 1^{st}$ solution iterate is updated as

$$\mathbf{U}^{i+1} = \mathbf{U}^i + \Delta \mathbf{U}^i. \tag{2.27}$$

The iterations given by Equation (2.22) continue until $\mathrm{r}^i_A = 0$ in an approximate sense. In the above, $\mathbf{K}^i$ is the consistent tangent matrix evaluated at the ith solution iterate. The use of consistent tangent matrices ensures quadratic convergence rate of the Newton–Raphson iteration.

This general finite element framework presented here accommodates a number of well-known methods, in particular, the *Bubnov–Galerkin Method*. The Bubnov–Galerkin method (or Galerkin's method) is the oldest and the most well-known finite element method. In fact, for many years, and to this day, the term "Galerkin's method" is synonymous with the FEM. The method is obtained by choosing $\mathcal{S}^h \subset \mathcal{S}$, $\mathcal{V}^h \subset \mathcal{V}$, and setting

$$\eta^w = \eta^s - \eta^s_g, \tag{2.28}$$

$$Q_A(\mathbf{x}) = N_A(\mathbf{x}) \; \forall A \in \eta^w, \tag{2.29}$$

$$B^h(w^h, u^h) = B(w^h, u^h), \tag{2.30}$$

$$F^h(w^h) = F(w^h). \tag{2.31}$$

That is, the discrete trial and test function sets are subsets of their continuous counterparts, they make use of the same basis functions, and the semilinear and linear forms remain unchanged from the continuous case. The Bubnov–Galerkin method is well suited for elliptic BVPs that are governed by symmetric operators, which explains its popularity and excellent performance in applications to heat conduction and structural mechanics. Furthermore, discretization of symmetric operators leads to symmetric left-hand-side matrices. The methods that do not use the same basis functions for the trial and test function sets are called *Petrov–Galerkin* methods.

2.3 Construction of Finite Element Basis Functions

In this section we look at the construction of the finite element basis functions. The typical construction begins at the level of an individual element. The element-level interpolation functions are constructed first and then the elements are put together into a finite element mesh, which also defines the global basis functions. Individual elements differ in size and shape, however every element in the mesh is an image of a simple geometrical shape that is defined in the parametric domain. While there may be millions of elements in the finite element mesh, there is only a handful of parametric shapes, which has important implications in the programming of the FEM. In this book, we will only focus on *triangular* and *quadrilateral* elements in 2D, and *tetrahedral* and *hexahedral* elements in 3D. We begin with the construction of element-level shape functions and then discuss the global case. We conclude the section by giving examples of finite elements that employ Lagrange polynomials as interpolation functions, the most common in the FEM.

2.3.1 Construction of Element Shape Functions

Let $\boldsymbol{\xi} \in \mathbb{R}^{n_{sd}}$ denote the coordinates of the parametric element $\hat{\Omega}^e$. We define a set of interpolation functions $N_a(\boldsymbol{\xi})$, $a = 1, \ldots, n_{en}$, where n_{en} is the number of element nodes. The functions $N_a(\boldsymbol{\xi})$'s are also known as the parametric finite element shape functions. Let $\mathbf{x}(\boldsymbol{\xi}) : \hat{\Omega}^e \to \Omega^e$ be a function that maps the parametric element to the physical element. This mapping is given by

$$\mathbf{x}(\boldsymbol{\xi}) = \sum_{a=1}^{n_{en}} \mathbf{x}_a N_a(\boldsymbol{\xi}), \qquad (2.32)$$

where $\mathbf{x}_a$'s are the coordinates of the finite element nodes in the physical space. We assume that the mapping $\mathbf{x}(\boldsymbol{\xi})$ is invertible and continuously differentiable on $\hat{\Omega}^e$, and we let $\boldsymbol{\xi}(\mathbf{x})$ denote its inverse. We also require that $N_a(\boldsymbol{\xi})$'s satisfy a *partition of unity* property,

$$\sum_{a=1}^{n_{en}} N_a(\boldsymbol{\xi}) = 1, \qquad (2.33)$$

that is, the parametric shape functions sum to one at every $\boldsymbol{\xi} \in \hat{\Omega}^e$. It is also desirable (but not strictly necessary) that $N_a(\boldsymbol{\xi})$'s satisfy the *interpolation* property, namely,

$$N_a(\boldsymbol{\xi}_b) = \delta_{ab}, \qquad (2.34)$$

where $\boldsymbol{\xi}_b$'s are the coordinates of the parametric element nodes. In this case, $\mathbf{x}_a$'s coincide with the coordinates of the mesh nodes. This can be seen as follows:

$$\mathbf{x}(\boldsymbol{\xi}_b) = \sum_{a=1}^{n_{en}} \mathbf{x}_a N_a(\boldsymbol{\xi}_b) = \sum_{a=1}^{n_{en}} \mathbf{x}_a \delta_{ab} = \mathbf{x}_b. \qquad (2.35)$$

Equation (2.35) states that the nodes of the parametric element are mapped to the nodes of the physical element. The above result and the invertibility of the finite element mapping imply

$$\boldsymbol{\xi}(\mathbf{x}_b) = \boldsymbol{\xi}_b. \qquad (2.36)$$

If the interpolation property given by Equation (2.34) is not fulfilled, $\mathbf{x}_a$'s may be thought of as generalized element coordinates.

Finite element formulations require the definition of a solution function over the physical element. To construct such a function, we first define the finite element solution over a parametric element as

$$u^h(\boldsymbol{\xi}) = \sum_{a=1}^{n_{en}} u_a N_a(\boldsymbol{\xi}), \qquad (2.37)$$

where u_a's are the nodal solution coefficients or degrees-of-freedom. Given the solution on the parametric domain, its physical-domain counterpart may be defined as

$$u^h(\mathbf{x}) = u^h(\boldsymbol{\xi}(\mathbf{x})). \qquad (2.38)$$

This definition should be understood as follows: if one wants to obtain a value of the finite element solution at a location $\mathbf{x}$, one finds $\boldsymbol{\xi}$, the preimage of $\mathbf{x}$ in the parametric domain, and evaluates the finite element solution in this location. Combining Equations (2.37) and (2.38), we obtain

$$u^h(\mathbf{x}) = \sum_{a=1}^{n_{en}} u_a N_a(\boldsymbol{\xi}(\mathbf{x})). \qquad (2.39)$$

Equation (2.39) leads to a natural definition of shape functions $N_a(\mathbf{x})$'s on a physical element:

$$N_a(\mathbf{x}) = N_a(\boldsymbol{\xi}(\mathbf{x})). \qquad (2.40)$$

If the element shape functions are constructed to satisfy the interpolation property given by Equation (2.34), a simple computation yields

$$u^h(\mathbf{x}_a) = \sum_{b=1}^{n_{en}} u_b N_b(\mathbf{x}_a) = \sum_{b=1}^{n_{en}} u_b N_b(\boldsymbol{\xi}(\mathbf{x}_a))$$
$$= \sum_{b=1}^{n_{en}} u_b N_b(\boldsymbol{\xi}_a) = \sum_{b=1}^{n_{en}} u_b \delta_{ba} = u_a, \qquad (2.41)$$

which implies that the solution coefficients are the nodal values of the solution. This property is particularly convenient for setting essential boundary conditions in finite element simulations.

Comparing equations for the geometry (2.32) and solution (2.37) interpolation on the parametric element, we note that the same parametric shape functions are employed for both. This is the well-known *isoparametric* finite element construction, a concept attributed to Irons (1966). The partition of unity property given by Equation (2.33) and the isoparametric construction ensure that the finite element shape functions are able to represent an arbitrary linear polynomial on every element in the physical domain. This may be seen as follows. Let the finite element solution function take the form

$$u^h(\mathbf{x}) = c_0 + c_1 (\mathbf{x})_1 + c_2 (\mathbf{x})_2 + c_3 (\mathbf{x})_3, \qquad (2.42)$$

where $(\mathbf{x})_i$ is the ith component of the position vector $\mathbf{x}$, and c_0–c_3 are the arbitrary real constants. The right-hand-side of Equation (2.42) represents an arbitrary linear polynomial in 3D.

Basics of the Finite Element Method for Nonmoving-Domain Problems

Introducing the geometrical mapping (2.32) and the partition of unity property (2.33) into Equation (2.42), and rearranging the terms, gives

$$u^h(\mathbf{x}) = c_0 \sum_{a=1}^{n_{en}} N_a(\mathbf{x}) + c_1 \sum_{a=1}^{n_{en}} (\mathbf{x}_a)_1 N_a(\mathbf{x})$$

$$+ c_2 \sum_{a=1}^{n_{en}} (\mathbf{x}_a)_2 N_a(\mathbf{x}) + c_3 \sum_{a=1}^{n_{en}} (\mathbf{x}_a)_3 N_a(\mathbf{x}) \quad (2.43)$$

$$= \sum_{a=1}^{n_{en}} (c_0 + c_1 (\mathbf{x}_a)_1 + c_2 (\mathbf{x}_a)_2 + c_3 (\mathbf{x}_a)_3) N_a(\mathbf{x}). \quad (2.44)$$

Equation (2.44) implies that by setting the nodal coefficients as

$$u_a = c_0 + c_1(\mathbf{x}_a)_1 + c_2(\mathbf{x}_a)_2 + c_3(\mathbf{x}_a)_3, \quad (2.45)$$

we recover an arbitrary linear polynomial on a physical element. This important result serves as the basic "convergence proof" for isoparametric elements employing shape functions that are not necessarily polynomial (for example, as in the case of mesh-free methods Belytschko et al., 1994 and isogeometric analysis Hughes et al., 2005). Furthermore, the ability to represent arbitrary global linear polynomials ensures that the rigid-body modes and constant-strain states in structural mechanics applications may be reproduced exactly.

The Jacobian matrix of the element mapping given by Equation (2.32) may be computed by a direct differentiation

$$\frac{\partial \mathbf{x}(\boldsymbol{\xi})}{\partial \boldsymbol{\xi}} = \sum_{a=1}^{n_{en}} \mathbf{x}_a \frac{\partial N_a(\boldsymbol{\xi})}{\partial \boldsymbol{\xi}}, \quad (2.46)$$

and its determinant is given by

$$J_{x\xi} = \det \frac{\partial \mathbf{x}}{\partial \boldsymbol{\xi}}. \quad (2.47)$$

The invertibility of the element mapping is guaranteed if $J_{x\xi} > 0$ at every point in the interior of the parametric element. Note that this condition depends on the definition of the element shape functions as well as the spatial locations of the element nodes. For better element approximation properties it is desirable to keep the variation of the Jacobian matrix over the element to a minimum. The Jacobian matrix is used in the computation of element shape function derivatives with respect to the physical coordinates as follows:

$$\frac{\partial N_a}{\partial \mathbf{x}} = \left(\frac{\partial \mathbf{x}}{\partial \boldsymbol{\xi}} \right)^{-T} \frac{\partial N_a}{\partial \boldsymbol{\xi}}. \quad (2.48)$$

Shape functions and their gradients with respect to physical coordinates are employed in the computation of left-hand-side element matrices and right-hand-side element vectors, which we present in the later sections.

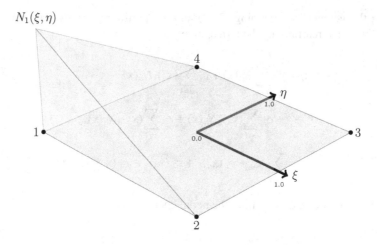

Figure 2.3 The quadrilateral parametric element and one of the shape functions

2.3.2 Finite Elements Based on Lagrange Interpolation Functions

In this section we present 2D and 3D finite elements that make use of Lagrange polynomial basis functions. In all cases the construction of the shape functions guarantees *completeness* to polynomial order p on the parametric element, and satisfaction of the interpolation and partition-of-unity properties given by Equations (2.34) and (2.33), respectively.

Quadrilateral Element. Let $\boldsymbol{\xi} = (\xi, \eta)$ be the element parametric coordinates, as shown in Figure 2.3. For this 2D element the parametric domain corresponds to a bi-unit square centered at the origin of the parametric coordinate system. To generate an element that is complete to polynomial order p in each direction, we place $p + 1$ equispaced nodes in each parametric interval, $\xi_1, \xi_2, \ldots, \xi_{p+1}$ and $\eta_1, \eta_2, \ldots, \eta_{p+1}$, and define 1D Lagrange polynomials in the usual way:

$$\ell_i^p(\xi) = \prod_{j=1, j \neq i}^{p+1} \frac{(\xi - \xi_j)}{(\xi_i - \xi_j)}, \qquad \text{for } i = 1, 2, \ldots, p+1. \tag{2.49}$$

Here the symbol $\prod$ denotes the product operation. The 2D parametric shape functions are defined as tensor products of 1D Lagrange polynomials,

$$N_a(\boldsymbol{\xi}) = \ell_i^p(\xi)\ell_j^p(\eta), \tag{2.50}$$

$$a = a(i, j), \qquad \text{for } i, j = 1, 2, \ldots, p+1, \tag{2.51}$$

where $a(i, j)$ gives the relationship between the indices and defines the local numbering of the parametric shape functions. A particular choice of the local shape function ordering is not important as long as it is consistent throughout the finite element mesh. In the case of a *bilinear quadrilateral* element, the parametric shape functions and their derivatives become

$$N_1(\boldsymbol{\xi}) = \frac{(1-\xi)(1-\eta)}{4}, \quad \frac{\partial N_1(\boldsymbol{\xi})}{\partial \xi} = -\frac{(1-\eta)}{4}, \quad \frac{\partial N_1(\boldsymbol{\xi})}{\partial \eta} = -\frac{(1-\xi)}{4}; \tag{2.52}$$

$$N_2(\boldsymbol{\xi}) = \frac{(1+\xi)(1-\eta)}{4}, \quad \frac{\partial N_2(\boldsymbol{\xi})}{\partial \xi} = +\frac{(1-\eta)}{4}, \quad \frac{\partial N_2(\boldsymbol{\xi})}{\partial \eta} = -\frac{(1+\xi)}{4}; \tag{2.53}$$

$$N_3(\boldsymbol{\xi}) = \frac{(1+\xi)(1+\eta)}{4}, \quad \frac{\partial N_3(\boldsymbol{\xi})}{\partial \xi} = +\frac{(1+\eta)}{4}, \quad \frac{\partial N_3(\boldsymbol{\xi})}{\partial \eta} = +\frac{(1+\xi)}{4}; \tag{2.54}$$

$$N_4(\boldsymbol{\xi}) = \frac{(1-\xi)(1+\eta)}{4}, \quad \frac{\partial N_4(\boldsymbol{\xi})}{\partial \xi} = -\frac{(1+\eta)}{4}, \quad \frac{\partial N_4(\boldsymbol{\xi})}{\partial \eta} = +\frac{(1-\xi)}{4}. \tag{2.55}$$

The determinant of the Jacobian matrix, $J_{x\xi}$, may be computed using Equation (2.47). The positivity of $J_{x\xi}$ for a bilinear quadrilateral element is guaranteed if all its angles in the physical domain are strictly less than 180°. In case this angle condition is violated, the Jacobian determinant goes to zero or even changes sign in the element interior, and the element is invalid for analysis.

Hexahedral Element. For this 3D element the parametric domain is a bi-unit cube centered at the origin of the parametric coordinate system, and the parametric coordinates are $\boldsymbol{\xi} = (\xi, \eta, \zeta)$. As in the case of the quadrilateral element, we place $p + 1$ equispaced nodes in each parametric interval and define the 3D parametric shape functions as tensor products of 1D Lagrange polynomials:

$$N_a(\boldsymbol{\xi}) = \ell_i^p(\xi)\ell_j^p(\eta)\ell_k^p(\zeta), \tag{2.56}$$

$$a = a(i, j, k), \quad \text{for } i, j, k = 1, 2, \ldots, p+1. \tag{2.57}$$

For a *trilinear hexahedron* the parametric shape functions and their derivatives are

$$N_1(\boldsymbol{\xi}) = \frac{(1-\xi)(1-\eta)(1-\zeta)}{8}, \quad \frac{\partial N_1(\boldsymbol{\xi})}{\partial \xi} = -\frac{(1-\eta)(1-\zeta)}{8}, \tag{2.58}$$

$$\frac{\partial N_1(\boldsymbol{\xi})}{\partial \eta} = -\frac{(1-\xi)(1-\zeta)}{8}, \quad \frac{\partial N_1(\boldsymbol{\xi})}{\partial \zeta} = -\frac{(1-\xi)(1-\eta)}{8}; \tag{2.59}$$

$$N_2(\boldsymbol{\xi}) = \frac{(1+\xi)(1-\eta)(1-\zeta)}{8}, \quad \frac{\partial N_2(\boldsymbol{\xi})}{\partial \xi} = +\frac{(1-\eta)(1-\zeta)}{8}, \tag{2.60}$$

$$\frac{\partial N_2(\boldsymbol{\xi})}{\partial \eta} = -\frac{(1+\xi)(1-\zeta)}{8}, \quad \frac{\partial N_2(\boldsymbol{\xi})}{\partial \zeta} = -\frac{(1+\xi)(1-\eta)}{8}; \tag{2.61}$$

$$N_3(\boldsymbol{\xi}) = \frac{(1+\xi)(1+\eta)(1-\zeta)}{8}, \quad \frac{\partial N_3(\boldsymbol{\xi})}{\partial \xi} = +\frac{(1+\eta)(1-\zeta)}{8}, \tag{2.62}$$

$$\frac{\partial N_3(\boldsymbol{\xi})}{\partial \eta} = +\frac{(1+\xi)(1-\zeta)}{8}, \quad \frac{\partial N_3(\boldsymbol{\xi})}{\partial \zeta} = -\frac{(1+\xi)(1+\eta)}{8}; \tag{2.63}$$

$$N_4(\boldsymbol{\xi}) = \frac{(1-\xi)(1+\eta)(1-\zeta)}{8}, \quad \frac{\partial N_4(\boldsymbol{\xi})}{\partial \xi} = -\frac{(1+\eta)(1-\zeta)}{8}, \tag{2.64}$$

$$\frac{\partial N_4(\boldsymbol{\xi})}{\partial \eta} = +\frac{(1-\xi)(1-\zeta)}{8}, \quad \frac{\partial N_4(\boldsymbol{\xi})}{\partial \zeta} = -\frac{(1-\xi)(1+\eta)}{8}; \tag{2.65}$$

$$N_5(\boldsymbol{\xi}) = \frac{(1-\xi)(1-\eta)(1+\zeta)}{8}, \quad \frac{\partial N_5(\boldsymbol{\xi})}{\partial \xi} = -\frac{(1-\eta)(1+\zeta)}{8}, \tag{2.66}$$

$$\frac{\partial N_5(\boldsymbol{\xi})}{\partial \eta} = -\frac{(1-\xi)(1+\zeta)}{8}, \quad \frac{\partial N_5(\boldsymbol{\xi})}{\partial \zeta} = +\frac{(1-\xi)(1-\eta)}{8}; \tag{2.67}$$

$$N_6(\boldsymbol{\xi}) = \frac{(1+\xi)(1-\eta)(1+\zeta)}{8}, \qquad \frac{\partial N_6(\boldsymbol{\xi})}{\partial \xi} = +\frac{(1-\eta)(1+\zeta)}{8}, \qquad (2.68)$$

$$\frac{\partial N_6(\boldsymbol{\xi})}{\partial \eta} = -\frac{(1+\xi)(1+\zeta)}{8}, \qquad \frac{\partial N_6(\boldsymbol{\xi})}{\partial \zeta} = +\frac{(1+\xi)(1-\eta)}{8}; \qquad (2.69)$$

$$N_7(\boldsymbol{\xi}) = \frac{(1+\xi)(1+\eta)(1+\zeta)}{8}, \qquad \frac{\partial N_7(\boldsymbol{\xi})}{\partial \xi} = +\frac{(1+\eta)(1+\zeta)}{8}, \qquad (2.70)$$

$$\frac{\partial N_7(\boldsymbol{\xi})}{\partial \eta} = +\frac{(1+\xi)(1+\zeta)}{8}, \qquad \frac{\partial N_7(\boldsymbol{\xi})}{\partial \zeta} = +\frac{(1+\xi)(1+\eta)}{8}; \qquad (2.71)$$

$$N_8(\boldsymbol{\xi}) = \frac{(1-\xi)(1+\eta)(1+\zeta)}{8}, \qquad \frac{\partial N_8(\boldsymbol{\xi})}{\partial \xi} = -\frac{(1+\eta)(1+\zeta)}{8}, \qquad (2.72)$$

$$\frac{\partial N_8(\boldsymbol{\xi})}{\partial \eta} = +\frac{(1-\xi)(1+\zeta)}{8}, \qquad \frac{\partial N_8(\boldsymbol{\xi})}{\partial \zeta} = +\frac{(1-\xi)(1+\eta)}{8}. \qquad (2.73)$$

Unlike in the case of a bilinear quadrilateral, no simple angle condition exists for a 3D trilinear hexahedron. However, one should avoid excessive mesh distortion, which tends to reduce the overall accuracy of the computations.

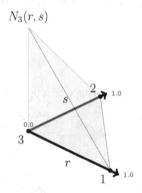

Figure 2.4 One of the shape functions for the triangular element

Triangular Element. Instead of the bi-unit square, the parametric domain for a triangular element is a right triangle (see Figure 2.4). In this case, we denote by r and s the independent natural coordinates with the range $0 \leq r \leq 1$, $0 \leq s \leq 1$, and $r + s \leq 1$. We also define a third, dependent coordinate t as

$$t = 1 - r - s. \qquad (2.74)$$

The triple $\mathbf{r} = (r, s, t)$ is known as the *triangular coordinates* (see Hughes, 2000). The 2D parametric shape functions for a triangular element of order p are defined as follows:

$$N_a(\mathbf{r}) = T_i(r) T_j(s) T_k(t), \qquad (2.75)$$
$$a = a(i, j, k), \qquad \text{for } i, j, k = 1, 2, \ldots, p+1, \qquad (2.76)$$

where

$$T_i(r) = \begin{cases} \ell_i^{i-1}(2r/r_i - 1), & \text{if } i > 1, \\ 1, & \text{if } i = 1, \end{cases} \qquad (2.77)$$

Basics of the Finite Element Method for Nonmoving-Domain Problems

and
$$r_i = (i-1)/p, \quad \text{for } i = 1, 2, \ldots, p+1. \tag{2.78}$$

In the case of the *linear triangle*, the shape functions and their derivatives become

$$N_1(\mathbf{r}) = r, \qquad \frac{\partial N_1(\mathbf{r})}{\partial r} = 1, \qquad \frac{\partial N_1(\mathbf{r})}{\partial s} = 0; \tag{2.79}$$

$$N_2(\mathbf{r}) = s, \qquad \frac{\partial N_2(\mathbf{r})}{\partial r} = 0, \qquad \frac{\partial N_2(\mathbf{r})}{\partial s} = 1; \tag{2.80}$$

$$N_3(\mathbf{r}) = t, \qquad \frac{\partial N_3(\mathbf{r})}{\partial r} = -1, \qquad \frac{\partial N_3(\mathbf{r})}{\partial s} = -1. \tag{2.81}$$

In the case of a linear triangle, the Jacobian matrix of the element mapping and its determinant are constant over the entire element. The Jacobian determinant is identically zero if the three nodal coordinates of the triangle in the physical domain are colinear. In this case the element has zero area and is invalid for analysis.

Tetrahedral Element. The parametric domain for a tetrahedral element is a tetrahedron with r, s, t as the independent natural coordinates. The shape functions for the tetrahedral element of order p may be obtained by extending the triangular element definition as

$$N_a(\mathbf{r}) = T_i(r)T_j(s)T_k(t)T_l(u), \tag{2.82}$$
$$a = a(i, j, k, l), \quad \text{for } i, j, k, l = 1, 2, \ldots, p+1, \tag{2.83}$$

where the fourth, dependent coordinate is defined as

$$u = 1 - r - s - t, \tag{2.84}$$

and the $T_i(r)$'s are defined as in Equation (2.77). In the case of the *linear tetrahedron* the shape functions and their derivatives become

$$N_1(\mathbf{r}) = r, \qquad \frac{\partial N_1(\mathbf{r})}{\partial r} = 1, \qquad \frac{\partial N_1(\mathbf{r})}{\partial s} = 0, \qquad \frac{\partial N_1(\mathbf{r})}{\partial t} = 0; \tag{2.85}$$

$$N_2(\mathbf{r}) = s, \qquad \frac{\partial N_2(\mathbf{r})}{\partial r} = 0, \qquad \frac{\partial N_2(\mathbf{r})}{\partial s} = 1, \qquad \frac{\partial N_2(\mathbf{r})}{\partial t} = 0; \tag{2.86}$$

$$N_3(\mathbf{r}) = t, \qquad \frac{\partial N_3(\mathbf{r})}{\partial r} = 0, \qquad \frac{\partial N_3(\mathbf{r})}{\partial s} = 0, \qquad \frac{\partial N_3(\mathbf{r})}{\partial t} = 1; \tag{2.87}$$

$$N_4(\mathbf{r}) = u, \qquad \frac{\partial N_4(\mathbf{r})}{\partial r} = -1, \qquad \frac{\partial N_4(\mathbf{r})}{\partial s} = -1, \qquad \frac{\partial N_4(\mathbf{r})}{\partial t} = -1. \tag{2.88}$$

For a linear tetrahedron the Jacobian matrix of the element mapping and its determinant are constant over the entire element. The Jacobian determinant is identically zero in case all four nodes of the tetrahedron in the physical domain are coplanar. In this case the element has zero volume and is invalid for analysis.

2.3.3 Construction of Global Basis Functions

The global basis functions employed in the simulations are constructed from element-level shape functions. In what follows, we denote the global basis functions by $N_A(\mathbf{x})$, where $A \in \eta$,

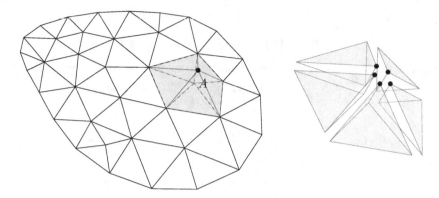

Figure 2.5 Illustration of the construction of the global basis function from the local shape functions

the index set of all basis functions defined on Ω^h. The construction is illustrated in Figure 2.5 on an example that uses linear triangles. Other element shapes are handled in an analogous fashion. For each global mesh node A, we find all the elements that share it. The union of such elements defines the *support* of N_A. Because only a few elements share a global node A, the support of N_A is local. The function $N_A(\mathbf{x})$ is constructed in a piecewise manner. On every element of its support, the global basis function coincides with its element-level counterpart corresponding to the local node with the same coordinates as the global node A. Outside of its support, the global basis function N_A is identically zero. This construction, in combination with the definition of the element-level shape functions using Lagrange polynomials, guarantees that the global basis functions are

- smooth on the element interiors,

- C^0-continuous at the inter-element boundaries.

The first property above follows from the fact that the element-level shape functions are obtained as a result of a combination of two smooth mappings (see Equation (2.40)). The second property may be argued as follows. In 2D, we consider two adjacent Lagrange elements that meet along an edge. Due to the interpolation property, the values of the local shape functions that contribute to the construction of the global basis function are coincident at the mesh nodes. The following happens between the mesh nodes. On the parametric domain, the local shape function restricted to the edge is a *unique* polynomial of order p that interpolates $p + 1$ mesh nodes. This follows directly from the fundamental properties of polynomials. Because of this uniqueness property, the shape functions in the parametric domain match at every point along the common edge. The continuity of the geometrical mapping also guarantees that the shape functions will match along the common edge in the physical domain. Note that the common edge may be straight or curved, without affecting the continuity of the global basis functions. The situation is illustrated in Figure 2.6 using two quadratic elements in 2D. The argument naturally extends to the 3D case. The C^0-continuity of the basis functions in the physical domain implies the function sets used in the finite element approximation are H^1-conforming, and are thus suitable for approximation of problems that involve second-order differential operators. Higher-order differential operators require more smoothness of

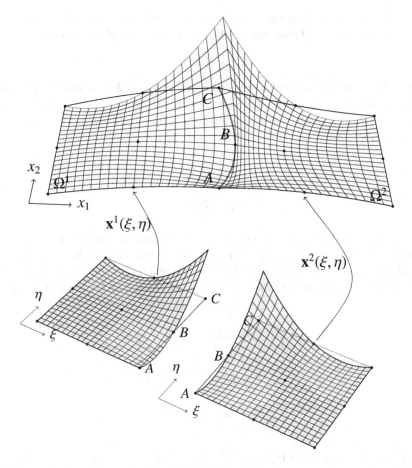

Figure 2.6 Illustration of the continuity of the global basis functions across element boundaries

the finite element functions. Higher-order smoothness may be achieved using polynomial or rational spline basis functions, as in the case of isogeometric analysis.

2.3.4 Element Matrices and Vectors and their Assembly into the Global Equation System

In a finite element computer code the relationship between the global basis functions and local shape functions (or the global and local node numbers) is represented by means of a connectivity array, typically called the *IEN array* (see, e.g., Hughes, 2000). The structure of the IEN array is as follows:

$$A = \text{IEN}(a, e), \tag{2.89}$$

where a is the local node or shape function number, e is the element number, and A is the corresponding global node or basis function number. The IEN array is, among others, a key finite element mesh data structure that is created at mesh generation and is typically read in by the analysis program. In addition to the IEN array, it is also convenient to define the *ID array*, which relates a global node number A to a corresponding global equation number P as

$$P = \text{ID}(A). \tag{2.90}$$

In the case A is a global node with a prescribed essential boundary condition, $\text{ID}(A)$ is set to 0 to indicate that there is no equation associated with A. Both the IEN and ID arrays are used in the assembly of global left-hand-side matrices and right-hand-side vectors. The assembly operation is expressed as

$$\mathbf{K} = \underset{e=1}{\overset{n_{\text{el}}}{\mathbf{A}}} \mathbf{k}^e, \tag{2.91}$$

$$\mathbf{F} = \underset{e=1}{\overset{n_{\text{el}}}{\mathbf{A}}} \mathbf{f}^e, \tag{2.92}$$

where $\mathbf{k}^e$ and $\mathbf{f}^e$ are the element-level left-hand-side matrices and right-hand-side vectors, respectively. They are obtained by restricting the semilinear and linear forms to each individual element e. Assuming the trial and test function sets are comprised of the same basis functions, in the linear FEM case (see Equations (2.17)–(2.21)), $\mathbf{k}^e$ and $\mathbf{f}^e$ are given by

$$\mathbf{k}^e = [\mathbf{k}^e_{ab}], \tag{2.93}$$

$$\mathbf{f}^e = [\mathbf{f}^e_a], \tag{2.94}$$

$$\mathbf{k}^e_{ab} = B^h(N_a, N_b), \tag{2.95}$$

$$\mathbf{f}^e_a = F^h(N_a) - \sum_c \mathbf{k}^e_{ac} g^e_c, \tag{2.96}$$

$$g^e_c = g_{\text{IEN}(c,e)}, \tag{2.97}$$

$$a, b, c = 1, \ldots, n_{\text{en}}. \tag{2.98}$$

In the nonlinear FEM case (see Equations (2.23)–(2.26)):

$$\mathbf{r}^e = [\mathbf{r}^e_a], \tag{2.99}$$

$$\mathbf{r}^e_a = B^h\left(N_a, \sum_c u^e_c N_c + \sum_c g^e_c N_c\right) - F^h(N_a), \tag{2.100}$$

$$\mathbf{k}^e_{ab} = \frac{\partial \mathbf{r}^e_a}{\partial u^e_b}, \tag{2.101}$$

$$u^e_c = u_{\text{IEN}(c,e)}, \tag{2.102}$$

where the iteration index i was omitted to simplify the exposition, and $\mathbf{r}^e$ plays the role of $\mathbf{f}^e$. The element-level matrices and vectors are obtained from numerical integration, which is presented in the following section. The assembly operation, denoted by $\underset{e=1}{\overset{n_{\text{el}}}{\mathbf{A}}}$, and used in Equations (2.91) and (2.92), is detailed in Algorithm 1.

REMARK 2.1 *Note that the developments in this section are applicable to BVPs with scalar-valued unknowns. For BVPs with vector-valued unknowns, all the FEM matrices and vectors are assigned an extra dimension corresponding to that of the BVP.*

Algorithm 1: Assembly of **K** and **F**.

Data: Element-level left-hand-side matrices $\mathbf{k}^e$, element-level right-hand-side vectors $\mathbf{f}^e$, and the IEN and ID arrays.
Result: Global left-hand-side matrix **K** and right-hand-side vector **F**.

```
// Initialize:
K = 0;
F = 0;

// Element loop:
for e = 1 to n_el do
    // Loop over element nodes:
    for a = 1 to n_en do
        A = ID(IEN(a, e));
        F_A = F_A + f^e_a;
        // Loop over element nodes:
        for b = 1 to n_en do
            B = ID(IEN(b, e));
            K_AB = K_AB + k^e_ab;
        end
    end
end
```

2.4 Finite Element Interpolation and Numerical Integration

In this section we discuss the finite element interpolation procedures and give interpolation error estimates in terms of the mesh size h and the polynomial order p. We conclude this section with a discussion of numerical integration, and, in particular, Gaussian quadrature, which is employed in the vast majority of the finite element programs.

2.4.1 Interpolation by Finite Elements

Let $u : \Omega \to \mathbb{R}$ be a function that we wish to interpolate using our finite element functions. We think of u as the exact solution to the continuous variational problem and we denote its finite element interpolant by $\Pi^h u$. The interpolant $\Pi^h u$ is the "best approximation" of u in our finite element space $\mathcal{S}^h$, that is,

$$\Pi^h u = \inf_{v^h \in \mathcal{S}^h} \left\| u - v^h \right\|_{A(\Omega)}. \tag{2.103}$$

Equation (2.103) states that $\Pi^h u$ minimizes the error between the exact solution and its interpolant in the norm $A(\Omega)$. The most common choices for $A(\Omega)$ is the $L^2(\Omega)$-norm or $H^1(\Omega)$ seminorm. The interpolant that minimizes the L^2-norm of the error may be obtained by solving the following variational problem: find $\Pi^h u \in \mathcal{S}^h$, such that $\forall w^h \in \mathcal{S}^h$:

$$\int_\Omega w^h \left(\Pi^h u - u \right) d\Omega = 0. \tag{2.104}$$

The solution $\Pi^h u$ of the above variational problem is called the L^2-projection of u onto the space of finite element functions. This projection is often employed in finite elements to impose initial conditions or to project the solution from one finite element mesh to another. The matrix form of the L^2-projection problem becomes

$$\mathbf{MU} = \mathbf{F}, \tag{2.105}$$

where

$$\mathbf{M} = [M_{AB}], \tag{2.106}$$

$$M_{AB} = \int_\Omega N_A N_B \, d\Omega, \tag{2.107}$$

$$\mathbf{F} = [F_A], \tag{2.108}$$

$$F_A = \int_\Omega N_A u \, d\Omega, \tag{2.109}$$

$$A, B \in \eta. \tag{2.110}$$

The solution of the matrix problem $\mathbf{U} = [u_B]$ is used to construct $\Pi^h u$ as

$$\Pi^h u(\mathbf{x}) = \sum_{A \in \eta} u_A N_A(\mathbf{x}). \tag{2.111}$$

Analogous to the L^2-projector, the H^1-projector may be obtained by solving the following variational problem: find $\Pi^h u \in \mathcal{S}^h$, such that $\forall \, w^h \in \mathcal{S}^h$:

$$\int_\Omega \nabla w^h \cdot \nabla \left(\Pi^h u - u \right) d\Omega = 0. \tag{2.112}$$

Essential boundary conditions are necessary for the above variational problem to be well-posed.

A projector that interpolates the function u at mesh nodes can also be constructed. For the basis functions that satisfy the interpolation property, the nodal interpolant of u is simply

$$\Pi^h u(\mathbf{x}) = \sum_{A \in \eta} u(\mathbf{x}_A) N_A(\mathbf{x}). \tag{2.113}$$

This projector in general does not have a minimization interpretation given by Equation (2.103). However, in 1D (and 1D only!), the nodal interpolant exactly coincides with the H^1-projector (see, e.g., Hughes and Sangalli, 2007). Given its implementation simplicity, the nodal interpolant is very convenient for setting essential boundary conditions.

The quality of the finite element interpolation depends on the mesh size h, the polynomial degree p, and the regularity of the function one is trying to interpolate. By regularity we mean the largest integer r, such that

$$\|u\|^2_{H^r(\Omega)} < \infty. \tag{2.114}$$

Here, $\|u\|_{H^r(\Omega)}$ denotes the rth norm of u given by

$$\|u\|^2_{H^r(\Omega)} = \sum_{0 \le |\alpha| \le r} \int_\Omega D^\alpha u \cdot D^\alpha u \, d\Omega, \tag{2.115}$$

with D^α being a short-hand notation for a differential operator $D^\alpha = D_1^{\alpha_1} D_2^{\alpha_2} \ldots D_d^{\alpha_d}$, where $\alpha = \{\alpha_1, \alpha_2, \ldots, \alpha_d\}$ is the multi-index, $|\alpha| = \sum_{i=1}^d \alpha_i$, and $D_i^j = \frac{\partial^j}{\partial x_i^j}$. The following finite element interpolation error estimate holds (see, e.g., Johnson, 1987; Brenner and Scott, 2002; Ern and Guermond, 2004):

$$\left\| u - \Pi^h u \right\|_{H^k(\Omega)} \leq Ch^{s-k} \|u\|_{H^s(\Omega)}, \tag{2.116}$$

where $s = \min(p+1, r)$, $k \leq s$, and C is a generic constant that depends on k, p, and Ω. Applying the estimate to the L^2-norm of the error and assuming full regularity of the solution we obtain

$$\left\| u - \Pi^h u \right\|_{L^2(\Omega)} \leq Ch^{p+1} \|u\|_{H^{p+1}(\Omega)}, \tag{2.117}$$

while in the case of the H^1-norm,

$$\left\| u - \Pi^h u \right\|_{H^1(\Omega)} \leq Ch^p \|u\|_{H^{p+1}(\Omega)}. \tag{2.118}$$

From the above interpolation error estimates we see that the accuracy of the finite element interpolation may be increased by decreasing the mesh size h or increasing the polynomial degree p, provided the function we are interpolating is smooth enough. We say that a solution to a given finite element formulation is optimally convergent if the error in the finite element solution is reduced at the same rate as the interpolation error.

2.4.2 Numerical Integration

In the previous section we saw that the finite element matrices and vectors are obtained as a result of integration of various quantities over the problem domain. In the FEM such an integration is performed numerically. In this section we briefly recall the basics of multidimensional Gaussian quadrature, which is the most common numerical integration method in finite elements.

We first note that that the integration over the finite element domain Ω is performed element-wise, namely,

$$\int_\Omega (\cdot) \, d\Omega = \sum_{e=1}^{n_{el}} \int_{\Omega^e} (\cdot) \, d\Omega. \tag{2.119}$$

To perform the element-level integral, we first change variables to the parametric element domain as

$$\int_{\Omega^e} f(\mathbf{x}) \, d\Omega = \int_{\hat{\Omega}^e} f(\mathbf{x}(\boldsymbol{\xi})) J_{x\xi}(\boldsymbol{\xi}) \, d\hat{\Omega}^e, \tag{2.120}$$

where $f(\mathbf{x})$ is a function we wish to integrate, $\hat{\Omega}^e$ is the domain of the parametric element (e.g., a brick or a tetrahedron in 3D), and $J_{x\xi}$ is the determinant of the Jacobian matrix (see Equation 2.47). Using the change of variables from the physical to the parametric element domain as in Equation (2.120) allows us to construct numerical-integration rules over simple, well-defined shapes.

To integrate the quantity given in Equation (2.120), we first define the function

$$g(\boldsymbol{\xi}) = f(\mathbf{x}(\boldsymbol{\xi})) J_{x\xi}(\boldsymbol{\xi}) \tag{2.121}$$

Table 2.1 Gaussian quadrature rules in 1D. One-point, two-point, and three-point quadrature rules are exact up to linear, cubic, and quintic polynomials, respectively. In general, the n-point rule is exact up to polynomials of degree $2n - 1$

n_{int}	γ	$\tilde{\xi}_\gamma$	W_γ
1	1	0	2
2	1	$-\dfrac{1}{\sqrt{3}}$	1
	2	$+\dfrac{1}{\sqrt{3}}$	1
3	1	$-\sqrt{\dfrac{3}{5}}$	$\dfrac{5}{9}$
	2	0	$\dfrac{8}{9}$
	3	$+\sqrt{\dfrac{3}{5}}$	$\dfrac{5}{9}$

and then *approximate* the integral over the parametric element with the sum

$$\int_{\hat{\Omega}^e} g(\xi)\mathrm{d}\hat{\Omega} \approx \sum_{\gamma=1}^{n_{int}} g(\tilde{\xi}_\gamma)W_\gamma. \qquad (2.122)$$

In Equation (2.122), $\tilde{\xi}_\gamma$'s are the quadrature points defined on the parametric domain, W_γ's are the quadrature weights, and n_{int} is their number. The choice of $\tilde{\xi}_\gamma$'s and W_γ's defines a quadrature rule. The quadrature weights are subject to the condition

$$\sum_{\gamma=1}^{n_{int}} W_\gamma = \int_{\hat{\Omega}^e} \mathrm{d}\hat{\Omega}. \qquad (2.123)$$

Gaussian quadrature rules are defined on a 1D parametric interval $[-1, 1]$. The locations of quadrature points and the corresponding weights are optimized to exactly integrate the maximum possible polynomial degree for the number of points employed. *The n-point Gaussian rule is able to exactly integrate an arbitrary polynomial of degree $2n - 1$.* Table 2.1 gives the locations of quadrature points and weights for one-, two-, and three-point Gaussian rules on the parametric interval.

In 2D and 3D, for rectangular and brick elements, respectively, Gaussian quadrature rule is defined by sequentially applying the 1D Gaussian quadrature rule in each tensor-product direction. For a 3D hexahedral element this leads to

$$\gamma = \gamma(i, j, k), \qquad (2.124)$$

$$\tilde{\xi}_\gamma = (\tilde{\xi}_i, \tilde{\eta}_j, \tilde{\zeta}_k), \qquad (2.125)$$

$$W_\gamma = W_{\xi i} W_{\eta j} W_{\zeta k}, \qquad (2.126)$$

Table 2.2 Quadrature point indices for the 3D hexahedron. The first four rows give the quadrature point indices for the 2D quadrilateral

γ	i	j	k
1	1	1	1
2	2	1	1
3	1	2	1
4	2	2	1
5	1	1	2
6	2	1	2
7	1	2	2
8	2	2	2

Table 2.3 Quadrature rules for the 2D triangle and 3D tetrahedron. In the case of the triangle, the one-point quadrature rule is exact up to linear polynomials, while the three-point quadrature rule is exact up to quadratic polynomials. In the case of the tetrahedron, the one-point quadrature rule is exact up to linear polynomials, while the four-point quadrature rule is exact up to quadratic polynomials

n_{int}	γ	$\tilde{r}_\gamma$	$\tilde{s}_\gamma$	w_γ
1	1	$\frac{1}{3}$	$\frac{1}{3}$	$\frac{1}{2}$
3	1	$\frac{1}{6}$	$\frac{1}{6}$	$\frac{1}{6}$
	2	$\frac{1}{6}$	$\frac{2}{3}$	$\frac{1}{6}$
	3	$\frac{2}{3}$	$\frac{1}{6}$	$\frac{1}{6}$

n_{int}	γ	$\tilde{r}_\gamma$	$\tilde{s}_\gamma$	$\tilde{t}_\gamma$	w_γ
1	1	$\frac{1}{4}$	$\frac{1}{4}$	$\frac{1}{4}$	$\frac{1}{6}$
4	1	$\frac{5+3\sqrt{5}}{20}$	$\frac{5-\sqrt{5}}{20}$	$\frac{5-\sqrt{5}}{20}$	$\frac{1}{24}$
	2	$\frac{5-\sqrt{5}}{20}$	$\frac{5+3\sqrt{5}}{20}$	$\frac{5-\sqrt{5}}{20}$	$\frac{1}{24}$
	3	$\frac{5-\sqrt{5}}{20}$	$\frac{5-\sqrt{5}}{20}$	$\frac{5+3\sqrt{5}}{20}$	$\frac{1}{24}$
	4	$\frac{5-\sqrt{5}}{20}$	$\frac{5-\sqrt{5}}{20}$	$\frac{5-\sqrt{5}}{20}$	$\frac{1}{24}$

where the single integer index γ is associated with a triple index (i, j, k) corresponding to the three tensor-product directions of a 3D hexahedron. The Gaussian quadrature points and weights for the case of $n_{int} = 8$ are summarized in Table 2.2, where it is assumed that the parametric domain is a bi-unit cube.

The choice of a quadrature rule for a given finite element formulation must be such that the errors due to numerical integration do not change the overall rate of convergence of a given finite element discretization. *As a rule of thumb, if p is the order of the finite element in a given tensor-product direction, one makes use of p + 1 quadrature points in that direction.* This choice is based on the requirement that mass and stiffness matrices for constant-Jacobian elements with uniform material properties are integrated exactly.

Quadrature rules for triangular and tetrahedral elements are defined directly on their multidimensional parametric domains in terms of triangular and tetrahedral coordinates, respectively. Table 2.3 summarizes those quadrature rules and their order for the 2D triangle and the 3D tetrahedron.

2.5 Examples of Finite Element Formulations

In this section we provide the finite element formulations of linear advection–diffusion and elastodynamics problems. Note that the examples provided include time-dependence of the solution. However, the spatial domain is still assumed to be fixed.

2.5.1 Galerkin Formulation of the Advection–Diffusion Equation

We begin with the time-dependent advection–diffusion equation given by Equation (1.27) in the weak form: find $\phi \in S$, such that $\forall w \in \mathcal{V}$:

$$\int_\Omega w \frac{\partial \phi}{\partial t} \, d\Omega + \int_\Omega w \mathbf{u} \cdot \nabla \phi \, d\Omega \\ + \int_\Omega \nabla w \cdot \nu \nabla \phi \, d\Omega - \int_{\Gamma_h} w h \, d\Gamma - \int_\Omega w f \, d\Omega = 0, \quad (2.127)$$

where f and h are the prescribed body force and diffusive flux, respectively. The Galerkin finite element formulation of the advection–diffusion equation is stated as follows: find $\phi^h \in S^h$, such that $\forall w^h \in \mathcal{V}^h$:

$$\int_\Omega w^h \frac{\partial \phi^h}{\partial t} \, d\Omega + \int_\Omega w^h \mathbf{u}^h \cdot \nabla \phi^h \, d\Omega \\ + \int_\Omega \nabla w^h \cdot \nu \nabla \phi^h \, d\Omega - \int_{\Gamma_h} w^h h^h d\Gamma - \int_\Omega w^h f^h d\Omega = 0. \quad (2.128)$$

The matrix form of the discrete advection–diffusion problem may be deduced directly from Equation (2.128):

$$\mathbf{M}\dot{\boldsymbol{\Phi}} + \mathbf{K}\boldsymbol{\Phi} = \mathbf{F}, \quad (2.129)$$

where

$$\mathbf{M} = [M_{AB}], \quad (2.130)$$

$$M_{AB} = \int_\Omega N_A N_B \, d\Omega, \quad (2.131)$$

$$\mathbf{K} = [K_{AB}], \tag{2.132}$$

$$K_{AB} = \int_{\Omega} N_A \mathbf{u}^h \cdot \nabla N_B \, d\Omega + \int_{\Omega} \nabla N_A \cdot \nu \nabla N_B \, d\Omega, \tag{2.133}$$

$$\mathbf{F} = [F_A], \tag{2.134}$$

$$F_A = \int_{\Gamma_h} N_A h^h d\Gamma + \int_{\Omega} N_A f^h d\Omega - \sum_{C \in \eta_g^s} K_{AC} g_C - \sum_{C \in \eta_g^s} M_{AC} \dot{g}_C, \tag{2.135}$$

$$\boldsymbol{\Phi} = \{\phi_B\}, \tag{2.136}$$

$$A, B \in \eta - \eta_g^s, \tag{2.137}$$

g_C's are the prescribed nodal values of the solution, and $(\dot{\cdot})$ denotes the time derivative. Equation (2.129) represents the *semi-discrete* formulation of the advection–diffusion equation, which is a system of ordinary differential equations. These equations may be integrated in time to obtain the evolution of the nodal solution coefficients. Time integration will be discussed in later sections. The steady case, which is often of interest, is obtained by neglecting the time derivative terms in the formulation.

The left-hand-side matrix $\mathbf{K}$ is nonsymmetric due to the presence of the convective term. Nevertheless, because ν is assumed to be strictly positive, for divergence-free convective velocity and essential boundary conditions set on all of Γ, the matrix $\mathbf{K}$ is positive definite. However, the positive-definiteness of $\mathbf{K}$ relies solely on the diffusion term. The convective term is skew-symmetric and does not contribute to the positive-definiteness property of $\mathbf{K}$. In practice, this situation leads to an instability in the discrete solution when the mesh Peclet number $\|\mathbf{u}^h\| h/\nu$ exceeds a critical value, which is $O(1)$. The instability manifests itself in an oscillatory behavior of the finite element solution. Furthermore, the magnitude of the oscillations grows as the mesh Peclet number increases. To ensure that in the Galerkin method the solution is free of oscillations, the mesh size h needs to be reduced such that the mesh Peclet number is sufficiently small. This is not practical in most cases.

2.5.2 Stabilized Formulation of the Advection–Diffusion Equation

To enhance the stability of Galerkin's method for the advection–diffusion equation without compromising the solution accuracy, stabilized methods have been proposed. The oldest and most common method is the Streamline-Upwind/Petrov–Galerkin (SUPG) formulation Brooks and Hughes, 1982; Hughes and Tezduyar, 1984. The SUPG formulation of the advection–diffusion equation is stated as follows: find $\phi^h \in \mathcal{S}^h$, such that $\forall w^h \in \mathcal{V}^h$:

$$\int_{\Omega} w^h \frac{\partial \phi^h}{\partial t} d\Omega + \int_{\Omega} w^h \mathbf{u}^h \cdot \nabla \phi^h d\Omega$$

$$+ \int_{\Omega} \nabla w^h \cdot \nu \nabla \phi^h d\Omega - \int_{\Gamma_h} w^h h^h d\Gamma - \int_{\Omega} w^h f^h d\Omega$$

$$+ \sum_{e=1}^{n_{el}} \int_{\Omega^e} \tau_{\text{SUPG}} \mathbf{u}^h \cdot \nabla w^h \left(\frac{\partial \phi^h}{\partial t} + \mathbf{u}^h \cdot \nabla \phi^h - \nabla \cdot \nu \nabla \phi^h - f^h \right) d\Omega = 0. \tag{2.138}$$

In the above formulation, the term in the parenthesis is the *residual* of the advection–diffusion equation. The residual is well defined on the interior of each finite element, and constitutes a function that is obtained by applying the original differential operator (in this case, the advection–diffusion operator) to the finite element solution. This design makes the formulation consistent, meaning, if ϕ^h is replaced by the exact solution ϕ, the semi-discrete variational formulation given by Equation (2.138) is identically satisfied. The residual is multiplied by $\mathbf{u}^h \cdot \nabla w^h$, a convective derivative of the weighting function, and τ_{SUPG}, a stabilization parameter. The parameter τ_{SUPG} is designed to render the discrete solution stable and optimally convergent under mesh refinement. It is $O(h/\|\mathbf{u}^h\|)$ in the convective limit (i.e., $\|\mathbf{u}^h\|h/\nu \to \infty$), $O(h^2/\nu)$ in the diffusive limit (i.e., $\|\mathbf{u}^h\|h/\nu \to 0$), and $O(\Delta t)$ when the time derivative term dominates, where Δt is the time-step size. A commonly used definition of the stabilization parameter, which respects these limits and is suitable for any element topology, is (see, e.g., Bazilevs et al., 2007a)

$$\tau_{\text{SUPG}} = \left(\frac{4}{\Delta t^2} + \mathbf{u}^h \cdot \mathbf{G} \mathbf{u}^h + C_I \nu^2 \mathbf{G} : \mathbf{G} \right)^{-1/2}, \tag{2.139}$$

where

$$\mathbf{G} = \frac{\partial \boldsymbol{\xi}}{\partial \mathbf{x}}^T \frac{\partial \boldsymbol{\xi}}{\partial \mathbf{x}} \tag{2.140}$$

is the element metric tensor and C_I is the constant of the element-wise inverse estimate, which depends on the element topology and polynomial order, but not the mesh size (see, e.g., Johnson, 1987).

REMARK 2.2 *The definition of τ_{SUPG} given by Equations (2.139) and (2.140) has the following attributes. The appropriate power of the mesh size is given implicitly by the use of the element metric tensor $\mathbf{G}$. Using $\mathbf{G}$ also unifies the definition of τ_{SUPG} for different element topology and is especially helpful in the case of higher-order elements with curved edges and faces. In the convective limit the definition "selects" h to be the mesh size in the direction of the convective velocity $\mathbf{u}^h$, while in the diffusive limit the size of the smallest element dimension is favored.*

Another commonly used definition of the stabilization parameter is as follows:

$$\tau_{\text{SUPG}} = \left(\frac{1}{\tau_{\text{SUGN1}}^2} + \frac{1}{\tau_{\text{SUGN2}}^2} + \frac{1}{\tau_{\text{SUGN3}}^2} \right)^{-\frac{1}{2}}, \tag{2.141}$$

$$\tau_{\text{SUGN1}} = \left(\sum_{a=1}^{n_{en}} |\mathbf{u}^h \cdot \nabla N_a| \right)^{-1}, \tag{2.142}$$

$$\tau_{\text{SUGN2}} = \frac{\Delta t}{2}, \tag{2.143}$$

$$\tau_{\text{SUGN3}} = \frac{h_{\text{RGN}}^2}{4\nu}, \tag{2.144}$$

$$h_{\text{RGN}} = 2 \left(\sum_{a=1}^{n_{en}} |\mathbf{r} \cdot \nabla N_a| \right)^{-1}, \tag{2.145}$$

$$\mathbf{r} = \frac{\nabla |\phi^h|}{\| \nabla |\phi^h| \|}. \tag{2.146}$$

We note that the unit vector defined in Equation (2.146) is different than the vector of triangular coordinates used in Equation (2.75) and the vector of tetrahedral coordinates used in Equation (2.82).

REMARK 2.3 *The definition of τ_{SUPG} given by Equations (2.141)–(2.146) comes from Tezduyar (2003a). These definitions sense, in addition to the element geometry, the order of the interpolation functions. The diffusive limit of τ_{SUPG} is solution dependent and the "element length" used as length scale is in the direction of the solution gradient. Near boundaries with steep solution gradients normal to the boundary, this becomes the "element length" in the normal direction.*

The matrix form of the equations in the case of the SUPG formulation becomes

$$\mathbf{M}\dot{\mathbf{\Phi}} + \mathbf{K}\mathbf{\Phi} = \mathbf{F}, \tag{2.147}$$

which is form-identical to the Galerkin formulation. However, the following definitions of the finite element matrices and vectors apply:

$$\mathbf{M} = [M_{AB}], \tag{2.148}$$

$$M_{AB} = \int_\Omega N_A N_B \, d\Omega + \sum_{e=1}^{n_{el}} \int_{\Omega^e} \tau_{SUPG} \mathbf{u}^h \cdot \nabla N_A N_B \, d\Omega, \tag{2.149}$$

$$\mathbf{K} = [K_{AB}], \tag{2.150}$$

$$K_{AB} = \int_\Omega N_A \mathbf{u}^h \cdot \nabla N_B \, d\Omega + \int_\Omega \nabla N_A \cdot \nu \nabla N_B \, d\Omega$$
$$+ \sum_{e=1}^{n_{el}} \int_{\Omega^e} \tau_{SUPG} \mathbf{u}^h \cdot \nabla N_A \left(\mathbf{u}^h \cdot \nabla N_B - \nabla \cdot \nu \nabla N_B \right) d\Omega, \tag{2.151}$$

$$\mathbf{F} = [F_A], \tag{2.152}$$

$$F_A = \int_\Omega N_A f^h \, d\Omega + \sum_{e=1}^{n_{el}} \int_{\Omega^e} \tau_{SUPG} \mathbf{u}^h \cdot \nabla N_A f^h \, d\Omega$$
$$+ \int_{\Gamma_h} N_A h^h \, d\Gamma - \sum_{C \in \eta_g^s} K_{AC} g_C - \sum_{C \in \eta_g^s} M_{AC} \dot{g}_C, \tag{2.153}$$

$$\mathbf{\Phi} = \{\phi_B\}, \tag{2.154}$$

$$A, B \in \eta - \eta_g^s. \tag{2.155}$$

The first stabilization term in the definition of K_{AB} has the structure of anisotropic diffusion, with the diffusivity tensor

$$\tau_{SUPG} \mathbf{u}^h \otimes \mathbf{u}^h = \left(\tau_{SUPG} \|\mathbf{u}^h\|^2 \right) \frac{\mathbf{u}^h}{\|\mathbf{u}^h\|} \otimes \frac{\mathbf{u}^h}{\|\mathbf{u}^h\|}. \tag{2.156}$$

The form of the diffusivity tensor given by Equation (2.156) suggests that the term provides extra diffusion of magnitude $\tau_{SUPG}\|\mathbf{u}^h\|^2$ in the direction of the velocity, thus adding the stability that is lacking in the Galerkin formulation. However, this term controls only the convective derivative of the solution and does not preclude oscillations in the directions orthogonal to the velocity. In order to further stabilize the solution and preclude "cross-wind" oscillations, additional dissipation mechanisms need to be built into the discrete formulation. Such dissipation mechanisms may be provided by adding the so-called discontinuity-capturing terms to Equation (2.138). The interested reader is referred to Hughes *et al.* (1986b), Tezduyar and Park (1986), Tezduyar *et al.* (2006d), Tezduyar and Sathe (2006), Corsini *et al.* (2006) and Bazilevs *et al.* (2007b) for examples of that.

2.5.3 Galerkin Formulation of Linear Elastodynamics

The variational formulation of linear elastodynamics may be stated as follows (see Equation (1.156)): find the structural displacement $\mathbf{y} \in \mathcal{S}_y$, such that $\forall \, \mathbf{w} \in \mathcal{V}_y$:

$$\int_\Omega \mathbf{w} \cdot \rho \frac{d^2 \mathbf{y}}{dt^2} \, d\Omega + \int_\Omega \boldsymbol{\varepsilon}(\mathbf{w}) : \boldsymbol{\sigma} \, d\Omega - \int_{\Gamma_h} \mathbf{w} \cdot \mathbf{h} \, d\Gamma - \int_\Omega \mathbf{w} \cdot \rho \mathbf{f} \, d\Omega = 0, \tag{2.157}$$

where, as before, the Cauchy stress $\boldsymbol{\sigma}$ is modeled as

$$\boldsymbol{\sigma} = \mathbb{C} \boldsymbol{\varepsilon}(\mathbf{y}), \tag{2.158}$$

where $\mathbb{C}$ is the fourth-rank tensor of elastic coefficients, and $\boldsymbol{\varepsilon}(\mathbf{y})$ is the infinitesimal strain. To state the Galerkin formulation, we first rewrite the stress term in Equation (2.157) using the Voigt notation:

$$\int_\Omega \boldsymbol{\varepsilon}(\mathbf{w}) : \boldsymbol{\sigma} \, d\Omega = \int_\Omega \boldsymbol{\epsilon}(\mathbf{w}) \cdot \mathbf{D} \boldsymbol{\epsilon}(\mathbf{y}) \, d\Omega, \tag{2.159}$$

where, in 3D,

$$\boldsymbol{\epsilon}(\mathbf{y}) = \begin{Bmatrix} y_{1,1} \\ y_{2,2} \\ y_{3,3} \\ y_{2,3} + y_{3,2} \\ y_{3,1} + y_{1,3} \\ y_{1,2} + y_{2,1} \end{Bmatrix} \tag{2.160}$$

is the strain vector,

$$\boldsymbol{\epsilon}(\mathbf{w}) = \begin{Bmatrix} w_{1,1} \\ w_{2,2} \\ w_{3,3} \\ w_{2,3} + w_{3,2} \\ w_{3,1} + w_{1,3} \\ w_{1,2} + w_{2,1} \end{Bmatrix} \tag{2.161}$$

is the "virtual strain" vector, and $\mathbf{D} = [D_{IJ}]$, $I, J = 1, \ldots, 6$ is the matrix representation of the tensor of elastic coefficients $\mathbb{C}_{ijkl}$, with the relationship between the indices given in Table 2.4.

Table 2.4 For the elastic tensor, indices i and j are collapsed into I, and k and l are collapsed into J. The value of I in the first column is obtained by reading i from the second column and j from the third column. The value of J from the first column is obtained by reading k from the second column and l from the third column

I/J	i/k	j/l
1	1	1
2	2	2
3	3	3
4	2	3
4	3	2
5	1	3
5	3	1
6	1	2
6	2	1

For an isotropic material the elastic tensor **D** takes on the form

$$\mathbf{D} = \begin{bmatrix} \lambda + 2\mu & \lambda & \lambda & 0 & 0 & 0 \\ \lambda & \lambda + 2\mu & \lambda & 0 & 0 & 0 \\ \lambda & \lambda & \lambda + 2\mu & 0 & 0 & 0 \\ 0 & 0 & 0 & \mu & 0 & 0 \\ 0 & 0 & 0 & 0 & \mu & 0 \\ 0 & 0 & 0 & 0 & 0 & \mu \end{bmatrix}. \quad (2.162)$$

The semi-discrete Galerkin formulation of the linear-elastodynamics problem is stated as follows: find $\mathbf{d}^h \in \mathcal{S}_y^h$, such that $\forall \, \mathbf{w}^h \in \mathcal{V}_y^h$:

$$\int_\Omega \mathbf{w}^h \cdot \rho \frac{d^2 \mathbf{y}^h}{dt^2} \, d\Omega + \int_\Omega \boldsymbol{\epsilon}(\mathbf{w}^h) \cdot \mathbf{D}\boldsymbol{\epsilon}(\mathbf{y}^h) \, d\Omega - \int_{\Gamma_h} \mathbf{w}^h \cdot \mathbf{h}^h \, d\Gamma - \int_\Omega \mathbf{w}^h \cdot \rho \mathbf{f}^h \, d\Omega = 0. \quad (2.163)$$

Note that the same finite element basis functions are typically used to approximate all the Cartesian components of the displacement vector.

The matrix form of the Galerkin formulation given by Equation (2.163) becomes

$$\mathbf{M}\ddot{\mathbf{Y}} + \mathbf{K}\mathbf{Y} = \mathbf{F}, \quad (2.164)$$

where

$$\mathbf{M} = [M_{AB}^{ij}], \quad (2.165)$$

$$M_{AB}^{ij} = \int_\Omega N_A \rho N_B \, d\Omega \, \delta_{ij}, \qquad (2.166)$$

$$\mathbf{K} = [K_{AB}^{ij}], \qquad (2.167)$$

$$K_{AB}^{ij} = \mathbf{e}_i \cdot \int_\Omega \mathbf{B}_A^T \mathbf{D} \mathbf{B}_B \, d\Omega \, \mathbf{e}_j, \qquad (2.168)$$

$$\mathbf{F} = [F_A^i], \qquad (2.169)$$

$$F_A^i = \int_\Omega N_A \mathbf{e}_i \cdot \rho \mathbf{f}^h \, d\Omega + \int_{\Gamma_h} N_A \mathbf{e}_i \cdot \mathbf{h}^h d\Gamma$$
$$- \sum_{C \in \eta_g^s} \sum_j K_{AC}^{ij} g_C^j - \sum_{C \in \eta_g^s} \sum_j M_{AC}^{ij} \ddot{g}_C^j, \qquad (2.170)$$

$$\mathbf{Y} = [y_B^j], \qquad (2.171)$$

$$A, B \in \eta^s - \eta_g^s, \qquad (2.172)$$

$$i, j = 1, 2, 3. \qquad (2.173)$$

In the above equations, $\mathbf{e}_i$ is the ith Cartesian basis vector, and $(\ddot{\cdot})$ denotes the second time derivative. The matrix $\mathbf{B}_A$ in Equation (2.168) is the *strain-displacement* matrix for the basis function N_A, which satisfies

$$\boldsymbol{\epsilon}(N_A \mathbf{e}_i) = \mathbf{B}_A \mathbf{e}_i, \qquad (2.174)$$

and may be explicitly computed from the above relationship as

$$\mathbf{B}_A = \begin{bmatrix} N_{A,1} & 0 & 0 \\ 0 & N_{A,2} & 0 \\ 0 & 0 & N_{A,3} \\ 0 & N_{A,3} & N_{A,2} \\ N_{A,3} & 0 & N_{A,1} \\ N_{A,2} & N_{A,1} & 0 \end{bmatrix}. \qquad (2.175)$$

REMARK 2.4 *In modeling the elastodynamic phenomena, one often needs to consider the structural damping effects. The structural damping is modeled as a linear combination of mass-proportional and stiffness-proportional damping. In this case the inertial term in Equation (2.157) is replaced as follows:*

$$\rho \frac{d^2 \mathbf{y}}{dt^2} \leftarrow \rho \frac{d^2 \mathbf{y}}{dt^2} + a\rho \frac{d\mathbf{y}}{dt} \qquad (2.176)$$

and the stress terms in Equation (2.157) are replaced as follows:

$$\boldsymbol{\sigma} = \mathbb{C}\boldsymbol{\varepsilon}(\mathbf{y}) \leftarrow \boldsymbol{\sigma} = \mathbb{C}(\boldsymbol{\varepsilon}(\mathbf{y}) + b\dot{\boldsymbol{\varepsilon}}(\mathbf{y})), \qquad (2.177)$$

where

$$\dot{\boldsymbol{\varepsilon}}(\mathbf{y}) = \frac{1}{2}\left(\nabla \frac{d\mathbf{y}}{dt} + \nabla \frac{d\mathbf{y}}{dt}^T\right) \qquad (2.178)$$

is the strain rate, and a and b are the damping parameters. This yields a new matrix form corresponding to the Galerkin formulation given by Equation (2.163):

$$\mathbf{M}\ddot{\mathbf{Y}} + \mathbf{C}\dot{\mathbf{Y}} + \mathbf{K}\mathbf{Y} = \mathbf{F}, \qquad (2.179)$$

where **C** is the damping matrix given by

$$\mathbf{C} = a\mathbf{M} + b\mathbf{K}, \qquad (2.180)$$

and the right-hand side vector is modified by the damping terms as

$$F_A^i \leftarrow F_A^i - \sum_{C \in \eta_g^s} \sum_j C_{AC}^{ij} \dot{g}_C^j. \qquad (2.181)$$

The linear-elastostatics formulation, which is often of interest in practice, may be obtained by neglecting all the time derivative terms in the developments presented in this section.

2.6 Finite Element Formulation of the Navier–Stokes Equations

In this section we present the stabilized and the residual-based variational multiscale (RBVMS) formulations of the Navier–Stokes equations of incompressible flows. We present the nonmoving-domain case to familiarize the reader with the essence of the stabilized and multiscale methods for this rather complex system of partial differential equations. We cover the formulations with both the standard and weakly enforced essential boundary conditions. The moving-domain ALE and space–time formulations of the Navier–Stokes equations will be presented in the later chapters.

2.6.1 Standard Essential Boundary Conditions

We take the variational formulation of the Navier–Stokes equations given by Equation (1.65) as the point of departure: find $\mathbf{u} \in \mathcal{S}_u$ and $p \in \mathcal{S}_p$, such that $\forall \, \mathbf{w} \in \mathcal{V}_u$ and $q \in \mathcal{V}_p$:

$$\int_\Omega \mathbf{w} \cdot \rho \left(\frac{\partial \mathbf{u}}{\partial t} + \mathbf{u} \cdot \nabla \mathbf{u} - \mathbf{f} \right) d\Omega + \int_\Omega \boldsymbol{\varepsilon}(\mathbf{w}) : \boldsymbol{\sigma}(\mathbf{u}, p) \, d\Omega \\ - \int_{\Gamma_h} \mathbf{w} \cdot \mathbf{h} \, d\Gamma + \int_\Omega q \nabla \cdot \mathbf{u} \, d\Omega = 0. \qquad (2.182)$$

To develop the numerical procedure, we make use of the VMS method (Hughes, 1995; Hughes et al., 1998, 2000) and split the weighting and solution function spaces into subspaces that contain coarse and fine scales. This is accomplished by a multiscale direct-sum decomposition:

$$\mathcal{S}_u = \mathcal{S}_u^h \oplus \mathcal{S}_u', \qquad (2.183)$$

$$\mathcal{S}_p = \mathcal{S}_p^h \oplus \mathcal{S}_p', \qquad (2.184)$$

$$\mathcal{V}_u = \mathcal{V}_u^h \oplus \mathcal{V}_u', \qquad (2.185)$$

$$\mathcal{V}_p = \mathcal{V}_p^h \oplus \mathcal{V}_p'. \qquad (2.186)$$

In the above, the spaces superscripted with h are the spaces of coarse scales. These are finite-dimensional function spaces associated with a given finite element discretization. The spaces superscripted with "/" are the spaces of fine scales, or subgrid scales. These infinite-dimensional function spaces contain the modes that are not represented in the discretization. Equations (2.183)–(2.186) imply that every member of S_u, S_p, $\mathcal{V}_u$, and $\mathcal{V}_p$ may be written as

$$\mathbf{u} = \mathbf{u}^h + \mathbf{u}', \tag{2.187}$$

$$p = p^h + p', \tag{2.188}$$

$$\mathbf{w} = \mathbf{w}^h + \mathbf{w}', \tag{2.189}$$

$$q = q^h + q'. \tag{2.190}$$

Following the developments in Bazilevs *et al.* (2007a), we choose $\mathbf{w} = \mathbf{w}^h$ and $q = q^h$, introduce Equations (2.187) and (2.188) into Equation (2.182), integrate by parts the fine-scale terms to move the derivatives onto the test functions, and model the fine-scale velocity and pressure fields as

$$\mathbf{u}' = -\frac{\tau_{\text{SUPS}}}{\rho} \mathbf{r}_M(\mathbf{u}^h, p^h), \tag{2.191}$$

$$p' = -\rho \nu_{\text{LSIC}} r_C(\mathbf{u}^h), \tag{2.192}$$

where $\mathbf{r}_M$ and r_C are the residuals of the Navier–Stokes linear-momentum and continuity equations:

$$\mathbf{r}_M(\mathbf{u}^h, p^h) = \rho \left(\frac{\partial \mathbf{u}^h}{\partial t} + \mathbf{u}^h \cdot \nabla \mathbf{u}^h - \mathbf{f}^h \right) - \nabla \cdot \boldsymbol{\sigma}\left(\mathbf{u}^h, p^h\right), \tag{2.193}$$

$$r_C(\mathbf{u}^h) = \nabla \cdot \mathbf{u}^h. \tag{2.194}$$

Note that the fine-scale velocity is proportional to the residual of the linear-momentum equations, while the fine-scale pressure is proportional to the residual of the continuity equation. This form of the fine-scale terms may be inferred from the fine-scale equations, which are obtained by choosing $\mathbf{w} = \mathbf{w}'$ and $q = q'$ in Equation (2.182).

With the above considerations, the RBVMS formulation is stated as follows: find $\mathbf{u}^h \in S_u^h$ and $p^h \in S_p^h$, such that $\forall \, \mathbf{w}^h \in \mathcal{V}_u^h$ and $q^h \in \mathcal{V}_p^h$:

$$\int_\Omega \mathbf{w}^h \cdot \rho \left(\frac{\partial \mathbf{u}^h}{\partial t} + \mathbf{u}^h \cdot \nabla \mathbf{u}^h - \mathbf{f}^h \right) d\Omega + \int_\Omega \boldsymbol{\varepsilon}(\mathbf{w}^h) : \boldsymbol{\sigma}(\mathbf{u}^h, p^h) \, d\Omega$$

$$- \int_{\Gamma_h} \mathbf{w}^h \cdot \mathbf{h}^h \, d\Gamma + \int_\Omega q^h \nabla \cdot \mathbf{u}^h \, d\Omega$$

$$+ \sum_{e=1}^{n_{\text{el}}} \int_{\Omega^e} \tau_{\text{SUPS}} \left(\mathbf{u}^h \cdot \nabla \mathbf{w}^h + \frac{\nabla q^h}{\rho} \right) \cdot \mathbf{r}_M(\mathbf{u}^h, p^h) \, d\Omega$$

$$+ \sum_{e=1}^{n_{\text{el}}} \int_{\Omega^e} \rho \nu_{\text{LSIC}} \nabla \cdot \mathbf{w}^h r_C(\mathbf{u}^h) \, d\Omega$$

$$-\sum_{e=1}^{n_{el}} \int_{\Omega^e} \tau_{SUPS} \mathbf{w}^h \cdot \left(\mathbf{r}_M\left(\mathbf{u}^h, p^h\right) \cdot \nabla \mathbf{u}^h\right) d\Omega$$

$$-\sum_{e=1}^{n_{el}} \int_{\Omega^e} \frac{\nabla \mathbf{w}^h}{\rho} : \left(\tau_{SUPS} \mathbf{r}_M\left(\mathbf{u}^h, p^h\right)\right) \otimes \left(\tau_{SUPS} \mathbf{r}_M\left(\mathbf{u}^h, p^h\right)\right) d\Omega = 0. \quad (2.195)$$

The stabilization parameters τ_{SUPS} and ν_{LSIC} are given by

$$\tau_{SUPS} = \left(\frac{4}{\Delta t^2} + \mathbf{u}^h \cdot \mathbf{G} \mathbf{u}^h + C_I \nu^2 \mathbf{G} : \mathbf{G}\right)^{-1/2} \quad (2.196)$$

and

$$\nu_{LSIC} = (\text{tr}\mathbf{G}\, \tau_{SUPS})^{-1}, \quad (2.197)$$

where

$$\text{tr}\mathbf{G} = \sum_{i=1}^{n_{sd}} G_{ii} \quad (2.198)$$

is the trace of the element metric tensor $\mathbf{G}$. Other commonly employed definitions of τ_{SUPS} and ν_{LSIC} are (see, e.g., Tezduyar, 2003a)

$$\tau_{SUPS} = \left(\frac{1}{\tau_{SUGN1}^2} + \frac{1}{\tau_{SUGN2}^2} + \frac{1}{\tau_{SUGN3}^2}\right)^{-\frac{1}{2}} \quad (2.199)$$

and

$$\nu_{LSIC} = \tau_{SUPS} \|\mathbf{u}^h\|^2, \quad (2.200)$$

where

$$\tau_{SUGN1} = \left(\sum_{a=1}^{n_{en}} |\mathbf{u}^h \cdot \nabla N_a|\right)^{-1}, \quad (2.201)$$

$$\tau_{SUGN2} = \frac{\Delta t}{2}, \quad (2.202)$$

$$\tau_{SUGN3} = \frac{h_{RGN}^2}{4\nu}, \quad (2.203)$$

$$h_{RGN} = 2 \left(\sum_{a=1}^{n_{en}} |\mathbf{r} \cdot \nabla N_a|\right)^{-1}, \quad (2.204)$$

$$\mathbf{r} = \frac{\nabla \|\mathbf{u}^h\|}{\|\nabla \|\mathbf{u}^h\|\|}. \quad (2.205)$$

Note that the stabilization parameter ν_{LSIC} has the dimensions of kinematic viscosity.

REMARK 2.5 *The stabilization parameters τ_{SUPS} and ν_{LSIC} in the above equations originate from stabilized finite element methods for fluid dynamics (see, e.g., Brooks and Hughes, 1982; Hughes and Tezduyar, 1984; Tezduyar and Park, 1986; Hughes et al., 1986a, 2004;*

Tezduyar and Osawa, 2000; Tezduyar, 2003a). The notation "SUPS," introduced in Takizawa and Tezduyar (2011), indicates that there is a single stabilization parameter for the SUPG and Pressure-Stabilizing/Petrov–Galerkin (PSPG) stabilizations, instead of two separate parameters. The notation "LSIC," introduced in Tezduyar and Osawa (2000), denotes the stabilization based on least-squares on the incompressibility constraint. The stabilization parameters were designed and studied extensively in the context of stabilized finite element formulations of linear model problems of direct relevance to fluid mechanics. These model problems include the advection–diffusion and Stokes equations. The design of τ_{SUPS} and ν_{LSIC} is such that optimal convergence with respect to the mesh size and polynomial order of discretization is attained for these cases (see, e.g., Hughes et al., 2004 and references therein). Furthermore, enhanced stability for advection-dominated flows and the ability to conveniently employ the same basis functions for velocity and pressure variables for incompressible flows are some of the attractive outcomes of this method. More recently, the stabilization parameters were derived in the context of the variational multiscale methods Hughes (1995) and Hughes et al. (1998) and were interpreted as the appropriate averages of the small-scale Green's function, a key mathematical object in the theory of VMS methods (see Hughes and Sangalli, 2007 for an elaboration).

REMARK 2.6 *To arrive at the RBVMS formulation given by Equation (2.195), we additionally assumed that the velocity fine scales are quasi-static (i.e., $\frac{\partial \mathbf{u}'}{\partial t} = 0$) and orthogonal to the velocity coarse scales in the H^1-seminorm. The same assumptions were employed in the original derivation of RBVMS. To improve on the RBVMS methodology, rather than assuming the velocity fine scales to be quasi-static, the authors in Codina et al. (2007) proposed to "track" them in time by solving an additional evolution equation at the quadrature points of the FEM discretization.*

The well-known SUPG/PSPG formulation of the Navier–Stokes equations of incompressible flows may be obtained from the RBVMS formulation by omitting the last two terms in Equation (2.195) and providing the option of using two different τ's in the momentum and continuity equations, namely: find $\mathbf{u}^h \in \mathcal{S}_u^h$ and $p^h \in \mathcal{S}_p^h$, such that $\forall \mathbf{w}^h \in \mathcal{V}_u^h$ and $q^h \in \mathcal{V}_p^h$:

$$\int_\Omega \mathbf{w}^h \cdot \rho \left(\frac{\partial \mathbf{u}^h}{\partial t} + \mathbf{u}^h \cdot \nabla \mathbf{u}^h - \mathbf{f}^h \right) d\Omega + \int_\Omega \boldsymbol{\varepsilon}(\mathbf{w}^h) : \boldsymbol{\sigma}(\mathbf{u}^h, p^h) \, d\Omega$$
$$- \int_{\Gamma_h} \mathbf{w}^h \cdot \mathbf{h}^h \, d\Gamma + \int_\Omega q^h \nabla \cdot \mathbf{u}^h \, d\Omega$$
$$+ \sum_{e=1}^{n_{\text{el}}} \int_{\Omega^e} \tau_{\text{SUPG}} \left(\mathbf{u}^h \cdot \nabla \mathbf{w}^h \right) \cdot \mathbf{r}_M \left(\mathbf{u}^h, p^h \right) d\Omega$$
$$+ \sum_{e=1}^{n_{\text{el}}} \int_{\Omega^e} \tau_{\text{PSPG}} \left(\frac{\nabla q^h}{\rho} \right) \cdot \mathbf{r}_M \left(\mathbf{u}^h, p^h \right) d\Omega$$
$$+ \sum_{e=1}^{n_{\text{el}}} \int_{\Omega^e} \rho \nu_{\text{LSIC}} \nabla \cdot \mathbf{w}^h r_C(\mathbf{u}^h) \, d\Omega = 0. \qquad (2.206)$$

Typically, $\tau_{\text{PSPG}} = \tau_{\text{SUPG}} = \tau_{\text{SUPS}}$. However, this is not always the case, one example being the Element-Matrix- and Element-Vector-Based definitions of τ's (Tezduyar and Osawa, 2000; Tezduyar, 2003a; Catabriga et al., 2005, 2006; Bazilevs et al., 2007b).

As we will see in the later chapters, these seemingly small differences between RBVMS and SUPG/PSPG formulations may produce notable differences in the computational results, especially for complex 3D flows.

REMARK 2.7 *In Bazilevs et al. (2007a), the RBVMS formulation was proposed in the context of Large-Eddy Simulation (LES) turbulence modeling. In contrast to classical LES turbulence modeling approaches, the proposed method does not explicitly rely on modeling turbulence phenomena by introducing ad hoc eddy-viscosity terms. Instead, a more mathematically rigorous approach is taken, in which a precise dependence of the large-scale equations on the fine-scale velocity and pressure solutions is derived. The modeling of the fine scales, which is the only approximation employed in the proposed approach, is motivated by the fine-scale equations and is supported by decades of experience with stabilized methods. The numerical results in Bazilevs et al. (2007a) and Akkerman et al. (2008) provide ample justification for using the RBVMS formulation in the LES regime. In Khurram and Masud (2006) the authors proposed a RBVMS formulation where the fine scales are modeled using element-level bubble functions.*

The vector form of the equations for the RBVMS formulation given by Equation (2.195) is written by starting with the following expressions for the space-discrete velocity and pressure trial and test functions:

$$\mathbf{u}^h(\mathbf{x}, t) = \sum_{\eta^s} \mathbf{u}_A(t) N_A(\mathbf{x}), \tag{2.207}$$

$$p^h(\mathbf{x}, t) = \sum_{\eta^s} p_A(t) N_A(\mathbf{x}), \tag{2.208}$$

$$\mathbf{w}^h(\mathbf{x}) = \sum_{\eta^w} \mathbf{w}_A N_A(\mathbf{x}), \tag{2.209}$$

$$q^h(\mathbf{x}) = \sum_{\eta^w} q_A N_A(\mathbf{x}). \tag{2.210}$$

Note that the spatial variation of the trial and test functions is accounted for by the basis functions N_A, while the time dependence of the trial functions is built into the nodal values of the velocity and pressure. Also note that the test functions have no time dependence.

Because we have separate test functions corresponding to the velocity and pressure, we define two discrete residual vectors corresponding to the linear-momentum and continuity equations. This is done by introducing Equations (2.209) and (2.210) into Equation (2.195) and assuming that $\mathbf{w}_A$'s and q_A's are arbitrary constants:

$$\mathbf{N}_M = [(N_M)_{A,i}], \tag{2.211}$$
$$\mathbf{N}_C = [(N_C)_A], \tag{2.212}$$

$$(N_M)_{A,i} = \int_\Omega N_A \mathbf{e}_i \cdot \rho \left(\frac{\partial \mathbf{u}^h}{\partial t} + \mathbf{u}^h \cdot \nabla \mathbf{u}^h - \mathbf{f}^h \right) d\Omega$$
$$+ \int_\Omega \boldsymbol{\varepsilon}(N_A \mathbf{e}_i) : \boldsymbol{\sigma}\left(\mathbf{u}^h, p^h\right) d\Omega - \int_{\Gamma_h} N_A \mathbf{e}_i \cdot \mathbf{h}^h \, d\Gamma$$
$$+ \sum_{e=1}^{n_{el}} \int_{\Omega^e} \tau_{\text{SUPS}} \left(\mathbf{u}^h \cdot \nabla N_A \mathbf{e}_i \right) \cdot \mathbf{r}_M \left(\mathbf{u}^h, p^h \right) d\Omega$$

$$+ \sum_{e=1}^{n_{el}} \int_{\Omega^e} \rho \nu_{LSIC} (\nabla \cdot N_A \mathbf{e}_i) r_C (\mathbf{u}^h) \, d\Omega$$

$$- \sum_{e=1}^{n_{el}} \int_{\Omega^e} \tau_{SUPS} N_A \mathbf{e}_i \cdot (\mathbf{r}_M (\mathbf{u}^h, p^h) \cdot \nabla \mathbf{u}^h) \, d\Omega$$

$$- \sum_{e=1}^{n_{el}} \int_{\Omega^e} \frac{\nabla N_A \mathbf{e}_i}{\rho} : (\tau_{SUPS} \mathbf{r}_M (\mathbf{u}^h, p^h)) \otimes (\tau_{SUPS} \mathbf{r}_M (\mathbf{u}^h, p^h)) \, d\Omega, \quad (2.213)$$

$$(N_C)_A = \int_{\Omega} N_A \nabla \cdot \mathbf{u}^h \, d\Omega + \sum_{e=1}^{n_{el}} \int_{\Omega^e} \tau_{SUPS} \frac{\nabla N_A}{\rho} \cdot \mathbf{r}_M (\mathbf{u}^h, p^h) \, d\Omega, \quad (2.214)$$

where $\mathbf{e}_i$ is the ith Cartesian basis vector.

Let $\mathbf{U} = [\mathbf{u}_B]$, $\dot{\mathbf{U}} = [\dot{\mathbf{u}}_B]$, and $\mathbf{P} = [p_B]$ denote the vectors of nodal degrees-of-freedom of velocity, velocity time derivative, and pressure, respectively. With this definition of the nodal degrees-of-freedom, the *vector form* of the semi-discrete equations corresponding to Equation (2.195) becomes: find $\mathbf{U}$, $\dot{\mathbf{U}}$, and $\mathbf{P}$, such that:

$$\mathbf{N}_M(\dot{\mathbf{U}}, \mathbf{U}, \mathbf{P}) = \mathbf{0}, \quad (2.215)$$

$$\mathbf{N}_C(\dot{\mathbf{U}}, \mathbf{U}, \mathbf{P}) = \mathbf{0}. \quad (2.216)$$

The *vector form* given by Equations (2.215) and (2.216) presents a system of ordinary differential-algebraic equations. We postpone the description of the time integration procedures, which will be presented in detail for the moving-domain case in Section 4.6.2. Note that because the discrete variational equations (2.215) and (2.216) are nonlinear, no natural matrix form exists at this stage.

2.6.2 Weakly Enforced Essential Boundary Conditions

In this section we state the formulation of the weakly enforced essential boundary conditions. These were first proposed for the advection–diffusion equation and for the Navier–Stokes equations of incompressible flows in Bazilevs and Hughes (2007) in an effort to improve the accuracy of the stabilized and multiscale formulations in the presence of unresolved boundary layers. In Bazilevs *et al.* (2007c, 2010e) and Bazilevs and Akkerman (2010), the weak boundary condition formulation was further refined and studied on a set of challenging wall-bounded turbulent flows, including aerodynamics of wind turbines in 3D at full spatial scale Hsu *et al.* (2011a).

To account for the weak enforcement of the essential boundary conditions, we remove them from the trial and test function sets $\mathcal{S}_u^h$ and $\mathcal{V}_u^h$, and add the following terms to the left-hand-side of Equations (2.195) and (2.206):

$$- \sum_{b=1}^{n_{eb}} \int_{\Gamma^b \cap \Gamma_g} \mathbf{w}^h \cdot \boldsymbol{\sigma} (\mathbf{u}^h, p^h) \mathbf{n} \, d\Gamma$$

$$- \sum_{b=1}^{n_{eb}} \int_{\Gamma^b \cap \Gamma_g} (2\mu \boldsymbol{\varepsilon} (\mathbf{w}^h) \mathbf{n} + q^h \mathbf{n}) \cdot (\mathbf{u}^h - \mathbf{g}^h) \, d\Gamma$$

$$- \sum_{b=1}^{n_{eb}} \int_{\Gamma^b \cap (\Gamma_g)^-} \mathbf{w}^h \cdot \rho (\mathbf{u}^h \cdot \mathbf{n}) (\mathbf{u}^h - \mathbf{g}^h) \, d\Gamma$$

$$+ \sum_{b=1}^{n_{\text{eb}}} \int_{\Gamma^b \cap \Gamma_g} \tau_{\text{TAN}}^B \left(\mathbf{w}^h - \left(\mathbf{w}^h \cdot \mathbf{n} \right) \mathbf{n} \right) \cdot \left(\left(\mathbf{u}^h - \mathbf{g}^h \right) - \left(\left(\mathbf{u}^h - \mathbf{g}^h \right) \cdot \mathbf{n} \right) \mathbf{n} \right) d\Gamma$$

$$+ \sum_{b=1}^{n_{\text{eb}}} \int_{\Gamma^b \cap \Gamma_g} \tau_{\text{NOR}}^B \left(\mathbf{w}^h \cdot \mathbf{n} \right) \left(\left(\mathbf{u}^h - \mathbf{g}^h \right) \cdot \mathbf{n} \right) d\Gamma, \tag{2.217}$$

where $\mathbf{g}^h$ is the prescribed velocity on Γ_g. The boundary Γ_g is decomposed into n_{eb} surface elements denoted by Γ^b, and Γ_g^- is defined as the "inflow" part of Γ_g:

$$\Gamma_g^- = \left\{ \mathbf{x} \mid \mathbf{u}^h \cdot \mathbf{n} < 0, \forall \mathbf{x} \subset \Gamma_g \right\}. \tag{2.218}$$

The term in the first line is the so-called consistency term. It is necessary to ensure that the discrete formulation is identically satisfied by the exact solution of the Navier–Stokes equations, which, in turn, has implications on the accuracy of the discrete formulation. Also note that this term cancels out with the contributions coming from the integration-by-parts of the stress terms in Equations (2.195) and (2.206), thus correctly removing traction boundary conditions from the no-slip boundary. The term in the second line is the adjoint consistency term. Its role is less intuitive, as it ensures that the analytical solution of the adjoint equations, when introduced in place of the linear-momentum and continuity equation test functions, also satisfies the discrete formulation. Adjoint consistency is linked to optimal convergence of the discrete solution in lower-order norms (see, e.g., Arnold et al., 2002). The term in the third line leads to better satisfaction of the inflow boundary conditions. The last two terms are penalty-like, in that they penalize the deviation of the discrete solution from its prescribed value at the boundary. These terms are necessary to ensure the stability (or coercivity) of the discrete formulation, which may be lost due to the introduction of the consistency and adjoint consistency terms.

The weak boundary condition formulation is numerically stable if

$$\tau_{\text{TAN}}^B = \tau_{\text{NOR}}^B = \frac{C_I^B \mu}{h_n}, \tag{2.219}$$

where h_n is the wall-normal element size, and C_I^B is a sufficiently large positive constant computed from an appropriate element-level inverse estimate (see, e.g., Johnson, 1987; Brenner and Scott, 2002; Ern and Guermond, 2004). The constant C_I^B depends on the space dimension n_{sd}, the element type (tetrahedron, hexahedron, etc.), and the polynomial order of the finite element approximation. For a linear tetrahedron, it is sufficient to take $4.0 \leq C_I^B \leq 8.0$ to obtain a stable discrete solution. The wall-normal element size may be computed from the element metric tensor:

$$h_n = (\mathbf{n} \cdot \mathbf{G} \mathbf{n})^{1/2}. \tag{2.220}$$

REMARK 2.8 *Rather than setting the no-slip boundary conditions exactly, the weak boundary condition formulation gives the no-slip solution only in the limit as $h_n \to 0$. As a result, coarse discretizations do not need to struggle to resolve the boundary layers; the flow simply slips on the solid boundary. Because of this added flexibility, the weak boundary condition enforcement tends to produce more accurate results on meshes that are too coarse to capture the boundary layer solution. However, as the mesh is refined to capture the boundary layer, the weak boundary condition formulation converges to the strong result (see Bazilevs et al., 2007c).*

REMARK 2.9 *Although the weak boundary condition formulation is also stable for very large values of C_I^B, we do not favor that. Large values of C_I^B place a heavy penalization on the no-slip condition, and the above mentioned flexibility of the method is lost together with the associated accuracy benefits. We favor using a C_I^B that is just large enough to guarantee the stability of the discrete formulation.*

REMARK 2.10 *In Bazilevs et al. (2007c), a connection was identified between the weakly-enforced boundary conditions and wall functions. The latter are commonly employed in conjunction with RANS formulations of turbulent flows (see, e.g., Launder and Spalding, 1974; Wilcox, 1998). In the case of wall function formulation, a no-slip boundary condition is replaced with a tangential traction boundary condition, where the traction direction is given by that of the local slip velocity, and the traction magnitude is computed by invoking the "law-of-the-wall." This is an empirical relationship between the flow speed and the normal distance to the wall, both appropriately normalized (see, e.g., Launder and Spalding, 1974). The penalty parameter τ_{TAN}^B may be defined as*

$$\tau_{\mathrm{TAN}}^B = \frac{\rho u^{*2}}{\|\mathbf{u}_{\mathrm{TAN}}^h\|}, \qquad (2.221)$$

where $\mathbf{u}_{\mathrm{TAN}}^h = \left(\left(\mathbf{u}^h - \mathbf{g}^h\right) - \left(\left(\mathbf{u}^h - \mathbf{g}^h\right) \cdot \mathbf{n}\right)\mathbf{n}\right)$ is the tangential slip velocity, and u^* is the so-called friction velocity, which, among other factors, depends on the magnitude of the slip velocity, and is computed from the law-of-the-wall formula by nonlinear iterations. It was shown in Bazilevs et al. (2007c), however, that when the boundary layer mesh is fine enough, τ_{TAN}^B from Equation (2.221) is independent of the local flow solution, and reverts to the definition given by Equation (2.219). This fact is remarkable in that Equation (2.219) is purely based on considerations of numerical stability, while Equation (2.221) derives from the physics of wall-bounded turbulent flows. In our limited experience, the "numerics-based" and "physics-based" definitions of the penalty parameter τ_{TAN}^B give very similar results.

REMARK 2.11 *In the case weakly-enforced boundary conditions are employed, Equations (2.213) and (2.214) will need to be augmented with the terms corresponding to Equation (2.217).*

3

Basics of the Isogeometric Analysis

The concept of Isogeometric Analysis (IGA) was introduced by Hughes *et al.* (2005). The motivation for introducing the IGA came from the need for tighter integration between engineering design, which is primarily done using Computer-Aided Design (CAD), and engineering simulation, which is primarily based on the FEM. The main idea behind the IGA is to focus on one, and only one, geometric model, which can be utilized directly as an analysis model, or from which geometrically precise analysis models can be automatically built. To instantiate such an idea requires a change from the classical FEM to an analysis procedure based on CAD representations.

There are several candidate technologies from computational geometry that may be used in IGA. The most widely used in engineering design are NURBS (non-uniform rational B-splines), the industry standard (see, Farin, 1995; Piegl and Tiller, 1997; Rogers, 2001). The major strengths of NURBS are that they are convenient for free-form surface modeling, can exactly represent all conic sections, and therefore circles, cylinders, spheres, ellipsoids, and other special geometries, and that there exist many efficient and numerically stable algorithms to generate NURBS objects. They also possess useful mathematical properties, such as good approximation, and the ability to be refined through knot insertion. Because NURBS are a CAD standard, representing many years of development, they were the natural starting point for IGA.

T-splines (Sederberg *et al.*, 2003, 2004) are a recently developed generalization of NURBS technology. They extend NURBS to permit local refinement (and coarsening). They are backward- and forward-compatible with NURBS, which makes them an attractive CAD technology. Preliminary investigations of T-splines as IGA technology may be found in Bazilevs *et al.* (2010a) and Dörfel *et al.* (2010). Recent results on linear independence and improved local refinement algorithms may be found in Li *et al.* (2012) and Scott *et al.* (2012). A recent attempt to construct solid T-Splines from existing hexahedral meshes may be found in Wang *et al.* (2012).

Computational Fluid–Structure Interaction: Methods and Applications, First Edition.
Yuri Bazilevs, Kenji Takizawa and Tayfun E. Tezduyar.
© 2013 John Wiley & Sons, Ltd. Published 2013 by John Wiley & Sons, Ltd.

Another computational geometry technology for IGA that warrants further investigation is subdivision surfaces, which use a limiting process to define a smooth surface from a mesh of triangles or quadrilaterals (see, e.g., Peters and Reif, 2008; Warren and Weimer, 2002). They have already been used in analysis of shell structures by Cirak *et al.* (2000, 2002) and Cirak and Ortiz (2001). The appeal of subdivision surfaces is that there is no restriction on the topology of the control grid. Like T-splines, they also create gap-free models. Subdivison surfaces were quickly adopted by the animation industry. However, they did not enjoy a wide-spread adoption in the CAD industry because they are not compatible with NURBS. Nevertheless, subdivision surfaces are playing an important role in IGA. Subdivision solids have been studied by Bajaj *et al.* (2002).

In this section we briefly review the main ideas behind the IGA, restricting the presentation to the NURBS-based analysis framework. Examples of IGA simulations for FSI problems will be shown in the later chapters. For more detailed mathematical developments, basis function research, geometry modeling, model quality assessment, and early applications, the reader is referred to Cottrell *et al.* (2009) and references therein.

3.1 B-Splines in 1D

NURBS are built from B-splines and so a discussion of B-splines is a natural starting point for their investigation. A *knot vector* in 1D is a non-decreasing set of coordinates (real numbers) in the parameter space, written $\Xi = \{\xi_1, \xi_2, \ldots, \xi_{n_k}\}$, where $\xi_i \in \mathbb{R}$ is the ith knot, i is the knot index, $i = 1, 2, \ldots, n_k$, $n_k = n_c + p + 1$, is the number of knots in the knot vector, p is the polynomial order, and n_c is the number of basis functions used in constructing the B-spline curve. The knots partition the parameter space into elements. Element boundaries in the physical space are the images of knot lines under the B-spline mapping.

Knot vectors may be uniform or nonuniform, depending on whether the knots are equally spaced or not. Knot values may be repeated, that is, more than one knot may take on the same value. The multiplicities of knot values have important implications for the properties of the basis. A knot vector is said to be *open* if its first and last knot values appear $p + 1$ times. Open knot vectors are the standard in the CAD literature. In 1D, basis functions formed from open knot vectors are interpolatory at the ends of the parameter space interval, $[\xi_1, \xi_{n_k}]$, and at the corners of patches in multiple dimensions, but they are not, in general, interpolatory at interior knots. A further consequence of the use of open knot vectors in multiple dimensions is that the boundary of a B-spline object with n_{sd} parametric dimensions is itself a B-spline object of dimension $n_{sd} - 1$. For example, each edge of a B-spline surface is itself a B-spline curve. This is an important property, which allows one to separate the basis function into two sets, one corresponding to the interior and the other to the boundary of the computational domain. This decomposition greatly simplifies imposition of essential and natural boundary conditions (see, e.g., Wang and Xuan, 2010).

For a given knot vector, the B-spline basis functions are defined recursively starting with piecewise constants ($p = 0$):

$$N_{i,0}(\xi) = \begin{cases} 1 & \text{if } \xi_i \leq \xi < \xi_{i+1}, \\ 0 & \text{otherwise.} \end{cases} \qquad (3.1)$$

Basics of the Isogeometric Analysis

For $p = 1, 2, 3, \ldots$, they are defined by

$$N_{i,p}(\xi) = \frac{\xi - \xi_i}{\xi_{i+p} - \xi_i} N_{i,p-1}(\xi) + \frac{\xi_{i+p+1} - \xi}{\xi_{i+p+1} - \xi_{i+1}} N_{i+1,p-1}(\xi), \qquad (3.2)$$

which is the *Cox-de Boor recursion formula* (see Cox, 1971; de Boor, 1972).

With the definition given by Equations (3.1) and (3.2), the B-spline basis constitutes a partition of unity, that is, $\forall \xi$

$$\sum_{i=1}^{n_c} N_{i,p}(\xi) = 1. \qquad (3.3)$$

Each basis function is pointwise nonnegative over the entire domain, that is, $N_{i,p}(\xi) \geq 0 \ \forall \ \xi$. Each pth order function has $p - 1$ continuous derivatives across the element boundaries (i.e., across the knots). This feature has many important implications for analysis and is one of the distinguishing features of IGA. The support of the B-spline functions of order p is always $p + 1$ knot spans. Thus higher-order functions have support over larger portions of the domain than the classical FEM functions. However, despite the larger support, the total number of B-spline basis functions that any given function shares support with (including itself) is $2p + 1$. As a result, the "bandwidth" of the left-hand-side matrix is the same for B-splines and C^0-continuous FEM functions of order p.

An example of a quadratic B-spline basis is presented in Figure 3.1a for the open, nonuniform knot vector $\Xi = \{0, 0, 0, 1, 2, 3, 4, 4, 5, 5, 5\}$. Note that the basis functions are interpolatory at the ends of the interval and also at $\xi = 4$, the location of a repeated knot where only C^0-continuity is attained. Elsewhere, the functions are C^1-continuous. In general, basis functions of order p have $p - m_i$ continuous derivatives across knot ξ_i, where m_i is the multiplicity of the value of ξ_i in the knot vector.

3.2 NURBS Basis Functions, Curves, Surfaces, and Solids

A NURBS entity in $\mathbb{R}^{n_{sd}}$ is obtained by the projective transformation of a B-spline entity in $\mathbb{R}^{n_{sd}+1}$. In particular, conic sections, such as circles and ellipses, can be *exactly* constructed by projective transformations of piecewise quadratic curves – one of the defining features of IGA. A good introduction to and discussion of projective geometry in the context of NURBS may be found in Farin (1995). Algebraically, the NURBS basis underlying projective transformations is

$$R_i^p(\xi) = \frac{N_{i,p}(\xi) w_i}{W(\xi)}, \qquad (3.4)$$

where $W(\xi)$ is the *weighting function* given as

$$W(\xi) = \sum_{i=1}^{n_c} N_{i,p}(\xi) w_i, \qquad (3.5)$$

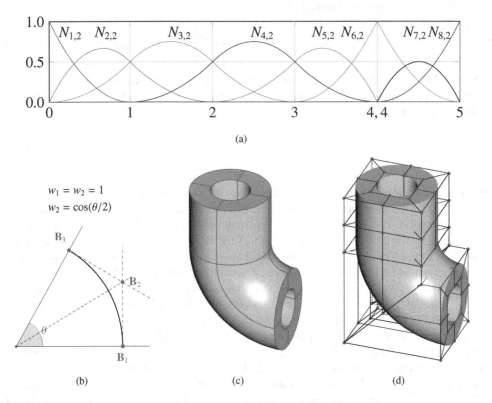

Figure 3.1 (a) Quadratic basis functions for open, nonuniform knot vector $\Xi = \{0,0,0,1,2,3,4,4,5,5,5\}$. (b) Quadratic NURBS description of a circular arc. Control points and weights are given in the figure and the underlying knot vector is $\Xi = \{0,0,0,1,1,1\}$. (c) Section of a hollow circular pipe represented as a NURBS solid: control mesh. (d) Section of a hollow circular pipe represented as a NURBS solid: quadratic NURBS mesh

$N_{i,p}(\xi)$ is the B-spline basis function, and w_i is a positive real *weight*. We conclude from Equations (3.4) and (3.5) that this new basis is no longer a piecewise polynomial, but piecewise rational.

NURBS curves in $\mathbb{R}^{n_{sd}}$ are constructed by taking a linear combination of NURBS basis functions. The vector-valued coefficients of the basis functions are called *control points*. They are analogous to nodal coordinates in the FEM. Given n_c basis functions, R_i^p, $i = 1, 2, \ldots, n_c$, and corresponding control points $\mathbf{B}_i \in \mathbb{R}^{n_{sd}}$, $i = 1, 2, \ldots, n_c$, and weights $w_i \in \mathbb{R}$, $i = 1, 2, \ldots, n_c$, a piecewise-rational NURBS curve is given by

$$\mathbf{C}(\xi) = \sum_{i=1}^{n_c} R_i^p(\xi) \mathbf{B}_i. \tag{3.6}$$

Piecewise linear interpolation of the control points gives the *control mesh*. Figure 3.1b gives an example of a quadratic NURBS curve corresponding to a piece of an exact circular arc.

We define the NURBS basis functions for multidimensional shapes, such as surfaces and solids, as follows:

$$R_{i,j}^{p,q}(\xi,\eta) = \frac{N_{i,p}(\xi)M_{j,q}(\eta)w_{i,j}}{\sum_{\hat{i}=1}^{n_c}\sum_{\hat{j}=1}^{m_c} N_{\hat{i},p}(\xi)M_{\hat{j},q}(\eta)w_{\hat{i},\hat{j}}}, \tag{3.7}$$

$$R_{i,j,k}^{p,q,r}(\xi,\eta,\zeta) = \frac{N_{i,p}(\xi)M_{j,q}(\eta)L_{k,r}(\zeta)w_{i,j,k}}{\sum_{\hat{i}=1}^{n_c}\sum_{\hat{j}=1}^{m_c}\sum_{\hat{k}=1}^{l_c} N_{\hat{i},p}(\xi)M_{\hat{j},q}(\eta)L_{\hat{k},r}(\zeta)w_{\hat{i},\hat{j},\hat{k}}}. \tag{3.8}$$

Here $\Xi = \{\xi_1, \xi_2, \ldots, \xi_{n_k}\}$, $\mathcal{H} = \{\eta_1, \eta_2, \ldots, \eta_{m_k}\}$, and $\mathcal{Z} = \{\zeta_1, \zeta_2, \ldots, \zeta_{l_k}\}$ are the knot vectors, and $w_{i,j}$ and $w_{i,j,k}$ are the weights for surfaces and solids, respectively. In addition, $m_k = m_c + q + 1$ and $l_k = l_c + r + 1$ denote the number of knots in $\mathcal{H}$ and $\mathcal{Z}$, respectively, where m_c and l_c denote the number of univariate B-spline basis functions of polynomial order q and r, respectively.

Given a control mesh $\{\mathbf{B}_{i,j}\}, i = 1, 2, \ldots, n_c, j = 1, 2, \ldots, m_c$, polynomial orders p and q, and knot vectors $\Xi = \{\xi_1, \xi_2, \ldots, \xi_{n_k}\}$ and $\mathcal{H} = \{\eta_1, \eta_2, \ldots, \eta_{m_k}\}$, a NURBS surface may be defined as

$$\mathbf{S}(\xi,\eta) = \sum_{i=1}^{n_c}\sum_{j=1}^{m_c} R_{i,j}^{p,q}(\xi,\eta)\mathbf{B}_{i,j}, \tag{3.9}$$

where $R_{i,j}^{p,q}(\xi,\eta)$'s are constructed according to Equation (3.7) using $N_{i,p}(\xi)$ and $M_{j,q}(\eta)$, the univariate B-spline basis functions of order p and q, corresponding to knot vectors Ξ and $\mathcal{H}$, respectively. Many of the properties of a B-spline surface are the result of its tensor-product nature. The basis is pointwise nonnegative, and forms a partition of unity. The number of continuous partial derivatives in a given parametric direction may be determined from the associated 1D knot vector and polynomial order.

Tensor product NURBS solids are defined in a fashion analogous to how the NURBS surfaces are defined. Given a control mesh $\{\mathbf{B}_{i,j,k}\}, i = 1, 2, \ldots, n_c, j = 1, 2, \ldots, m_c, k = 1, 2, \ldots, l_c$, polynomial orders p, q, and r, and knot vectors $\Xi = \{\xi_1, \xi_2, \ldots, \xi_{n_k}\}$, $\mathcal{H} = \{\eta_1, \eta_2, \ldots, \eta_{m_k}\}$, and $\mathcal{Z} = \{\zeta_1, \zeta_2, \ldots, \zeta_{l_k}\}$, a NURBS solid is defined as

$$\mathbf{S}(\xi,\eta,\zeta) = \sum_{i=1}^{n_c}\sum_{j=1}^{m_c}\sum_{k=1}^{l_c} R_{i,j,k}^{p,q,r}(\xi,\eta,\zeta)\mathbf{B}_{i,j,k}. \tag{3.10}$$

The properties of NURBS solids are trivariate generalizations of those for NURBS surfaces. An example of a NURBS solid is given in Figure 3.1c, where the object is depicted together with its control mesh.

REMARK 3.1 *If all weights are equal, the NURBS curves, surfaces, and solids reduce to their B-spline counterparts. This is evident from the definition of the NURBS basis functions given by Equations (3.4), (3.7) and (3.8) and the partition of unity property of the B-spline basis functions.*

3.3 *h-*, *p-*, and *k-*Refinement of NURBS Meshes

In this section we summarize various mesh refinement procedures that are available in IGA.

h-refinement. The NURBS curve (and basis) may be enriched by *knot insertion* as follows. Given a knot vector $\Xi = \{\xi_1, \xi_2, \ldots, \xi_{n_k}\}$, we introduce the notion of an *extended* knot vector $\bar{\Xi} = \{\bar{\xi}_1 = \xi_1, \bar{\xi}_2, \ldots, \bar{\xi}_{n_k+m} = \xi_{n_k}\}$, such that $\Xi \subset \bar{\Xi}$. The new $n_c + m$ basis functions are formed by applying the Cox–de Boor recursion given by Equations (3.1) and (3.2) to the new knot vector $\bar{\Xi}$. The new $n_c + m$ control points and weights are formed from linear combinations of the original control points and weights using a linear transformation given in Cottrell et al. (2007, 2009). This procedure ensures that the original curve stays unchanged geometrically or parametrically. The knot insertion procedure is analogous to *h*-refinement in the FEM.

REMARK 3.2 *Knot values already present in the knot vector may be repeated in this way, thereby increasing their multiplicity. As a result, the continuity of the NURBS basis will be reduced. However, as a result of the knot insertion procedure, the new control points will be formed such that the continuity of the NURBS curve will remain unchanged.*

p-refinement. The NURBS curve (and basis) may also be enriched by *degree elevation*, which is analogous to *p*-refinement in the FEM. The process involves raising the polynomial degree of the basis functions used to represent the geometry. Because the NURBS basis has $p - m_i$ continuous derivatives across element boundaries, it is clear that when the polynomial order is increased, the multiplicity of the knots must also be increased if we are to exactly preserve the continuity of the original curve. Elevating the order of the curve by one increases the multiplicity of each knot value by one, however, no new knot values are added. As with knot insertion, neither the geometry nor its parameterization are changed. The process for order elevation consists of thee steps. First, all the existing knots are repeated such that their multiplicity equals p, thus effectively subdividing the curve into many Bézier curves (see e.g., Rogers, 2001 or Farin, 1995 for a discussion of Bézier curves). Then, the polynomial order is increased on each one of the individual segments. Lastly, excess knots are removed to combine the segments into one, order-elevated, curve. For a thorough treatment and efficient algorithms (see Piegl and Tiller, 1997).

k-refinement. There is a more flexible higher-order refinement procedure in IGA, which stems from the fact that the processes of order elevation and knot insertion do not commute. If a unique knot value, $\bar{\xi}$, is inserted between two distinct knot values in a curve of order p, the number of continuous derivatives of the basis functions at $\bar{\xi}$ is $p - 1$. If we subsequently elevate the order to q, the multiplicity of every distinct knot value (including the knot just inserted) is increased so that discontinuities in the pth derivative of the basis are preserved. That is, the basis still has $p - 1$ continuous derivatives at $\bar{\xi}$, although the polynomial order is now q. If, instead, we elevated the order of the original, coarsest curve to q and only then inserted the unique knot value $\bar{\xi}$, the basis would have $q - 1$ continuous derivatives at $\bar{\xi}$. We refer to this latter procedure as *k*-refinement. There are no analogs of this procedure in the FEM. The concept of *k*-refinement is potentially a superior approach to high-precision analysis than *p*-refinement. In traditional *p*-refinement there is a proliferation in the number of nodes because C^0-continuity is maintained in the refinement process. In *k*-refinement, the growth in the number of control variables is limited (see Cottrell et al., 2009 for more details).

3.4 NURBS Analysis Framework

We have introduced NURBS (and, as a special case, B-spline) basis functions, curves, surfaces, and solids. We also introduced mesh refinement procedures. At this point, we consolidate the notation and terminology in order to introduce the IGA framework as an extension of the FEM framework presented in earlier sections.

We refer to "NURBS" even in the case we are dealing with "B-splines." As a result, we write $N(\xi)$ to refer to any basis function, which could be a univariate, bivariate, or trivariate, non-rational, or rational basis function.

We denote the physical domain (i.e., the geometry) by Ω, and the parametric domain by $\hat{\Omega}$. The parametric domain in 2D is a rectangle and in 3D is a cuboid. Both $\hat{\Omega}$ and Ω are referred to as the patch. We write $\mathbf{x} : \hat{\Omega} \to \Omega$ as the geometrical mapping, taking points in the parametric domain and returning the corresponding points in the physical domain. We assume that this mapping is invertible, and we denote the inverse mapping by $\mathbf{x}^{-1} : \Omega \to \hat{\Omega}$. The inverse mapping takes points in the physical domain and identifies their corresponding parameter values. While most geometries utilized for academic test cases can be modeled with a single patch, more complicated shapes are typically comprised of multiple patches. The patches may be merged with C^0 or higher continuity.

Each patch is decomposed into knot spans. Knots are points, lines, and surfaces in 1D, 2D, and 3D topologies, respectively. Knot spans are bounded by knots. These define element domains, where basis functions are smooth (i.e., C^∞). Across knots, basis functions will be C^{p-m}, where p is the degree of the polynomial and m is the multiplicity of the knot in question. In NURBS-based IGA, we consider elements to be the images of knot spans under the NURBS mapping. We denote these knot spans in the parameter space by $\hat{\Omega}^e$, and their image in the physical space as Ω^e, where $e = 1, \ldots, n_{el}$, with n_{el} being the total number of elements in the NURBS mesh. As a result, $\hat{\Omega} = \bigcup_{e=1}^{n_{el}} \hat{\Omega}^e$ and $\Omega = \bigcup_{e=1}^{n_{el}} \Omega^e$. An example of a NURBS mesh of a solid object is shown in Figure 3.1d.

In what follows, indices $A, B, C, \ldots$ identify the global basis functions. Their range is $1, \ldots, n_{np}$, where n_{np} is the total number of basis functions in the NURBS mesh. Likewise, indices $a, b, c, \ldots$ identify the local basis functions, or basis functions that are supported on a given NURBS element. Their range is $1, \ldots, n_{en}$, where n_{en} is the total number of basis functions supported on a given NURBS element. The double and triple sum over the indices, which is typically used in the definition of NURBS surfaces and solids (see, e.g., Equations (3.9) and (3.10)), are replaced with a single sum over the global indices $A, B, C, \ldots$ or the local indices $a, b, c, \ldots$, with the appropriate mapping between the multiply-indexed and singly-indexed numberings.

With this new notation, the geometrical mapping for a patch is

$$\mathbf{x}(\xi) = \sum_{A=1}^{n_{np}} \mathbf{x}_A N_A(\xi), \tag{3.11}$$

where $\mathbf{x}_A$'s are the coordinates of the control points. We *do not* require that the basis functions $N_A(\xi)$'s satisfy the interpolation property. The above mapping may be *restricted* to the element level as follows:

$$\mathbf{x}(\xi) = \sum_{a=1}^{n_{en}} \mathbf{x}_a N_a(\xi), \tag{3.12}$$

where $\mathbf{x}_a$'s and $N_a(\xi)$ are the local control points and basis functions extracted from their global counterparts. This extraction requires a data structure, which is completely analogous to the *IEN array* data structure used in the FEM and given by Equation (2.89).

To construct the IGA solution field over a patch, we first define it on a parametric domain as

$$u^h(\boldsymbol{\xi}) = \sum_{A=1}^{n_{np}} u_A N_A(\boldsymbol{\xi}). \qquad (3.13)$$

where u_A's are the control variables or degrees-of-freedom. Given the solution in the parametric domain, its physical-domain counterpart may be defined as

$$u^h(\mathbf{x}) = \sum_{A=1}^{n_{np}} u_A N_A(\boldsymbol{\xi}(\mathbf{x})), \qquad (3.14)$$

which leads to a natural definition of the NURBS basis functions on the physical domain:

$$N_A(\mathbf{x}) = N_A(\boldsymbol{\xi}(\mathbf{x})). \qquad (3.15)$$

This is precisely the isoparametric construction employed in the FEM.

The basis functions and the solution field may be restricted to each NURBS element as follows:

$$N_a(\mathbf{x}) = N_a(\boldsymbol{\xi}(\mathbf{x})), \qquad (3.16)$$

$$u^h(\mathbf{x}) = \sum_{a=1}^{n_{en}} u_a N_a(\mathbf{x}). \qquad (3.17)$$

The correspondence between the local and global quantities is given through the IEN array. This restriction of the basis functions and solution vectors to the element is necessary to construct element-level left-hand-side matrices and right-hand-side vectors. These are assembled into the global matrices and vectors in a fashion analogous to the standard FEM.

REMARK 3.3 *We note that, unlike in finite elements, where the parameter space is local to the element and every element in the mesh is an image of the same parametric element, in NURBS-based analysis the parameter space is local to the patch and each physical element is an image of its own parametric element.*

REMARK 3.4 *We also note that knot spans are convenient for numerical quadrature. In the current implementation of IGA, Gaussian quadrature rule is defined on each individual knot span in the parametric domain. While this gives accurate evaluation of integrals used in the construction of the left-hand-side matrices and right-hand-side vectors, one might argue that this approach is not very efficient because Gaussian quadrature rule is insensitive to the underlying continuity of the basis functions employed. Recently, more efficient quadrature rules were proposed for NURBS-based IGA, which take the continuity of the basis functions into account (see Hughes et al., 2010).*

REMARK 3.5 *The above construction of the approximation spaces for NURBS-based IGA, which relies on the isoparametric concept, is readily extendible to T-splines and subdivision surfaces. In fact, if one is looking for a concise and descriptive definition of IGA, then one may think of it as a combination of CAD basis functions and the isoparametric concept.*

The natural question is whether NURBS are capable of producing computational results that are convergent under mesh refinement. This question was answered in Bazilevs *et al.* (2006b), where the authors rigorously proved the following interpolation estimate for NURBS approximation in the physical domain: let k and l be integer indices such that $0 \leq k \leq l \leq p+1$, and let $u \in H^l(\Omega)$; then

$$\sum_{e=1}^{n_{el}} |u - \Pi^h u|^2_{H^k(\Omega^e)} \leq C \sum_{e=1}^{n_{el}} h_e^{2(l-k)} \sum_{i=0}^{l} \|\nabla_\xi \mathbf{x}\|^{2(i-l)}_{L^\infty(\mathbf{x}^{-1}(\Omega^e))} |u|^2_{H^i(\Omega^e)}. \tag{3.18}$$

Here $\Pi^h u$ is an appropriately constructed NURBS interpolant or projector (see Bazilevs *et al.*, 2006b for details), the constant C depends on p and the shape (but not size) of the domain Ω, as well as the shape regularity of the mesh. See da Veiga *et al.* (2012) for a recent extension of the above result, which removes the shape regularity requirement of the NURBS mesh. The factors involving the gradient of the geometrical mapping in Equation (3.18) render the estimate dimensionally consistent. The fact that the geometrical mapping is fixed at the coarsest level of discretization (i.e., it stays unchanged throughout the refinement process) renders the above interpolation estimate optimal with respect to the mesh size for a fixed polynomial order. Despite the fact that rational functions are employed for the computations, the error in isogeometric solution converges at the same rate as in the FEM. This makes NURBS-based IGA a realistic simulation technology.

4

ALE and Space–Time Methods for Moving Boundaries and Interfaces

In this chapter we show the construction of the basis functions for moving-domain problems. We present the ALE and space–time formulations of flows with moving boundaries and interfaces. We conclude this chapter with a discussion of basic mesh update methods for moving-domain problems.

4.1 Interface-Tracking (Moving-Mesh) and Interface-Capturing (Nonmoving-Mesh) Techniques

We can view a method for flow problems with moving boundaries and interfaces as an interface-tracking (moving-mesh) technique or an interface-capturing (nonmoving-mesh[1]) technique, or a combination of the two. In interface-tracking methods, as the interface moves and the spatial domain occupied by the fluid changes its shape, the mesh moves to accommodate this shape change and to follow (i.e., "track") the interface. Moving the fluid mesh to track a fluid–solid interface enables us, at least for interfaces with reasonable geometric complexity, to control the mesh resolution near that interface and obtain accurate solutions in such critical flow regions. Sometimes the geometric complexity of the interface may require a fluid mechanics mesh that is not affordable or not desirable or just not manageable in mesh moving, and this is one of the most common reasons given for favoring an interface-capturing method. This approach can be seen as a special case of interface representation techniques where the interface geometry is somehow represented over a nonmoving fluid mechanics mesh, the main point being that the fluid mesh does not move to track the interfaces. However, as pointed out in Tezduyar (2003a), a consequence of the mesh not moving to track the interface is that for fluid–solid interfaces, independent of

---

[1] We use the terminology "nonmoving mesh" as a generalization of "fixed mesh," where the mesh does not move to track the interface but does not have to remain fixed either and could change from time to time.

Computational Fluid–Structure Interaction: Methods and Applications, First Edition.
Yuri Bazilevs, Kenji Takizawa and Tayfun E. Tezduyar.
© 2013 John Wiley & Sons, Ltd. Published 2013 by John Wiley & Sons, Ltd.

how accurately the interface geometry is represented, the resolution of the boundary layer will be limited by the resolution of the fluid mesh where the interface is. Therefore, for interfaces with reasonable geometric complexity, if a moving-mesh method can be used with a reasonable remeshing (see Tezduyar and Sathe, 2007 for various remeshing options) cost, its fluid mechanics accuracy near the interface will be superior to that of an nonmoving-mesh method. Furthermore, while it is understandable that nonmoving-mesh methods become more favored when the interface geometric complexity appears to be too high for a moving-mesh method, we need to remember that there is a difference between making the problem computable and obtaining good fluid mechanics accuracy near the interface. In addition, when the problem includes interfaces with high geometric complexity, we can find a number of ways to make the problem computable also with moving-mesh methods. Yet, we can still expect to obtain good accuracy where the interfaces have reasonable geometric complexity. Examples of that were given by Tezduyar *et al.* (2008a).

4.2 Mixed Interface-Tracking/Interface-Capturing Technique (MITICT)

We of course recognize that certain classes of interfaces (such as free-surface or two-fluid flows with splashing) might be too complex to deal with an interface-tracking technique and therefore, for all practical purposes, require an interface-capturing technique. The Mixed Interface-Tracking/Interface-Capturing Technique (MITICT) Tezduyar (2001b) was introduced in 2001 for computation of flow problems that involve both interfaces that can be accurately tracked with a moving-mesh method and interfaces that are too complex to be tracked and therefore require falling back to an interface-capturing technique. The mesh moves to track the interfaces that can be tracked, and the interfaces that are too complex to track are captured over that moving mesh. The MITICT was successfully tested in Akin *et al.* (2007) and Cruchaga *et al.* (2007) and used in ship hydrodynamics computations in Akkerman *et al.* (2012).

4.3 ALE Methods

In Chapter 1 we derived the conservative and convective forms of the incompressible-flow momentum balance equations in the ALE form. These are given by Equations (1.220) and (1.223), respectively. One of the earliest ALE finite element formulations for flow problems was given, in the context of viscous incompressible flows, in Hughes *et al.* (1981). In ALE methods, the partial time and space derivatives employed in the formulation of the balance equations are taken with respect to different descriptions, namely, the referential and spatial descriptions, respectively. As it turns out, this engenders significant simplifications for the underlying numerics as shown in the sequel.

The following finite-dimensional representation of a time-dependent solution in the fixed referential domain $\hat{\Omega}$ is assumed:

$$\hat{u}(\hat{\mathbf{x}}, t) = \sum_{A \in \eta^s} u_A(t) \hat{N}_A(\hat{\mathbf{x}}), \qquad (4.1)$$

where $\hat{N}_A$'s are the fixed basis functions that result from a finite element discretization of $\hat{\Omega}$, and u_A's are the time-dependent solution coefficients. From the above equation we can easily compute the referential time derivative of $\hat{u}$ as

$$\left.\frac{\partial \hat{u}(\hat{\mathbf{x}}, t)}{\partial t}\right|_{\hat{\mathbf{x}}} = \sum_{A \in \eta^s} \frac{du_A(t)}{dt} \hat{N}_A(\hat{\mathbf{x}}). \tag{4.2}$$

The basis functions on the spatial domain Ω_t are defined through the following mapping:

$$N_A(\mathbf{x}, t) = \hat{N}_A(\boldsymbol{\phi}^{-1}(\mathbf{x}, t)), \tag{4.3}$$

which is a *push forward* of the referential-domain basis functions to the spatial domain by the ALE mapping given in Equation (1.199). In the discrete case, the ALE mapping is given as

$$\boldsymbol{\phi}(\hat{\mathbf{x}}, t) = \sum_{A \in \eta^s} (\hat{\mathbf{x}}_A + \hat{\mathbf{y}}_A(t)) \hat{N}_A(\hat{\mathbf{x}}), \tag{4.4}$$

where $\hat{\mathbf{x}}_A$'s are the nodal coordinates of the referential domain and $\hat{\mathbf{y}}_A(t)$'s are the time-dependent nodal displacements. Note that because the spatial domain is in motion, this construction produces the spatial-configuration basis functions that are time-dependent. The discrete solution on the spatial domain Ω_t is now given as

$$u(\mathbf{x}, t) = \hat{u}(\boldsymbol{\phi}^{-1}(\mathbf{x}, t), t) = \sum_{A \in \eta^s} u_A(t) \hat{N}_A(\boldsymbol{\phi}^{-1}(\mathbf{x}, t)) = \sum_{A \in \eta^s} u_A(t) N_A(\mathbf{x}, t), \tag{4.5}$$

where the last equality follows from Equation (4.3). Equation (4.5) implies that the solution spatial derivative is given as

$$\frac{\partial u(\mathbf{x}, t)}{\partial \mathbf{x}} = \sum_{A \in \eta^s} u_A(t) \frac{\partial N_A(\mathbf{x}, t)}{\partial \mathbf{x}}. \tag{4.6}$$

The solution referential time derivative pushed forward to the spatial configuration that is utilized in the ALE description may be computed as

$$\left.\frac{\partial u}{\partial t}\right|_{\hat{\mathbf{x}}} (\mathbf{x}, t) = \left.\frac{\partial \hat{u}}{\partial t}\right|_{\hat{\mathbf{x}}} (\boldsymbol{\phi}^{-1}(\mathbf{x}, t), t)$$

$$= \sum_{A \in \eta^s} \frac{du_A(t)}{dt} \hat{N}_A(\boldsymbol{\phi}^{-1}(\mathbf{x}, t)) = \sum_{A \in \eta^s} \frac{du_A(t)}{dt} N_A(\mathbf{x}, t). \tag{4.7}$$

Comparing expressions (4.6) and (4.7), we see that the spatial gradient affects the basis functions, while the time derivative only affects the solution coefficients. This separation of time and space derivatives is the essence of the discrete ALE approach. Because the referential time derivative of the solution in the spatial configuration may be obtained by simply taking a time derivative of its coefficients, the semi-discrete equations are amenable to finite difference treatment in time. As a result, we may think of an ALE method as an extension of the classical semi-discrete approach to moving-domain problems. This explains the popularity of ALE methods, since going from a stationary- to a moving-domain formulation requires little change to one's finite element program.

REMARK 4.1 *In the context of finite elements, the referential time derivative may be understood as the rate of change of the solution at a mesh node or integration point. It is for this reason that the $\left.\frac{\partial}{\partial t}\right|_{\hat{\mathbf{x}}}$ is often replaced with $\left.\frac{\partial}{\partial t}\right|_{\xi}$ to indicate the rate of change of the solution at a given element parametric location, which remains fixed as the mesh moves through space.*

4.4 Space–Time Methods

In the Deforming-Spatial-Domain/Stabilized Space–Time (DSD/SST) method (Tezduyar, 1992, 2003a; Tezduyar *et al.*, 1992a,c; Takizawa and Tezduyar, 2011), the finite element formulation of the governing equations is written over a sequence of N space–time slabs Q_n, where Q_n is the slice of the space–time domain between the time levels n and $n + 1$ (see Figure 4.1). At each time step, the integrations involved in the finite element formulation are performed over Q_n. The finite element interpolation functions are discontinuous across the space–time slabs.

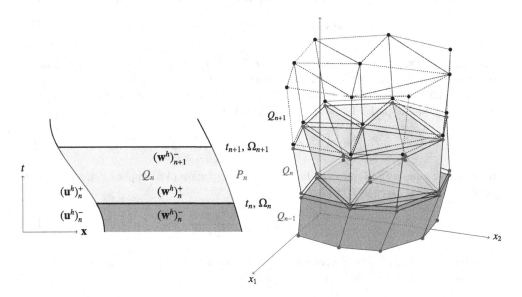

Figure 4.1 Space–time slab in an abstract representation (left) and in a 2D context (right)

Prior to the introduction of the DSD/SST formulation, the stabilized space–time formulations were introduced and tested in the context of problems with fixed spatial domains (see Hughes and Hulbert, 1988).

The space–time computations are carried out for one space–time slab at a time. This spares a 3D computational problem from becoming a 4D problem including the time dimension. In a typical space–time computation all the nodes of the space–time slab would be on the time planes n or $n + 1$ or in between, not necessarily on time planes parallel to the n and $n + 1$ planes but without requiring a space–time mesh that is unstructured in time. In other words, the spatial meshes of a space–time slab would simply be deformed versions of each other. With that understanding, the space–time shape functions can be written as follows:

$$N_a^\alpha = T^\alpha(\theta) N_a(\boldsymbol{\xi}), \qquad a = 1, 2, \ldots, n_{\text{ens}}, \qquad \alpha = 1, 2, \ldots, n_{\text{ent}}, \qquad (4.8)$$

where $\theta \in [-1, 1]$ is the temporal element coordinate (see Figure 4.2 for examples of temporal basis functions), and n_{ens} and n_{ent} are the number of spatial and temporal element nodes. The

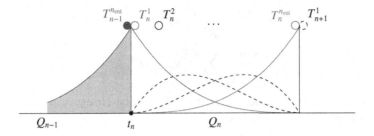

Figure 4.2 Temporal basis functions

number of space–time element nodes, $n_{en} = n_{ent} \times n_{ens}$. In general, the values

$$\phi_n^- = \lim_{t \to t_n^-} \phi^h(t) \tag{4.9}$$

and

$$\phi_n^+ = \lim_{t \to t_n^+} \phi^h(t) \tag{4.10}$$

do not need to be equal to $\phi_{n-1}^{n_{ent}}$ and ϕ_n^1 (coefficients of the basis functions $T_{n-1}^{n_{ent}}$ and T_n^1). However, for the cases we consider here the basis functions are interpolatory at $\theta = -1$ and $\theta = 1$ and therefore $\phi_n^- = \phi_{n-1}^{n_{ent}}$ and $\phi_n^+ = \phi_n^1$.

With shape functions given by Equation (4.8), the spatial-coordinate vector and time are interpolated as follows:

$$\mathbf{x}(\boldsymbol{\xi}, \theta) = \sum_{\alpha=1}^{n_{ent}} \sum_{a=1}^{n_{ens}} \mathbf{x}_a^\alpha T^\alpha(\theta) N_a(\boldsymbol{\xi}), \tag{4.11}$$

$$t(\boldsymbol{\xi}, \theta) = \sum_{\alpha=1}^{n_{ent}} \sum_{a=1}^{n_{ens}} t_a^\alpha T^\alpha(\theta) N_a(\boldsymbol{\xi}), \tag{4.12}$$

where $\mathbf{x}_a^\alpha$ and t_a^α are the spatial and temporal coordinates of the finite element nodes in the physical space–time domain. The Jacobian matrix for the space–time element mapping, $\mathbf{Q}^{ST}$, may be computed by a direct differentiation:

$$\mathbf{Q}^{ST} \equiv \begin{Bmatrix} \frac{\partial t}{\partial \theta} & \frac{\partial t}{\partial \boldsymbol{\xi}} \\ \frac{\partial \mathbf{x}}{\partial \theta} & \frac{\partial \mathbf{x}}{\partial \boldsymbol{\xi}} \end{Bmatrix}. \tag{4.13}$$

Each component of Equation (4.13) is evaluated by using Equations (4.11) and (4.12) as follows:

$$\frac{\partial t}{\partial \theta} = \sum_{\alpha=1}^{n_{ent}} \sum_{a=1}^{n_{ens}} t_a^\alpha \frac{dT^\alpha}{d\theta} N_a, \tag{4.14}$$

$$\frac{\partial t}{\partial \boldsymbol{\xi}} = \sum_{\alpha=1}^{n_{ent}} \sum_{a=1}^{n_{ens}} t_a^\alpha T^\alpha \frac{\partial N_a}{\partial \boldsymbol{\xi}}, \tag{4.15}$$

$$\frac{\partial \mathbf{x}}{\partial \theta} = \sum_{\alpha=1}^{n_{ent}} \sum_{a=1}^{n_{ens}} \mathbf{x}_a^\alpha \frac{dT^\alpha}{d\theta} N_a, \tag{4.16}$$

$$\frac{\partial \mathbf{x}}{\partial \boldsymbol{\xi}} = \sum_{\alpha=1}^{n_{ent}} \sum_{a=1}^{n_{ens}} \mathbf{x}_a^\alpha \frac{\partial N_a}{\partial \boldsymbol{\xi}} T^\alpha. \tag{4.17}$$

To perform the space–time element-level integral, we first change variables to the parent space–time element domain as follows:

$$\int_{Q^e} f(\mathbf{x}, t) dQ = \int_{-1}^{1} \int_{\hat{\Omega}^e} f(\mathbf{x}(\boldsymbol{\xi}, \theta), t(\theta)) J_{x\xi}^{ST}(\boldsymbol{\xi}, \theta) d\hat{\Omega} d\theta, \tag{4.18}$$

where $f(\mathbf{x}, t)$ is a function we wish to integrate over the space–time element Q^e, $\hat{\Omega}^e$ is the "cross-section" of the space–time parent element at θ, and $J_{x\xi}^{ST}$ is the determinant of $\mathbf{Q}^{ST}$:

$$J_{x\xi}^{ST} = \det \mathbf{Q}^{ST}. \tag{4.19}$$

In the numerical integration, we first define the following function:

$$g(\boldsymbol{\xi}, \theta) = f(\mathbf{x}(\boldsymbol{\xi}, \theta), t(\theta)) J_{x\xi}^{ST}(\boldsymbol{\xi}, \theta), \tag{4.20}$$

and then approximate the integration of that over the parametric element:

$$\int_{-1}^{1} \int_{\hat{\Omega}^e} g(\mathbf{x}(\boldsymbol{\xi}, \theta), t(\theta)) d\hat{\Omega} d\theta \approx \sum_{\gamma_t}^{n_{intt}} \sum_{\gamma}^{n_{int}} g\left(\tilde{\boldsymbol{\xi}}_\gamma, \tilde{\theta}_{\gamma_t}\right) W_\gamma W_{\gamma_t}. \tag{4.21}$$

Here n_{intt} is the number of integration points in time. The Jacobian matrix $\mathbf{Q}^{ST}$ is used in the computation of the element shape function derivatives with respect to the physical space and time coordinates as follows:

$$\left\{ \begin{array}{c} \frac{\partial T^\alpha N_a}{\partial t} \\ \frac{\partial T^\alpha N_a}{\partial \mathbf{x}} \end{array} \right\} = \left(\mathbf{Q}^{ST}\right)^{-T} \left\{ \begin{array}{c} N_a \frac{dT^\alpha}{d\theta} \\ T^\alpha \frac{\partial N_a}{\partial \boldsymbol{\xi}} \end{array} \right\}. \tag{4.22}$$

As mentioned earlier, in a typical space–time computation, all the nodes of the space–time slab associated with a given value of α would be on the same time plane; that is, $t_a^\alpha \equiv t^\alpha$. In such cases, we can further rearrange Equations (4.14) and (4.15) and rewrite Equation (4.13) as follows:

$$\mathbf{Q}^{ST} = \left\{ \begin{array}{cc} \frac{dt}{d\theta} & \mathbf{0}^T \\ \hat{\mathbf{u}} \frac{dt}{d\theta} & \frac{\partial \mathbf{x}}{\partial \boldsymbol{\xi}} \end{array} \right\}, \tag{4.23}$$

where

$$\frac{dt}{d\theta} = \sum_{\alpha=1}^{n_{ent}} t^\alpha \frac{dT^\alpha}{d\theta}, \tag{4.24}$$

$$\hat{\mathbf{u}} = \sum_{\alpha=1}^{n_{ent}} \sum_{a=1}^{n_{ens}} \mathbf{x}_a^\alpha \frac{dT^\alpha}{dt} N_a. \tag{4.25}$$

Thus, Equation (4.19) and the inverse matrix can be simplified as follows:

$$J^{ST}_{x\xi} = \frac{dt}{d\theta} \det \mathbf{Q}, \tag{4.26}$$

where

$$\mathbf{Q} = \sum_{a=1}^{n_{ens}} \mathbf{x}_a \frac{\partial N_a}{\partial \xi}, \tag{4.27}$$

and

$$\left(\mathbf{Q}^{ST}\right)^{-T} = \begin{Bmatrix} \frac{d\theta}{dt} & -\hat{\mathbf{u}}^T \mathbf{Q}^{-T} \\ \mathbf{0} & \mathbf{Q}^{-T} \end{Bmatrix}. \tag{4.28}$$

Therefore we need to invert only a $n_{sd} \times n_{sd}$ matrix.

4.5 Advection–Diffusion Equation

In this section we provide a formulation for the advection–diffusion equation on a moving domain. Both the ALE and space–time formulations are presented. We directly present the stabilized formulation of the advection–diffusion because we already discussed that the Galerkin formulation is unstable when the advection dominates locally (see discussion in Section 2.5).

4.5.1 ALE Formulation

Following the developments of Section 1.3, the convective form of the advection–diffusion equation in the ALE description becomes

$$\left.\frac{\partial \phi}{\partial t}\right|_{\hat{x}} + (\mathbf{u} - \hat{\mathbf{u}}) \cdot \nabla \phi - \nabla \cdot \nu \nabla \phi - f = 0. \tag{4.29}$$

The above equation holds in the spatial domain Ω_t for all $0 < t < T$. To define the SUPG-stabilized formulation of this equation, we recognize that the convective term is modified by the spatial-domain velocity $\hat{\mathbf{u}}$. This observation leads to the following straightforward extension of the SUPG formulation of the advection–diffusion equation to the moving-domain case: find $\phi^h \in S^h$, such that $\forall w^h \in \mathcal{V}^h$:

$$\int_{\Omega_t} w^h \left.\frac{\partial \phi^h}{\partial t}\right|_{\hat{x}} d\Omega + \int_{\Omega_t} w^h \left(\mathbf{u}^h - \hat{\mathbf{u}}^h\right) \cdot \nabla \phi^h d\Omega$$
$$+ \int_{\Omega_t} \nabla w^h \cdot \nu \nabla \phi^h d\Omega - \int_{(\Gamma_t)_h} w^h h^h d\Gamma - \int_{\Omega_t} w^h f^h d\Omega$$
$$+ \sum_{e=1}^{n_{el}} \int_{\Omega_t^e} \tau_{SUPG} \left(\mathbf{u}^h - \hat{\mathbf{u}}^h\right) \cdot \nabla w^h \left(\left.\frac{\partial \phi^h}{\partial t}\right|_{\hat{x}} + \left(\mathbf{u}^h - \hat{\mathbf{u}}^h\right) \cdot \nabla \phi^h - \nabla \cdot \nu \nabla \phi^h - f^h\right) d\Omega = 0. \tag{4.30}$$

The consistency of the formulation is retained. The stabilization parameter τ_{SUPG} is modified as

$$\tau_{SUPG} = \left(\frac{4}{\Delta t^2} + \left(\mathbf{u}^h - \hat{\mathbf{u}}^h\right) \cdot \mathbf{G} \left(\mathbf{u}^h - \hat{\mathbf{u}}^h\right) + C_I \nu^2 \mathbf{G} : \mathbf{G}\right)^{-1/2} \tag{4.31}$$

to account for the new structure of the convective term. The other commonly employed definition of τ_{SUPG} becomes

$$\tau_{\text{SUPG}} = \left(\frac{1}{\tau_{\text{SUGN1}}^2} + \frac{1}{\tau_{\text{SUGN2}}^2} + \frac{1}{\tau_{\text{SUGN3}}^2} \right)^{-\frac{1}{2}}, \qquad (4.32)$$

where

$$\tau_{\text{SUGN1}} = \left(\sum_{a=1}^{n_{\text{en}}} \left| \left(\mathbf{u}^h - \hat{\mathbf{u}}^h \right) \cdot \nabla N_a \right| \right)^{-1}, \qquad (4.33)$$

and τ_{SUGN2} and τ_{SUGN3} are given by Equations (2.143) and (2.144), respectively. The matrix form of the stabilized ALE formulation is form-identical to the nonmoving-domain case and is given as

$$\mathbf{M}\dot{\boldsymbol{\Phi}} + \mathbf{K}\boldsymbol{\Phi} = \mathbf{F}, \qquad (4.34)$$

where

$$\mathbf{M} = [M_{AB}], \qquad (4.35)$$

$$M_{AB} = \int_{\Omega_t} N_A N_B \, d\Omega + \sum_{e=1}^{n_{\text{el}}} \int_{\Omega_t^e} \tau_{\text{SUPG}} \left(\mathbf{u}^h - \hat{\mathbf{u}}^h \right) \cdot \nabla N_A N_B \, d\Omega, \qquad (4.36)$$

$$\mathbf{K} = [K_{AB}], \qquad (4.37)$$

$$K_{AB} = \int_{\Omega_t} N_A \left(\mathbf{u}^h - \hat{\mathbf{u}}^h \right) \cdot \nabla N_B \, d\Omega + \int_{\Omega_t} \nabla N_A \cdot \nu \nabla N_B \, d\Omega$$

$$+ \sum_{e=1}^{n_{\text{el}}} \int_{\Omega_t^e} \tau_{\text{SUPG}} \left(\mathbf{u}^h - \hat{\mathbf{u}}^h \right) \cdot \nabla N_A \left(\left(\mathbf{u}^h - \hat{\mathbf{u}}^h \right) \cdot \nabla N_B - \nabla \cdot \nu \nabla N_B \right) d\Omega, \qquad (4.38)$$

$$\mathbf{F} = [F_A], \qquad (4.39)$$

$$F_A = \int_{\Omega_t} N_A f^h \, d\Omega + \sum_{e=1}^{n_{\text{el}}} \int_{\Omega_t^e} \tau_{\text{SUPG}} \left(\mathbf{u}^h - \hat{\mathbf{u}}^h \right) \cdot \nabla N_A f^h \, d\Omega$$

$$+ \int_{(\Gamma_t)_h} N_A h^h \, d\Gamma - \sum_{C \in \eta_g^s} K_{AC} g_C - \sum_{C \in \eta_g^s} M_{AC} \dot{g}_C, \qquad (4.40)$$

$$\boldsymbol{\Phi} = [\phi_B], \qquad (4.41)$$

$$A, B \in \eta - \eta_g^s. \qquad (4.42)$$

The ALE construction allows us to integrate Equation (4.34) using any standard finite-difference time marching method. Note, however, that the integrals in the left-hand-side matrix and right-hand-side vector definitions are taken over a time-dependent domain Ω_t. As a result, the left-hand-side matrices and the right-hand-side vectors in this formulation are time-dependent and need to be recomputed as the equations are integrated in time. This is a major difference between the moving- and nonmoving- domain cases.

4.5.2 Space–Time Formulation

Let us assume that we have constructed some suitably-defined finite-dimensional trial and test function spaces S_n^h and $\mathcal{V}_n^h$. The subscript n implies that corresponding to different space–time slabs we might have different discretizations. The DSD/SST formulation of Equation (1.27) can be written as follows: given $(\phi^h)_n^-$, find $\phi^h \in S_n^h$, such that $\forall\, w^h \in \mathcal{V}_n^h$:

$$\int_{Q_n} w^h \left(\frac{\partial \phi^h}{\partial t} + \mathbf{u}^h \cdot \nabla \phi^h - f^h \right) dQ + \int_{Q_n} \nabla w^h \cdot \nu \nabla \phi^h dQ$$
$$- \int_{(P_n)_h} w^h h^h dP + \int_{\Omega_n} (w^h)_n^+ \left((\phi^h)_n^+ - (\phi^h)_n^- \right) d\Omega$$
$$+ \sum_{e=1}^{(n_{el})_n} \int_{Q_n^e} \tau_{\text{SUPG}} \left(\frac{\partial w^h}{\partial t} + \mathbf{u}^h \cdot \nabla w^h \right) \left(\frac{\partial \phi^h}{\partial t} + \mathbf{u}^h \cdot \nabla \phi^h - \nabla \cdot \nu \nabla \phi^h - f^h \right) dQ = 0. \quad (4.43)$$

We start with $(\phi^h)_0^- = \phi_0^h$, and apply this formulation sequentially to all space–time slabs $Q_0, Q_1, Q_2, \ldots, Q_{N-1}$. For an earlier, detailed reference on the DSD/SST formulation, see (Tezduyar, 1992; Tezduyar et al., 1992a,c). The stabilization parameter τ_{SUPG} is defined as

$$\tau_{\text{SUPG}} = \left(\frac{1}{\tau_{\text{SUGN12}}^2} + \frac{1}{\tau_{\text{SUGN3}}^2} \right)^{-\frac{1}{2}}, \quad (4.44)$$

$$\tau_{\text{SUGN12}} = \left(\sum_{\alpha=1}^{n_{\text{ent}}} \sum_{a=1}^{n_{\text{ens}}} \left| \frac{\partial N_a^\alpha}{\partial t} + \mathbf{u}^h \cdot \nabla N_a^\alpha \right| \right)^{-1}, \quad (4.45)$$

$$\tau_{\text{SUGN3}} = \frac{h_{\text{RGN}}^2}{4\nu}, \quad (4.46)$$

$$h_{\text{RGN}} = 2 \left(\sum_{\alpha=1}^{n_{\text{ent}}} \sum_{a=1}^{n_{\text{ens}}} |\mathbf{r} \cdot \nabla N_a^\alpha| \right)^{-1}, \quad (4.47)$$

$$\mathbf{r} = \frac{\nabla |\phi|}{\| \nabla |\phi| \|}. \quad (4.48)$$

As an alternative to the construction of τ_{SUPG} as given by Equations (4.44)–(4.48), it was proposed in Tezduyar and Sathe (2007) the option of constructing τ_{SUPG} based on separate definitions for the advection-dominated and transient-dominated limits, as given by

$$\tau_{\text{SUPG}} = \left(\frac{1}{\tau_{\text{SUGN1}}^2} + \frac{1}{\tau_{\text{SUGN2}}^2} + \frac{1}{\tau_{\text{SUGN3}}^2} \right)^{-\frac{1}{2}}, \quad (4.49)$$

$$\tau_{\text{SUGN1}} = \left(\sum_{\alpha=1}^{n_{\text{ent}}} \sum_{a=1}^{n_{\text{ens}}} \left| (\mathbf{u}^h - \hat{\mathbf{u}}^h) \cdot \nabla N_a^\alpha \right| \right)^{-1}, \quad (4.50)$$

$$\tau_{\text{SUGN2}} = \frac{\Delta t}{2}, \quad (4.51)$$

and Equations (4.46)–(4.48).

4.6 Navier–Stokes Equations

In this section we provide the ALE and space–time formulations of the Navier–Stokes equations on a moving domain. We begin by presenting the RBVMS formulation and then show how to reduce it to a more standard stabilized formulation. Weak enforcement of the essential boundary conditions is also presented.

4.6.1 ALE Formulation

To give the RBVMS formulation of the Navier–Stokes equations in the ALE form, we begin with the convective form of the Navier–Stokes equations of incompressible flows:

$$\rho \left(\left. \frac{\partial \mathbf{u}}{\partial t} \right|_{\hat{\mathbf{x}}} + (\mathbf{u} - \hat{\mathbf{u}}) \cdot \nabla \mathbf{u} - \mathbf{f} \right) - \nabla \cdot \boldsymbol{\sigma}(\mathbf{u}, p) = \mathbf{0}, \tag{4.52}$$

$$\nabla \cdot \mathbf{u} = 0, \tag{4.53}$$

which holds for every $\mathbf{x} \in \Omega_t$, $t \in (0, T)$. Note that the form of the incompressibility constraint is unchanged for the moving-domain case. The variational formulation of the above equations may be stated as follows: find $\mathbf{u} \in \mathcal{S}_u$ and $p \in \mathcal{S}_p$, such that $\forall \, \mathbf{w} \in \mathcal{V}_u$ and $q \in \mathcal{V}_p$:

$$\int_{\Omega_t} \mathbf{w} \cdot \rho \left(\left. \frac{\partial \mathbf{u}}{\partial t} \right|_{\hat{\mathbf{x}}} + (\mathbf{u} - \hat{\mathbf{u}}) \cdot \nabla \mathbf{u} - \mathbf{f} \right) d\Omega + \int_{\Omega_t} \boldsymbol{\varepsilon}(\mathbf{w}) : \boldsymbol{\sigma}(\mathbf{u}, p) \, d\Omega$$

$$- \int_{(\Gamma_t)_h} \mathbf{w} \cdot \mathbf{h} \, d\Gamma + \int_{\Omega_t} q \nabla \cdot \mathbf{u} \, d\Omega = 0. \tag{4.54}$$

We extend the semi-discrete RBVMS formulation of incompressible flows given by Equation (2.195) to the moving-domain case as follows: find $\mathbf{u}^h \in \mathcal{S}_u^h$ and $p^h \in \mathcal{S}_p^h$, such that $\forall \, \mathbf{w}^h \in \mathcal{V}_u^h$ and $q^h \in \mathcal{V}_p^h$:

$$\int_{\Omega_t} \mathbf{w}^h \cdot \rho \left(\left. \frac{\partial \mathbf{u}^h}{\partial t} \right|_{\hat{\mathbf{x}}} + \left(\mathbf{u}^h - \hat{\mathbf{u}}^h\right) \cdot \nabla \mathbf{u}^h - \mathbf{f}^h \right) d\Omega + \int_{\Omega_t} \boldsymbol{\varepsilon}\left(\mathbf{w}^h\right) : \boldsymbol{\sigma}\left(\mathbf{u}^h, p^h\right) d\Omega$$

$$- \int_{(\Gamma_t)_h} \mathbf{w}^h \cdot \mathbf{h}^h \, d\Gamma + \int_{\Omega_t} q^h \nabla \cdot \mathbf{u}^h \, d\Omega$$

$$+ \sum_{e=1}^{n_{el}} \int_{\Omega_t^e} \tau_{\text{SUPS}} \left(\left(\mathbf{u}^h - \hat{\mathbf{u}}^h\right) \cdot \nabla \mathbf{w}^h + \frac{\nabla q^h}{\rho} \right) \cdot \mathbf{r}_M \left(\mathbf{u}^h, p^h\right) d\Omega$$

$$+ \sum_{e=1}^{n_{el}} \int_{\Omega_t^e} \rho \nu_{\text{LSIC}} \nabla \cdot \mathbf{w}^h r_C(\mathbf{u}^h) \, d\Omega$$

$$- \sum_{e=1}^{n_{el}} \int_{\Omega_t^e} \tau_{\text{SUPS}} \mathbf{w}^h \cdot \left(\mathbf{r}_M \left(\mathbf{u}^h, p^h\right) \cdot \nabla \mathbf{u}^h \right) d\Omega$$

$$- \sum_{e=1}^{n_{el}} \int_{\Omega_t^e} \frac{\nabla \mathbf{w}^h}{\rho} : \left(\tau_{\text{SUPS}} \mathbf{r}_M \left(\mathbf{u}^h, p^h\right) \right) \otimes \left(\tau_{\text{SUPS}} \mathbf{r}_M \left(\mathbf{u}^h, p^h\right) \right) d\Omega = 0. \tag{4.55}$$

We refer to the formulation given by Equation (4.55) as *ALE-VMS*. In Equation (4.55) $\hat{\mathbf{u}}^h$ is the discrete counterpart of the fluid-domain velocity and the integrals are taken over the fluid domain and its boundary in a time-dependent spatial configuration. The definitions of τ_{SUPG} and ν_{LSIC} may be obtained from Equations (2.196)–(2.205) by replacing the fluid velocity relative to a stationary observer with the fluid velocity relative to that of the fluid domain as $\mathbf{u}^h \leftarrow \mathbf{u}^h - \hat{\mathbf{u}}^h$. The definition of the linear-momentum equation residual, $\mathbf{r}_M(\mathbf{u}^h, p^h)$, is also modified to maintain the consistency of the discrete formulation:

$$\mathbf{r}_M(\mathbf{u}^h, p^h) = \rho \left(\left. \frac{\partial \mathbf{u}^h}{\partial t} \right|_{\hat{x}} + (\mathbf{u}^h - \hat{\mathbf{u}}^h) \cdot \nabla \mathbf{u}^h - \mathbf{f}^h \right) - \nabla \cdot \boldsymbol{\sigma}(\mathbf{u}^h, p^h). \tag{4.56}$$

The SUPG/PSPG counterpart of the ALE-VMS formulation may be obtained by omitting the last two terms on the left-hand-side of Equation (4.55) and providing the option of using two different τ's in the momentum and continuity equations:

$$\int_{\Omega_t} \mathbf{w}^h \cdot \rho \left(\left. \frac{\partial \mathbf{u}^h}{\partial t} \right|_{\hat{x}} + (\mathbf{u}^h - \hat{\mathbf{u}}^h) \cdot \nabla \mathbf{u}^h - \mathbf{f}^h \right) d\Omega + \int_{\Omega_t} \boldsymbol{\varepsilon}(\mathbf{w}^h) : \boldsymbol{\sigma}(\mathbf{u}^h, p^h) d\Omega$$

$$- \int_{(\Gamma_t)_h} \mathbf{w}^h \cdot \mathbf{h}^h \, d\Gamma + \int_{\Omega_t} q^h \nabla \cdot \mathbf{u}^h \, d\Omega$$

$$+ \sum_{e=1}^{n_{\text{el}}} \int_{\Omega_t^e} \tau_{\text{SUPG}} \left((\mathbf{u}^h - \hat{\mathbf{u}}^h) \cdot \nabla \mathbf{w}^h \right) \cdot \mathbf{r}_M(\mathbf{u}^h, p^h) \, d\Omega$$

$$+ \sum_{e=1}^{n_{\text{el}}} \int_{\Omega_t^e} \tau_{\text{PSPG}} \left(\frac{\nabla q^h}{\rho} \right) \cdot \mathbf{r}_M(\mathbf{u}^h, p^h) \, d\Omega$$

$$+ \sum_{e=1}^{n_{\text{el}}} \int_{\Omega_t^e} \rho \nu_{\text{LSIC}} \nabla \cdot \mathbf{w}^h r_C(\mathbf{u}^h) \, d\Omega = 0. \tag{4.57}$$

To account for the weak enforcement of the essential boundary conditions (see Equation (2.217)), we remove them from the trial and test function sets, and add the following terms to the left-hand-side of Equations (4.55) and (4.57):

$$- \sum_{b=1}^{n_{\text{eb}}} \int_{\Gamma^b \cap (\Gamma_t)_g} \mathbf{w}^h \cdot \boldsymbol{\sigma}(\mathbf{u}^h, p^h) \mathbf{n} \, d\Gamma$$

$$- \sum_{b=1}^{n_{\text{eb}}} \int_{\Gamma^b \cap (\Gamma_t)_g} \left(2\mu \boldsymbol{\varepsilon}(\mathbf{w}^h) \mathbf{n} + q^h \mathbf{n} \right) \cdot (\mathbf{u}^h - \mathbf{g}^h) \, d\Gamma$$

$$- \sum_{b=1}^{n_{\text{eb}}} \int_{\Gamma^b \cap (\Gamma_t)_g^-} \mathbf{w}^h \cdot \rho \left((\mathbf{u}^h - \hat{\mathbf{u}}^h) \cdot \mathbf{n} \right) (\mathbf{u}^h - \mathbf{g}^h) \, d\Gamma$$

$$+ \sum_{b=1}^{n_{\text{eb}}} \int_{\Gamma^b \cap (\Gamma_t)_g} \tau_{\text{TAN}}^B \left(\mathbf{w}^h - (\mathbf{w}^h \cdot \mathbf{n}) \mathbf{n} \right) \cdot \left((\mathbf{u}^h - \mathbf{g}^h) - ((\mathbf{u}^h - \mathbf{g}^h) \cdot \mathbf{n}) \mathbf{n} \right) d\Gamma$$

$$+ \sum_{b=1}^{n_{\text{eb}}} \int_{\Gamma^b \cap (\Gamma_t)_g} \tau_{\text{NOR}}^B (\mathbf{w}^h \cdot \mathbf{n}) \left((\mathbf{u}^h - \mathbf{g}^h) \cdot \mathbf{n} \right) d\Gamma. \tag{4.58}$$

The "inflow" part of the essential boundary, $(\Gamma_t)_g^-$, is now defined as

$$(\Gamma_t)_g^- = \left\{ \mathbf{x} \mid \left(\mathbf{u}^h - \hat{\mathbf{u}}^h\right) \cdot \mathbf{n} < 0, \forall \mathbf{x} \subset (\Gamma_t)_g \right\}, \tag{4.59}$$

which is the extension of Equation (2.218) to the moving-domain case. Also note that if $(\Gamma_t)_g$ coincides with a moving wall, then $\mathbf{g}^h$ is the wall velocity prescribed as the essential boundary condition for the fluid mechanics equations.

The vector form of the equations for the ALE-VMS formulation given by Equation (4.55) is written, as we did in Section 2.6.1, by starting with the following expressions for the space-discrete velocity and pressure trial and test functions:

$$\mathbf{u}^h(\mathbf{x}, t) = \sum_{\eta^s} \mathbf{u}_A(t) N_A(\mathbf{x}, t), \tag{4.60}$$

$$p^h(\mathbf{x}, t) = \sum_{\eta^s} p_A(t) N_A(\mathbf{x}, t), \tag{4.61}$$

$$\mathbf{w}^h(\mathbf{x}, t) = \sum_{\eta^w} \mathbf{w}_A N_A(\mathbf{x}, t), \tag{4.62}$$

$$q^h(\mathbf{x}, t) = \sum_{\eta^w} q_A N_A(\mathbf{x}, t). \tag{4.63}$$

The basis functions N_A's on the spatial domain are defined in Equation (4.3), and, because the spatial domain is in motion, they are time-dependent. We define two discrete residual vectors corresponding to the momentum and continuity equations by introducing Equations (4.62) and (4.63) into Equation (4.55) and assuming that $\mathbf{w}_A$'s and q_A's are arbitrary constants:

$$\mathbf{N}_M = [(N_M)_{A,i}], \tag{4.64}$$
$$\mathbf{N}_C = [(N_C)_A], \tag{4.65}$$

$$(N_M)_{A,i} = \int_{\Omega_t} N_A \mathbf{e}_i \cdot \rho \left(\left. \frac{\partial \mathbf{u}^h}{\partial t} \right|_{\hat{x}} + \left(\mathbf{u}^h - \hat{\mathbf{u}}^h\right) \cdot \nabla \mathbf{u}^h - \mathbf{f}^h \right) d\Omega$$

$$+ \int_{\Omega_t} \varepsilon(N_A \mathbf{e}_i) : \sigma\left(\mathbf{u}^h, p^h\right) d\Omega - \int_{(\Gamma_t)_h} N_A \mathbf{e}_i \cdot \mathbf{h}^h \, d\Gamma$$

$$+ \sum_{e=1}^{n_{el}} \int_{\Omega_t^e} \tau_{\text{SUPS}} \left(\left(\mathbf{u}^h - \hat{\mathbf{u}}^h\right) \cdot \nabla N_A \mathbf{e}_i \right) \cdot \mathbf{r}_M\left(\mathbf{u}^h, p^h\right) d\Omega$$

$$+ \sum_{e=1}^{n_{el}} \int_{\Omega_t^e} \rho \nu_{\text{LSIC}} \left(\nabla \cdot N_A \mathbf{e}_i\right) r_C\left(\mathbf{u}^h\right) d\Omega$$

$$- \sum_{e=1}^{n_{el}} \int_{\Omega_t^e} \tau_{\text{SUPS}} N_A \mathbf{e}_i \cdot \left(\mathbf{r}_M\left(\mathbf{u}^h, p^h\right) \cdot \nabla \mathbf{u}^h\right) d\Omega$$

$$- \sum_{e=1}^{n_{el}} \int_{\Omega_t^e} \frac{\nabla N_A \mathbf{e}_i}{\rho} : \left(\tau_{\text{SUPS}} \mathbf{r}_M\left(\mathbf{u}^h, p^h\right)\right) \otimes \left(\tau_{\text{SUPS}} \mathbf{r}_M\left(\mathbf{u}^h, p^h\right)\right) d\Omega, \tag{4.66}$$

$$(N_C)_A = \int_{\Omega_t} N_A \nabla \cdot \mathbf{u}^h \, d\Omega + \sum_{e=1}^{n_{el}} \int_{\Omega_t^e} \tau_{\text{SUPS}} \frac{\nabla N_A}{\rho} \cdot \mathbf{r}_M\left(\mathbf{u}^h, p^h\right) d\Omega. \tag{4.67}$$

Let, as before, $\mathbf{U}$, $\dot{\mathbf{U}}$, and $\mathbf{P}$ denote the vectors of nodal degrees-of-freedom of velocity, velocity time derivative, and pressure, respectively. The vector form of the semi-discrete equations corresponding to Equation (4.55) becomes: find $\mathbf{U}$, $\dot{\mathbf{U}}$, and $\mathbf{P}$ such that:

$$\mathbf{N}_M(\dot{\mathbf{U}}, \mathbf{U}, \mathbf{P}) = \mathbf{0}, \tag{4.68}$$

$$\mathbf{N}_C(\dot{\mathbf{U}}, \mathbf{U}, \mathbf{P}) = \mathbf{0}. \tag{4.69}$$

4.6.2 Generalized-α Time Integration of the ALE Equations

In what follows, we present the application of the generalized-α time integration method to the above equation system. The generalized-α method for the Navier–Stokes equations of incompressible flows was first proposed in Jansen et al. (2000), where the spatial domain was not moving.

We begin by partitioning the time interval of interest, $[0, T]$, into subintervals, or time steps. The time levels t_n and t_{n+1} define the endpoints of the nth time step, and $\Delta t_n = t_{n+1} - t_n$ is the size of the nth time step. In general, the time-step size may vary from step to step.

The equations of the generalized-α time integration method are: given the nodal solutions at n, $\mathbf{U}_n$, $\dot{\mathbf{U}}_n$, and $\mathbf{P}_n$, find the solutions at $n+1$, $\mathbf{U}_{n+1}$, $\dot{\mathbf{U}}_{n+1}$ and $\mathbf{P}_{n+1}$, such that

$$\mathbf{N}_M(\dot{\mathbf{U}}_{n+\alpha_m}, \mathbf{U}_{n+\alpha_f}, \mathbf{P}_{n+1}) = \mathbf{0}, \tag{4.70}$$

$$\mathbf{N}_C(\dot{\mathbf{U}}_{n+\alpha_m}, \mathbf{U}_{n+\alpha_f}, \mathbf{P}_{n+1}) = \mathbf{0}, \tag{4.71}$$

where

$$\dot{\mathbf{U}}_{n+\alpha_m} = \dot{\mathbf{U}}_n + \alpha_m(\dot{\mathbf{U}}_{n+1} - \dot{\mathbf{U}}_n), \tag{4.72}$$

$$\mathbf{U}_{n+\alpha_f} = \mathbf{U}_n + \alpha_f(\mathbf{U}_{n+1} - \mathbf{U}_n) \tag{4.73}$$

are the intermediate values of the nodal velocity vector and its time derivative. In the generalized-α method the discrete momentum and continuity equations are collocated at intermediate values of the solution at every time step. The relationship between the nodal velocity degrees-of-freedom and their time derivatives is approximated by the discrete Newmark formula (see, e.g., Hughes, 2000):

$$\mathbf{U}_{n+1} = \mathbf{U}_n + \Delta t_n \left((1-\gamma)\dot{\mathbf{U}}_n + \gamma \dot{\mathbf{U}}_{n+1}\right). \tag{4.74}$$

Here α_m, α_f, and γ are the real-valued parameters chosen based on the stability and accuracy considerations. It was shown in Jansen et al. (2000) that second-order accuracy in time is achieved provided that

$$\gamma = 1/2 + \alpha_m - \alpha_f, \tag{4.75}$$

while unconditional stability is attained provided that

$$\alpha_m \geq \alpha_f \geq 1/2. \tag{4.76}$$

A *one-parameter* family of second-order accurate and unconditionally stable time integration schemes may be obtained by setting γ according to Equation (4.75) and employing the following parameterization of the intermediate time levels:

$$\alpha_m = \frac{1}{2}\left(\frac{3-\rho_\infty}{1+\rho_\infty}\right) \quad \text{and} \quad \alpha_f = \frac{1}{1+\rho_\infty}. \tag{4.77}$$

In Equation (4.77), the parameter ρ_∞ is the spectral radius of the amplification matrix as $\Delta t_n \to \infty$, which controls the high-frequency dissipation (see Hughes, 2000).

To solve the nonlinear system of equations (4.70)–(4.74), we employ a Newton–Raphson method, which results in a two-stage predictor–multicorrector algorithm.

Predictor stage. At this stage, given the solution at time level n, we "predict" the solution at time level $n + 1$:

$$\dot{\mathbf{U}}_{n+1}^0 = \frac{(\gamma - 1)}{\gamma}\dot{\mathbf{U}}_n, \tag{4.78}$$

$$\mathbf{U}_{n+1}^0 = \mathbf{U}_n, \tag{4.79}$$

$$\mathbf{P}_{n+1}^0 = \mathbf{P}_n, \tag{4.80}$$

where the superscript 0 represents the zeroth value of the iteration counter. This choice corresponds to the *same velocity* predictor, meaning that the initial guess for the fluid velocity at time level $n + 1$ is the velocity from time level n. To maintain consistency with the Newmark formula given by Equation (4.74), the velocity time derivative must be initialized as in Equation (4.78). Other predictor choices are possible.

Multicorrector stage. At this stage, we iterate on the solution until Equations (4.70) and (4.71) are satisfied. For this, we repeat the following steps for $i = 0, 1, \ldots, (i_{\max} - 1)$, where i is the iteration counter and $i_{\max}$ is the maximum number of nonlinear iterations specified for the current time step.

1. Evaluate the iterates at the intermediate time levels:

$$\dot{\mathbf{U}}_{n+\alpha_m}^i = \dot{\mathbf{U}}_n + \alpha_m(\dot{\mathbf{U}}_{n+1}^i - \dot{\mathbf{U}}_n), \tag{4.81}$$

$$\mathbf{U}_{n+\alpha_f}^i = \mathbf{U}_n + \alpha_f(\mathbf{U}_{n+1}^i - \mathbf{U}_n), \tag{4.82}$$

$$\mathbf{P}_{n+1}^i = \mathbf{P}_{n+1}^i. \tag{4.83}$$

2. Use the intermediate values to assemble the linear system of equations corresponding to the linearization of Equations (4.70) and (4.71) with respect to the nodal unknowns $\dot{\mathbf{U}}_{n+1}$ and $\mathbf{P}_{n+1}$:

$$\left.\frac{\partial \mathbf{N}_\mathrm{M}}{\partial \dot{\mathbf{U}}_{n+1}}\right|_i \Delta\dot{\mathbf{U}}_{n+1}^i + \left.\frac{\partial \mathbf{N}_\mathrm{M}}{\partial \mathbf{P}_{n+1}}\right|_i \Delta\mathbf{P}_{n+1}^i = -\mathbf{N}_\mathrm{M}^i, \tag{4.84}$$

$$\left.\frac{\partial \mathbf{N}_\mathrm{C}}{\partial \dot{\mathbf{U}}_{n+1}}\right|_i \Delta\dot{\mathbf{U}}_{n+1}^i + \left.\frac{\partial \mathbf{N}_\mathrm{C}}{\partial \mathbf{P}_{n+1}}\right|_i \Delta\mathbf{P}_{n+1}^i = -\mathbf{N}_\mathrm{C}^i. \tag{4.85}$$

Solve this linear system using a preconditioned GMRES algorithm (see Saad and Schultz, 1986) to a specified tolerance.

3. Update the solution:

$$\dot{\mathbf{U}}_{n+1}^{i+1} = \dot{\mathbf{U}}_{n+1}^i + \Delta\dot{\mathbf{U}}_{n+1}^i, \tag{4.86}$$

$$\mathbf{U}_{n+1}^{i+1} = \mathbf{U}_{n+1}^i + \gamma\Delta t_n\Delta\dot{\mathbf{U}}_{n+1}^i, \tag{4.87}$$

$$\mathbf{P}_{n+1}^{i+1} = \mathbf{P}_{n+1}^i + \Delta\mathbf{P}_{n+1}^i. \tag{4.88}$$

To evaluate the integrals in the definition of the discrete residual vectors given by Equations (4.64) and (4.65), we use the fluid domain at $t = t_{(n+\alpha_f)}$, namely,

$$\int_{\Omega_t} (\cdot) \, d\Omega = \int_{\Omega_{t_{(n+\alpha_f)}}} (\cdot) \, d\Omega, \tag{4.89}$$

where

$$\Omega_{t_{(n+\alpha_f)}} = \left\{ \mathbf{x} \mid \mathbf{x}(\hat{\mathbf{x}}, t_{(n+\alpha_f)}) = \hat{\mathbf{x}} + \hat{\mathbf{y}}^h(\hat{\mathbf{x}}, t_{(n+\alpha_f)}) \right\}. \tag{4.90}$$

If the fluid-domain position is known only at discrete time levels, then we use the definition

$$\Omega_{t_{(n+\alpha_f)}} = \left\{ \mathbf{x} \mid \mathbf{x}(\hat{\mathbf{x}}, t_{(n+\alpha_f)}) = \hat{\mathbf{x}} + \hat{\mathbf{y}}^h(\hat{\mathbf{x}}, t_n) + \alpha_f \left(\hat{\mathbf{y}}^h(\hat{\mathbf{x}}, t_{n+1}) - \hat{\mathbf{y}}^h(\hat{\mathbf{x}}, t_n) \right) \right\}. \tag{4.91}$$

In what follows, we provide explicit formulas for the left-hand-side matrices involved in Step 2 of the multicorrector stage:

$$\frac{\partial \mathbf{N}_M}{\partial \dot{\mathbf{U}}_{n+1}} = \left[K_{AB}^{ij} \right], \tag{4.92}$$

$$K_{AB}^{ij} = \alpha_m \int_{\Omega_{t_{(n+\alpha_f)}}} N_A \rho N_B \, d\Omega \, \delta_{ij}$$

$$+ \alpha_m \int_{\Omega_{t_{(n+\alpha_f)}}} \tau_{\text{SUPS}} \left(\mathbf{u}^h - \hat{\mathbf{u}}^h \right) \cdot \nabla N_A \rho N_B \, d\Omega \, \delta_{ij}$$

$$+ \alpha_f \gamma \Delta t_n \int_{\Omega_{t_{(n+\alpha_f)}}} N_A \rho \left(\mathbf{u}^h - \hat{\mathbf{u}}^h \right) \cdot \nabla N_B \, d\Omega \, \delta_{ij}$$

$$+ \alpha_f \gamma \Delta t_n \int_{\Omega_{t_{(n+\alpha_f)}}} \nabla N_A \cdot \mu \nabla N_B \, d\Omega \, \delta_{ij}$$

$$+ \alpha_f \gamma \Delta t_n \int_{\Omega_{t_{(n+\alpha_f)}}} \nabla N_A \cdot \mathbf{e}_j \mu \nabla N_B \cdot \mathbf{e}_i \, d\Omega$$

$$+ \alpha_f \gamma \Delta t_n \int_{\Omega_{t_{(n+\alpha_f)}}} \tau_{\text{SUPS}} \left(\mathbf{u}^h - \hat{\mathbf{u}}^h \right) \cdot \nabla N_A \rho \left(\mathbf{u}^h - \hat{\mathbf{u}}^h \right) \cdot \nabla N_B \, d\Omega \, \delta_{ij}$$

$$+ \alpha_f \gamma \Delta t_n \int_{\Omega_{t_{(n+\alpha_f)}}} \rho \nu_{\text{LSIC}} \nabla N_A \cdot \mathbf{e}_i \nabla N_B \cdot \mathbf{e}_j \, d\Omega, \tag{4.93}$$

$$\frac{\partial \mathbf{N}_M}{\partial \mathbf{P}_{n+1}} = \left[G_{AB}^i \right], \tag{4.94}$$

$$G_{AB}^i = - \int_{\Omega_{t_{(n+\alpha_f)}}} \nabla N_A \cdot \mathbf{e}_i N_B \, d\Omega$$

$$+ \int_{\Omega_{t_{(n+\alpha_f)}}} \tau_{\text{SUPS}} \left(\mathbf{u}^h - \hat{\mathbf{u}}^h \right) \cdot \nabla N_A \nabla N_B \cdot \mathbf{e}_i \, d\Omega, \tag{4.95}$$

$$\frac{\partial \mathbf{N}_C}{\partial \dot{\mathbf{U}}_{n+1}} = \left[D_{AB}^j \right], \tag{4.96}$$

$$D^j_{AB} = \alpha_f \gamma \Delta t_n \int_{\Omega_{t_{(n+\alpha_f)}}} N_A \nabla N_B \cdot \mathbf{e}_j \, d\Omega$$

$$+ \alpha_f \gamma \Delta t_n \int_{\Omega_{t_{(n+\alpha_f)}}} \tau_{\text{SUPS}} \nabla N_A \cdot \mathbf{e}_j \left(\mathbf{u}^h - \hat{\mathbf{u}}^h\right) \cdot \nabla N_B \, d\Omega$$

$$+ \alpha_m \int_{\Omega_{t_{(n+\alpha_f)}}} \tau_{\text{SUPS}} \nabla N_A \cdot \mathbf{e}_j N_B \, d\Omega, \tag{4.97}$$

and

$$\frac{\partial \mathbf{N}_C}{\partial \mathbf{P}_{n+1}} = [L_{AB}], \tag{4.98}$$

$$L_{AB} = \int_{\Omega_{t_{(n+\alpha_f)}}} \frac{\tau_{\text{SUPS}}}{\rho} \nabla N_A \cdot \nabla N_B \, d\Omega. \tag{4.99}$$

These left-hand-side matrices depend on the solution and are typically evaluated at every Newton–Raphson iteration (the iteration index i has been omitted to simplify the notation). The convective velocity and the stabilization parameters are "lagged" by one iteration and their derivatives are not accounted for in the derivation of these matrices.

REMARK 4.2 *To keep the length of mathematical expressions to a manageable level, we did not present the contributions of the weakly enforced essential boundary conditions to the discrete residual vectors and left-hand-side matrices. These may be inferred directly from the terms given by Equation (4.58).*

4.6.3 Space–Time Formulation

To write the space–time variational formulation of incompressible flows (see Tezduyar, 1992, 2003a; Tezduyar *et al.*, 1992a,c; Takizawa and Tezduyar, 2011) we use again the space–time slab Q_n defined in Section 4.4. We denote the trial and test functions spaces for the velocity and pressure as $\mathbf{u} \in S_u$, $p \in S_p$, $\mathbf{w} \in \mathcal{V}_u$ and $q \in \mathcal{V}_p$. In deriving the variational formulation, we multiply Equations (1.1) and (1.2) with the corresponding test functions and integrate them over Q_n:

$$\int_{Q_n} \mathbf{w} \cdot \rho \left(\frac{\partial \mathbf{u}}{\partial t} + \nabla \cdot (\mathbf{u} \otimes \mathbf{u}) - \mathbf{f} \right) dQ - \int_{Q_n} \mathbf{w} \cdot \nabla \cdot \boldsymbol{\sigma} \, dQ + \int_{Q_n} q \nabla \cdot \mathbf{u} \, dQ = 0. \tag{4.100}$$

We integrate by parts all the terms except for the external force and enforce the essential (i.e., strong Dirichlet) and natural boundary conditions over $(P_n)_g$ and $(P_n)_h$, the complementary subsets of P_n. That gives us the following variational formulation: find $\mathbf{u} \in S_u$ and $p \in S_p$, such that $\forall \, \mathbf{w} \in \mathcal{V}_u$ and $q \in \mathcal{V}_p$:

$$\int_{\Omega_{n+1}} \mathbf{w}^-_{n+1} \cdot \rho \mathbf{u}^-_{n+1} d\Omega - \int_{\Omega_n} \mathbf{w}^+_n \cdot \rho \mathbf{u}^-_n d\Omega - \int_{Q_n} \frac{\partial \mathbf{w}}{\partial t} \cdot \rho \mathbf{u} \, dQ$$

$$- \int_{(P_n)_h} (\mathbf{w} \cdot \rho \mathbf{u})(\mathbf{n} \cdot \hat{\mathbf{u}}) \, dP + \int_{(P_n)_h} (\mathbf{w} \cdot \rho \mathbf{u})(\mathbf{n} \cdot \mathbf{u}) \, dP - \int_{Q_n} \nabla \mathbf{w} : \rho \mathbf{u} \otimes \mathbf{u} \, dQ$$

$$- \int_{Q_n} \mathbf{w} \cdot \rho \mathbf{f} dQ - \int_{(P_n)_h} \mathbf{w} \cdot \mathbf{h} dP + \int_{Q_n} \varepsilon(\mathbf{w}) : \sigma(\mathbf{u}, p) dQ$$
$$+ \int_{P_n} q\mathbf{n} \cdot \mathbf{u} dP - \int_{Q_n} \nabla q \cdot \mathbf{u} dQ = 0. \quad (4.101)$$

We apply the scale separation given by Equations (2.187)–(2.190) at the continuous level, where $\overline{(\ )}$ denotes the coarse scales. The coarse-scale part of Equation (4.101) is written as follows:

$$\int_{\Omega_{n+1}} \overline{\mathbf{w}}_{n+1}^- \cdot \rho \overline{\mathbf{u}}_{n+1}^- d\Omega - \int_{\Omega_n} \overline{\mathbf{w}}_n^+ \cdot \rho \overline{\mathbf{u}}_n^- d\Omega - \int_{Q_n} \frac{\partial \overline{\mathbf{w}}}{\partial t} \cdot \rho \mathbf{u} dQ$$
$$- \int_{(P_n)_h} (\overline{\mathbf{w}} \cdot \rho \mathbf{u})(\mathbf{n} \cdot \hat{\mathbf{u}}) dP + \int_{(P_n)_h} (\overline{\mathbf{w}} \cdot \rho \mathbf{u})(\mathbf{n} \cdot \mathbf{u}) dP - \int_{Q_n} \nabla \overline{\mathbf{w}} : \rho \mathbf{u} \otimes \mathbf{u} dQ$$
$$- \int_{Q_n} \overline{\mathbf{w}} \cdot \rho \mathbf{f} dQ - \int_{(P_n)_h} \overline{\mathbf{w}} \cdot \mathbf{h} dP + \int_{Q_n} \varepsilon(\overline{\mathbf{w}}) : \sigma dQ$$
$$+ \int_{P_n} \overline{q}\mathbf{n} \cdot \mathbf{u} dP - \int_{Q_n} \nabla \overline{q} \cdot \mathbf{u} dQ = 0. \quad (4.102)$$

The fine-scale solutions are represented as

$$\mathbf{u}' = -\frac{\tau_{\text{SUPS}}}{\rho} \mathbf{r}_M(\overline{\mathbf{u}}, \overline{p}), \qquad p' = -\rho \nu_{\text{LSIC}} r_C(\overline{\mathbf{u}}). \quad (4.103)$$

Using scale separation in Equation (4.102) for velocity and pressure, we obtain:

$$\int_{\Omega_{n+1}} (\overline{\mathbf{w}})_{n+1}^- \cdot \rho \left((\overline{\mathbf{u}})_{n+1}^- + (\mathbf{u}')_{n+1}^- \right) d\Omega - \int_{\Omega_n} (\overline{\mathbf{w}})_n^+ \cdot \rho \left((\overline{\mathbf{u}})_n^- + (\mathbf{u}')_n^- \right) d\Omega$$
$$- \int_{Q_n} \frac{\partial \overline{\mathbf{w}}}{\partial t} \cdot \rho(\overline{\mathbf{u}} + \mathbf{u}') dQ + \int_{(P_n)_h} (\overline{\mathbf{w}} \cdot \rho(\overline{\mathbf{u}} + \mathbf{u}'))(\mathbf{n} \cdot (\overline{\mathbf{u}} + \mathbf{u}' - \hat{\mathbf{u}})) dP$$
$$- \int_{Q_n} \nabla \overline{\mathbf{w}} : \rho(\overline{\mathbf{u}} + \mathbf{u}') \otimes (\overline{\mathbf{u}} + \mathbf{u}') dQ - \int_{Q_n} \overline{\mathbf{w}} \cdot \rho \mathbf{f} dQ - \int_{(P_n)_h} \overline{\mathbf{w}} \cdot \mathbf{h} dP$$
$$+ \int_{Q_n} \varepsilon(\overline{\mathbf{w}}) : (\sigma(\overline{\mathbf{u}}, \overline{p}) + \sigma') dQ + \int_{P_n} \overline{q}\mathbf{n} \cdot (\overline{\mathbf{u}} + \mathbf{u}') dP - \int_{Q_n} \nabla \overline{q} \cdot (\overline{\mathbf{u}} + \mathbf{u}') dQ = 0. \quad (4.104)$$

Here $\sigma' \equiv \sigma - \sigma^h$ is introduced temporarily. We set the fine-scale solution to zero at the spatial and temporal boundaries, use the assumption $\varepsilon(\mathbf{w}^h) : 2\mu \nabla \mathbf{u}' = 0$ (see Hughes and Sangalli, 2007; Hughes and Oberai, 2003), and obtain the following form:

$$\int_{\Omega_{n+1}} (\overline{\mathbf{w}})_{n+1}^- \cdot \rho(\overline{\mathbf{u}})_{n+1}^- d\Omega - \int_{\Omega_n} (\overline{\mathbf{w}})_n^+ \cdot \rho(\overline{\mathbf{u}})_n^- d\Omega$$
$$- \int_{Q_n} \frac{\partial \overline{\mathbf{w}}}{\partial t} \cdot \rho(\overline{\mathbf{u}} + \mathbf{u}') dQ + \int_{(P_n)_h} (\overline{\mathbf{w}} \cdot \rho \overline{\mathbf{u}})(\mathbf{n} \cdot (\overline{\mathbf{u}} - \hat{\mathbf{u}})) dP$$
$$- \int_{Q_n} \nabla \overline{\mathbf{w}} : \rho(\overline{\mathbf{u}} + \mathbf{u}') \otimes (\overline{\mathbf{u}} + \mathbf{u}') dQ - \int_{Q_n} \overline{\mathbf{w}} \cdot \rho \mathbf{f} dQ - \int_{(P_n)_h} \overline{\mathbf{w}} \cdot \mathbf{h} dP$$
$$+ \int_{Q_n} \varepsilon(\overline{\mathbf{w}}) : (\sigma(\overline{\mathbf{u}}, \overline{p}) - p'\mathbf{I}) dQ + \int_{P_n} \overline{q}\mathbf{n} \cdot \overline{\mathbf{u}} dP - \int_{Q_n} \nabla \overline{q} \cdot (\overline{\mathbf{u}} + \mathbf{u}') dQ = 0. \quad (4.105)$$

We collect the fine-scale terms in one place and write the global integrations as the sum of element-level integrals:

$$\int_{\Omega_{n+1}} (\overline{\mathbf{w}})^-_{n+1} \cdot \rho(\overline{\mathbf{u}})^-_{n+1} d\Omega - \int_{\Omega_n} (\overline{\mathbf{w}})^+_n \cdot \rho(\overline{\mathbf{u}})^-_n d\Omega - \int_{Q_n} \frac{\partial \overline{\mathbf{w}}}{\partial t} \cdot \rho \overline{\mathbf{u}} dQ$$
$$+ \int_{(P_n)_h} (\overline{\mathbf{w}} \cdot \rho\overline{\mathbf{u}}) (\mathbf{n} \cdot (\overline{\mathbf{u}} - \hat{\mathbf{u}})) dP - \int_{Q_n} \nabla\overline{\mathbf{w}} : \rho\overline{\mathbf{u}} \otimes \overline{\mathbf{u}} dQ - \int_{Q_n} \overline{\mathbf{w}} \cdot \rho\mathbf{f} dQ$$
$$- \int_{(P_n)_h} \overline{\mathbf{w}} \cdot \mathbf{h} dP + \int_{Q_n} \boldsymbol{\varepsilon}(\overline{\mathbf{w}}) : \boldsymbol{\sigma}(\overline{\mathbf{u}}, \overline{p}) dQ + \int_{P_n} \overline{q}\mathbf{n} \cdot \overline{\mathbf{u}} dP - \int_{Q_n} \nabla\overline{q} \cdot \overline{\mathbf{u}} dQ$$
$$- \sum_{e=1}^{(n_{el})_n} \int_{Q_n^e} \left[\left(\rho \frac{\partial \overline{\mathbf{w}}}{\partial t} + \nabla\overline{q} \right) \cdot \mathbf{u}' + \nabla\overline{\mathbf{w}} : (\rho(\mathbf{u}' \otimes \overline{\mathbf{u}} + \overline{\mathbf{u}} \otimes \mathbf{u}' + \mathbf{u}' \otimes \mathbf{u}') + p'\mathbf{I}) \right] dQ = 0. \quad (4.106)$$

Here each Q_n is decomposed into elements Q_n^e, where $e = 1, 2, \ldots, (n_{el})_n$. The subscript n used with n_{el} is for the general case where the number of space–time elements may change from one space–time slab to another. The spatially discretized version of Equation (4.106) is written as follows: find $\mathbf{u}^h \in (\mathcal{S}_u^h)_n$ and $p^h \in (\mathcal{S}_p^h)_n$, such that $\forall \; \mathbf{w}^h \in (\mathcal{V}_u^h)_n$ and $q^h \in (\mathcal{V}_p^h)_n$:

$$\int_{\Omega_{n+1}} (\mathbf{w}^h)^-_{n+1} \cdot \rho(\mathbf{u}^h)^-_{n+1} d\Omega - \int_{\Omega_n} (\mathbf{w}^h)^+_n \cdot \rho(\mathbf{u}^h)^-_n d\Omega - \int_{Q_n} \frac{\partial \mathbf{w}^h}{\partial t} \cdot \rho \mathbf{u}^h dQ$$
$$+ \int_{(P_n)_h} \left(\mathbf{w}^h \cdot \rho\mathbf{u}^h \right) \left(\mathbf{n}^h \cdot (\mathbf{u}^h - \hat{\mathbf{u}}^h) \right) dP - \int_{Q_n} \nabla\mathbf{w}^h : \rho\mathbf{u}^h \otimes \mathbf{u}^h dQ - \int_{Q_n} \mathbf{w}^h \cdot \rho\mathbf{f}^h dQ$$
$$- \int_{(P_n)_h} \mathbf{w}^h \cdot \mathbf{h}^h dP + \int_{Q_n} \boldsymbol{\varepsilon}(\mathbf{w}^h) : \boldsymbol{\sigma}(\mathbf{u}^h, p^h) dQ + \int_{P_n} q^h \mathbf{n}^h \cdot \mathbf{u}^h dP - \int_{Q_n} \nabla q^h \cdot \mathbf{u}^h dQ$$
$$- \sum_{e=1}^{(n_{el})_n} \int_{Q_n^e} \left[\left(\rho \frac{\partial \mathbf{w}^h}{\partial t} + \nabla q^h \right) \cdot \mathbf{u}' + \nabla\mathbf{w}^h : \left(\rho(\mathbf{u}' \otimes \mathbf{u}^h + \mathbf{u}^h \otimes \mathbf{u}' + \mathbf{u}' \otimes \mathbf{u}') + p'\mathbf{I} \right) \right] dQ = 0.$$
$$(4.107)$$

We further rearrange the terms in Equation (4.107):

$$\int_{Q_n} \mathbf{w}^h \cdot \rho \left(\frac{\partial \mathbf{u}^h}{\partial t} + \nabla \cdot (\mathbf{u}^h \otimes \mathbf{u}^h) - \mathbf{f}^h \right) dQ + \int_{Q_n} \boldsymbol{\varepsilon}(\mathbf{w}^h) : \boldsymbol{\sigma}(\mathbf{u}^h, p^h) dQ$$
$$- \int_{(P_n)_h} \mathbf{w}^h \cdot \mathbf{h}^h dP + \int_{Q_n} q^h \nabla \cdot \mathbf{u}^h dQ + \int_{\Omega_n} (\mathbf{w}^h)^+_n \cdot \rho \left((\mathbf{u}^h)^+_n - (\mathbf{u}^h)^-_n \right) d\Omega$$
$$- \sum_{e=1}^{(n_{el})_n} \int_{Q_n^e} \left[\rho \left(\frac{\partial \mathbf{w}^h}{\partial t} + \mathbf{u}^h \cdot \nabla \mathbf{w}^h \right) + \nabla q^h \right] \cdot \mathbf{u}' dQ - \sum_{e=1}^{(n_{el})_n} \int_{Q_n^e} \nabla \cdot \mathbf{w}^h p' dQ$$
$$- \sum_{e=1}^{(n_{el})_n} \int_{Q_n^e} \rho \left(\nabla\mathbf{w}^h \right) : \mathbf{u}^h \otimes \mathbf{u}' dQ - \sum_{e=1}^{(n_{el})_n} \int_{Q_n^e} \rho \left(\nabla\mathbf{w}^h \right) : \mathbf{u}' \otimes \mathbf{u}' dQ = 0, \quad (4.108)$$

and expand the fine-scale terms with the expressions given by Equations (2.191) and (2.192):

$$\int_{Q_n} \mathbf{w}^h \cdot \rho \left(\frac{\partial \mathbf{u}^h}{\partial t} + \nabla \cdot (\mathbf{u}^h \otimes \mathbf{u}^h) - \mathbf{f}^h \right) dQ + \int_{Q_n} \boldsymbol{\varepsilon}(\mathbf{w}^h) : \boldsymbol{\sigma}(\mathbf{u}^h, p^h) dQ$$

$$- \int_{(P_n)_h} \mathbf{w}^h \cdot \mathbf{h}^h dP + \int_{Q_n} q^h \nabla \cdot \mathbf{u}^h dQ + \int_{\Omega_n} (\mathbf{w}^h)_n^+ \cdot \rho \left((\mathbf{u}^h)_n^+ - (\mathbf{u}^h)_n^- \right) d\Omega$$

$$+ \sum_{e=1}^{(n_{\mathrm{el}})_n} \int_{Q_n^e} \frac{\tau_{\mathrm{SUPS}}}{\rho} \left[\rho \left(\frac{\partial \mathbf{w}^h}{\partial t} + \mathbf{u}^h \cdot \nabla \mathbf{w}^h \right) + \nabla q^h \right] \cdot \mathbf{r}_{\mathrm{M}}(\mathbf{u}^h, p^h) dQ$$

$$+ \sum_{e=1}^{(n_{\mathrm{el}})_n} \int_{Q_n^e} \rho \nu_{\mathrm{LSIC}} \nabla \cdot \mathbf{w}^h r_{\mathrm{C}}(\mathbf{u}^h) dQ$$

$$+ \sum_{e=1}^{(n_{\mathrm{el}})_n} \int_{Q_n^e} \tau_{\mathrm{SUPS}} \left(\nabla \mathbf{w}^h \right) : \mathbf{u}^h \otimes \mathbf{r}_{\mathrm{M}}(\mathbf{u}^h, p^h) dQ$$

$$- \sum_{e=1}^{(n_{\mathrm{el}})_n} \int_{Q_n^e} \frac{\tau_{\mathrm{SUPS}}^2}{\rho} \left(\nabla \mathbf{w}^h \right) : \mathbf{r}_{\mathrm{M}}(\mathbf{u}^h, p^h) \otimes \mathbf{r}_{\mathrm{M}}(\mathbf{u}^h, p^h) dQ = 0. \quad (4.109)$$

This was named the DSD/SST-VMST formulation in Takizawa and Tezduyar (2011) (i.e., the version with the VMS turbulence model). The shorter acronym "ST-VMS" (meaning "Space–Time VMS") was introduced in Takizawa and Tezduyar (2012b) and is used interchangeably with DSD/SST-VMST. This is the "conservative form" of the DSD/SST-VMST formulation.

We now provide, from Takizawa and Tezduyar (2012b), a shortcut derivation for the "convective form" of the DSD/SST-VMST formulation. We integrate by parts two of the pieces in the 5th term of Equation (4.105):

$$- \int_{Q_n} \nabla \overline{\mathbf{w}} : \rho \overline{\mathbf{u}} \otimes \overline{\mathbf{u}} dQ = - \int_{(P_n)_h} (\overline{\mathbf{w}} \cdot \rho \overline{\mathbf{u}})(\mathbf{n} \cdot \overline{\mathbf{u}}) dP + \int_{Q_n} \rho \overline{\mathbf{w}} \otimes \overline{\mathbf{u}} : (\nabla \overline{\mathbf{u}}) dQ$$

$$+ \int_{Q_n} (\overline{\mathbf{w}} \cdot \rho \overline{\mathbf{u}}) \nabla \cdot \overline{\mathbf{u}} dQ, \quad (4.110)$$

$$- \int_{Q_n} \nabla \overline{\mathbf{w}} : \rho \overline{\mathbf{u}} \otimes \mathbf{u}' dQ = - \int_{(P_n)_h} (\overline{\mathbf{w}} \cdot \rho \overline{\mathbf{u}})(\mathbf{n} \cdot \mathbf{u}') dP + \int_{Q_n} \rho \overline{\mathbf{w}} \otimes \mathbf{u}' : (\nabla \overline{\mathbf{u}}) dQ$$

$$+ \int_{Q_n} (\overline{\mathbf{w}} \cdot \rho \overline{\mathbf{u}}) \nabla \cdot \mathbf{u}' dQ. \quad (4.111)$$

The first term on the right-hand-side of Equation (4.110) cancels the same term with opposite sign in Equation (4.105). The first term on the right-hand-side of Equation (4.111) vanishes because the fine-scale solution is zero at the spatial boundaries. The sum of the last terms of Equations (4.110) and (4.111) is zero. The discrete version of the second term on the right-hand-side of Equation (4.110) replaces in the first term of Equation (4.108) the advection piece in the conservation-law form. The discrete version of the second term on the right-hand-side of Equation (4.111) replaces the 8th term of Equation (4.108). With all that, we obtain:

$$\int_{Q_n} \mathbf{w}^h \cdot \rho \left(\frac{\partial \mathbf{u}^h}{\partial t} + \mathbf{u}^h \cdot \nabla \mathbf{u}^h - \mathbf{f}^h \right) dQ + \int_{Q_n} \boldsymbol{\varepsilon}(\mathbf{w}^h) : \boldsymbol{\sigma}(\mathbf{u}^h, p^h) dQ$$
$$- \int_{(P_n)_h} \mathbf{w}^h \cdot \mathbf{h}^h dP + \int_{Q_n} q^h \nabla \cdot \mathbf{u}^h dQ + \int_{\Omega_n} (\mathbf{w}^h)_n^+ \cdot \rho \left((\mathbf{u}^h)_n^+ - (\mathbf{u}^h)_n^- \right) d\Omega$$
$$- \sum_{e=1}^{(n_{el})_n} \int_{Q_n^e} \left[\rho \left(\frac{\partial \mathbf{w}^h}{\partial t} + \mathbf{u}^h \cdot \nabla \mathbf{w}^h \right) + \nabla q^h \right] \cdot \mathbf{u}' dQ - \sum_{e=1}^{(n_{el})_n} \int_{Q_n^e} \nabla \cdot \mathbf{w}^h p' dQ$$
$$+ \sum_{e=1}^{(n_{el})_n} \int_{Q_n^e} \rho \mathbf{w}^h \otimes \mathbf{u}' : (\nabla \mathbf{u}^h) dQ - \sum_{e=1}^{(n_{el})_n} \int_{Q_n^e} (\nabla \mathbf{w}^h) : \rho \mathbf{u}' \otimes \mathbf{u}' dQ = 0. \qquad (4.112)$$

We again expand the fine-scale terms with the expressions given by Equations (2.191) and (2.192) and obtain the convective form of the DSD/SST-VMST formulation:

$$\int_{Q_n} \mathbf{w}^h \cdot \rho \left(\frac{\partial \mathbf{u}^h}{\partial t} + \mathbf{u}^h \cdot \nabla \mathbf{u}^h - \mathbf{f}^h \right) dQ + \int_{Q_n} \boldsymbol{\varepsilon}(\mathbf{w}^h) : \boldsymbol{\sigma}(\mathbf{u}^h, p^h) dQ$$
$$- \int_{(P_n)_h} \mathbf{w}^h \cdot \mathbf{h}^h dP + \int_{Q_n} q^h \nabla \cdot \mathbf{u}^h dQ + \int_{\Omega_n} (\mathbf{w}^h)_n^+ \cdot \rho \left((\mathbf{u}^h)_n^+ - (\mathbf{u}^h)_n^- \right) d\Omega$$
$$+ \sum_{e=1}^{(n_{el})_n} \int_{Q_n^e} \frac{\tau_{\text{SUPS}}}{\rho} \left[\rho \left(\frac{\partial \mathbf{w}^h}{\partial t} + \mathbf{u}^h \cdot \nabla \mathbf{w}^h \right) + \nabla q^h \right] \cdot \mathbf{r}_{\text{M}}(\mathbf{u}^h, p^h) dQ$$
$$+ \sum_{e=1}^{(n_{el})_n} \int_{Q_n^e} \rho \nu_{\text{LSIC}} \nabla \cdot \mathbf{w}^h r_{\text{C}}(\mathbf{u}^h) dQ$$
$$- \sum_{e=1}^{(n_{el})_n} \int_{Q_n^e} \tau_{\text{SUPS}} \mathbf{w}^h \otimes \mathbf{r}_{\text{M}} : (\nabla \mathbf{u}^h) dQ$$
$$- \sum_{e=1}^{(n_{el})_n} \int_{Q_n^e} \frac{\tau_{\text{SUPS}}^2}{\rho} (\nabla \mathbf{w}^h) : \mathbf{r}_{\text{M}}(\mathbf{u}^h, p^h) \otimes \mathbf{r}_{\text{M}}(\mathbf{u}^h, p^h) dQ = 0. \qquad (4.113)$$

REMARK 4.3 *One of the main differences between the ALE and DSD/SST forms of the VMS method is that the DSD/SST formulation retains the fine-scale time derivative term $\frac{\partial \mathbf{u}'}{\partial t}|_\xi$. Dropping this term is called the "quasi-static" assumption (see Bazilevs et al., 2011b for the terminology). This is the same as the WTSE option in the DSD/SST formulation (see Remark 2 of Tezduyar and Sathe, 2007). We believe that this makes a significant difference, especially when the polynomial orders in space or time are higher (see Takizawa and Tezduyar, 2011).*

The original DSD/SST formulation (Tezduyar, 1992, 2003a; Tezduyar et al., 1992a,c) can be obtained by dropping the last two terms in Equation (4.113). The original formulation was named "DSD/SST-SUPS" in Takizawa and Tezduyar (2012b) (i.e., the version with the SUPG/PSPG stabilization). The short acronym "ST-SUPS" will be used interchangeably with DSD/SST-SUPS. For backward compatibility, the acronym DSD/SST, when written without any of the two option indicators "-SUPS" and "-VMST", will imply DSD/SST-SUPS. For completeness, we also provide here the DSD/SST-SUPS formulation from Tezduyar (2003a):

$$\int_{Q_n} \mathbf{w}^h \cdot \rho \left(\frac{\partial \mathbf{u}^h}{\partial t} + \mathbf{u}^h \cdot \nabla \mathbf{u}^h - \mathbf{f}^h \right) dQ + \int_{Q_n} \boldsymbol{\varepsilon}(\mathbf{w}^h) : \boldsymbol{\sigma}(\mathbf{u}^h, p^h) dQ$$

$$- \int_{(P_n)_h} \mathbf{w}^h \cdot \mathbf{h}^h dP + \int_{Q_n} q^h \nabla \cdot \mathbf{u}^h dQ + \int_{\Omega_n} (\mathbf{w}^h)_n^+ \cdot \rho \left((\mathbf{u}^h)_n^+ - (\mathbf{u}^h)_n^- \right) d\Omega$$

$$+ \sum_{e=1}^{(n_{el})_n} \int_{Q_n^e} \frac{1}{\rho} \left[\tau_{\text{SUPG}} \rho \left(\frac{\partial \mathbf{w}^h}{\partial t} + \mathbf{u}^h \cdot \nabla \mathbf{w}^h \right) + \tau_{\text{PSPG}} \nabla q^h \right] \cdot \mathbf{r}_{\text{M}}(\mathbf{u}^h, p^h) dQ$$

$$+ \sum_{e=1}^{(n_{el})_n} \int_{Q_n^e} \rho \nu_{\text{LSIC}} \nabla \cdot \mathbf{w}^h r_{\text{C}}(\mathbf{u}^h) dQ = 0. \quad (4.114)$$

We note that one can also have the conservative form of the DSD/SST-SUPS formulation, which can be obtained by dropping the last two terms in Equation (4.109).

We now provide the complete set of choices for the stabilization parameters used with the DSD/SST formulations. We start with

$$\tau_{\text{SUPS}} = \left(\frac{1}{\tau_{\text{SUGN12}}^2} + \frac{1}{\tau_{\text{SUGN3}}^2} \right)^{-\frac{1}{2}}, \quad (4.115)$$

$$\tau_{\text{SUGN12}} = \left(\sum_{\alpha=1}^{n_{\text{ent}}} \sum_{a=1}^{n_{\text{ens}}} \left| \frac{\partial N_a^\alpha}{\partial t} + \mathbf{u}^h \cdot \nabla N_a^\alpha \right| \right)^{-1}, \quad (4.116)$$

$$\tau_{\text{SUGN3}} = \frac{h_{\text{RGN}}^2}{4\nu}, \quad (4.117)$$

$$h_{\text{RGN}} = 2 \left(\sum_{\alpha=1}^{n_{\text{ent}}} \sum_{a=1}^{n_{\text{ens}}} |\mathbf{r} \cdot \nabla N_a^\alpha| \right)^{-1}, \quad (4.118)$$

$$\mathbf{r} = \frac{\nabla \|\mathbf{u}^h\|}{\|\nabla \|\mathbf{u}^h\|\|}, \quad (4.119)$$

$$\tau_{\text{PSPG}} = \tau_{\text{SUPG}} = \tau_{\text{SUPS}}. \quad (4.120)$$

When we construct τ_{SUPS} based on separate definitions for the advection-dominated and transient-dominated limits, we have

$$\tau_{\text{SUPS}} = \left(\frac{1}{\tau_{\text{SUGN1}}^2} + \frac{1}{\tau_{\text{SUGN2}}^2} + \frac{1}{\tau_{\text{SUGN3}}^2} \right)^{-\frac{1}{2}}, \quad (4.121)$$

$$\tau_{\text{SUGN1}} = \left(\sum_{\alpha=1}^{n_{\text{ent}}} \sum_{a=1}^{n_{\text{ens}}} \left| (\mathbf{u}^h - \hat{\mathbf{u}}^h) \cdot \nabla N_a^\alpha \right| \right)^{-1}, \quad (4.122)$$

$$\tau_{\text{SUGN2}} = \frac{\Delta t}{2}. \quad (4.123)$$

It was noted in Tezduyar and Sathe (2007) that separating τ_{SUGN12} into its advection- and transient-dominated components as given by Equations (4.122) and (4.123) is equivalent to excluding the $\frac{\partial N_a}{\partial t}\big|_\xi$ part of $\frac{\partial N_a}{\partial t}$ in Equation (4.116), making that the definition for τ_{SUGN1}, and accounting for $\frac{\partial N_a}{\partial t}\big|_\xi$ in the definition for τ_{SUGN2} given by Equation (4.123). Here ξ is the vector of element coordinates, and $\frac{\partial}{\partial t}\big|_\xi$ is equivalent to $\frac{\partial}{\partial t}\big|_{\hat{x}}$. Both notations for the same partial derivative are kept in the interest of backward compatibility with prior articles.

The LSIC parameter was defined in Tezduyar (2003a) as

$$\nu_{\text{LSIC}} = \tau_{\text{SUPS}} \|\mathbf{u}^h\|^2, \qquad (4.124)$$

and in Tezduyar and Sathe (2007) as

$$\nu_{\text{LSIC-TC2}} = \tau_{\text{SUPS}} \|\mathbf{u}^h - \hat{\mathbf{u}}^h\|^2, \qquad (4.125)$$

which is called "TC2." There are two more LSIC parameter definitions used in conjunction with the DSD/SST-VMST formulation. First one was introduced in Takizawa and Tezduyar (2011) and is defined as follows:

$$\nu_{\text{LSIC-TGI}} = \left(\tau_{\text{SUPS}} \sum_{i=1}^{n_{\text{sd}}} G_{ii}\right)^{-1}, \qquad (4.126)$$

which was motivated from Bazilevs and Akkerman (2010). This is called "TGI." We note that $\sum_{i=1}^{n_{\text{sd}}} G_{ii}$ is the trace of $\mathbf{G}$, which is expressed by Equation (2.140). The second option, called "LHC," was introduced in Takizawa et al. (2011a) and is defined as follows:

$$\nu_{\text{LSIC-LHC}} = \left(\nu_{\text{LSIC-TC2}}^{-2} + \nu_{\text{LSIC-HRGN}}^{-2}\right)^{-\frac{1}{2}}, \qquad (4.127)$$

$$\nu_{\text{LSIC-HRGN}} = \frac{h_{\text{RGN}}^2}{\tau_{\text{SUPS}}}. \qquad (4.128)$$

For more ways of calculating τ_{SUPG}, τ_{PSPG}, and ν_{LSIC}, see (Tezduyar and Osawa, 2000; Tezduyar, 2003a, 2007b; Akin et al., 2003; Akin and Tezduyar, 2004; Catabriga et al., 2005, 2006; Tezduyar et al., 2006d; Tezduyar and Sathe, 2006; Onate et al., 2006; Corsini et al., 2006, 2010, 2011; Rispoli et al., 2007; Hsu et al., 2010). References Tezduyar (2003a, 2007b) also include the Discontinuity-Capturing Directional Dissipation (DCDD) stabilization, which was introduced as an alternative to the LSIC stabilization.

REMARK 4.4 *As an alternative to how the SUPG test function is defined in Equation (4.114), another option was proposed in Tezduyar and Sathe (2007), where the SUPG test function* $\left(\frac{\partial \mathbf{w}^h}{\partial t} + \mathbf{u}^h \cdot \nabla \mathbf{w}^h\right)$ *is replaced with* $\left((\mathbf{u}^h - \hat{\mathbf{u}}^h) \cdot \nabla \mathbf{w}^h\right)$. *This replacement is equivalent to excluding the* $\frac{\partial \mathbf{w}^h}{\partial t}\big|_\xi$ *part of* $\frac{\partial \mathbf{w}^h}{\partial t}$. *In Tezduyar and Sathe (2007), this option was called "WTSE," and the option where the* $\frac{\partial \mathbf{w}^h}{\partial t}\big|_\xi$ *term is active, "WTSA." If the SUPG test function option is not explicitly specified, it will imply WTSA.*

REMARK 4.5 *The stability and accuracy analysis reported in Takizawa et al. (2011f) and Takizawa and Tezduyar (2011) for the DSD/SST formulation of the advection equation shows for linear functions in space and time that the WTSA option yields higher-order accuracy than the WTSE option.*

REMARK 4.6 *The τ_{SUGN12} component of the τ_{SUPG} definition given by Equations (107)–(109) in Tezduyar (2003a) is the space–time version of the original definition in Tezduyar and Park (1986). These definitions sense, in addition to the element geometry, the order of the interpolation functions. Some τ definitions do that and some do not. The definitions in Sections 3.3.1 and 3.3.2 of Shakib et al. (1991), for example, are among those that do not.*

REMARK 4.7 *Remark 4.6 is applicable also when the interpolation functions are NURBS functions. This includes classical p-refinement and also k-refinement, except when used in conjunction with periodic B-splines.*

REMARK 4.8 *For each space–time slab, velocity and pressure assume double unknown values at each spatial node. One value corresponds to the lower end of the slab, and the other one the upper end. In Tezduyar and Sathe (2007), the option of using double unknown values at a spatial node is called "DV" for velocity and "DP" for pressure. In this case, as pointed out in Tezduyar and Sathe (2007), we use two integration points over the time interval of the space–time slab, and this time integration option is called "TIP2." This version of the DSD/SST-SUPS formulation, with the options set DV, DP, and TIP2, is called "DSD/SST-DP."*

REMARK 4.9 *In Tezduyar and Sathe (2007), the option of using, for each space–time slab, a single unknown pressure value at each spatial node was proposed with the option name "SP." With this, another version of the DSD/SST-SUPS formulation was proposed in Tezduyar and Sathe (2007), where the options set is DV, SP, and TIP2. This version is called "DSD/SST-SP." Because the number of pressure unknowns is halved, the computational cost is reduced.*

REMARK 4.10 *To reduce the computational cost further, the option of using only one integration point over the time interval of the space–time slab was proposed in Tezduyar and Sathe (2007). This time integration option is called "TIP1." With that, a third version of the DSD/SST-SUPS formulation was proposed in Tezduyar and Sathe (2007), where the options set is DV, SP, and TIP1. This version is called "DSD/SST-TIP1."*

REMARK 4.11 *As a third way of reducing the computational cost, the option of using, for each space–time slab, a single unknown velocity value at each spatial node was proposed in Tezduyar and Sathe (2007) with the option name "SV." In the SV option, of the two parts of Equation (4.114), the one generated by $(\mathbf{w}^h)_n^+$ is removed, and we explicitly set $(\mathbf{u}^h)_n^+ = (\mathbf{u}^h)_n^-$, which makes the velocity field continuous in time. Based on the SV option, a fourth version of the DSD/SST-SUPS formulation was proposed in Tezduyar and Sathe (2007), where the options set is SV, SP, and TIP1. This version is called "DSD/SST-SV." In this version of the DSD/SST-SUPS formulation, as it was proposed in Tezduyar and Sathe (2007), one can use the SUPG test function option WTSE.*

REMARK 4.12 *The DSD/SST-SV and DSD/SST-TIP1 versions were introduced to reduce the computational cost per time step, so that the DSD/SST formulation offers versions that are competitive with the ALE formulation based on that measure. However, as the stability and accuracy analysis reported in Takizawa et al. (2011f) and Takizawa and Tezduyar (2011) for the DSD/SST formulation of the advection equation shows, the DSD/SST-SP version (and therefore the DSD/SST-DP version) has higher-order time accuracy than the DSD/SST-SV and DSD/SST-TIP1 versions. Consequently, unless there are other reasons to use smaller time* steps, with the DSD/SST-SP and DSD/SST-DP versions the desired accuracy can be obtained with larger time steps. This, in some cases, might make the DSD/SST-SP and DSD/SST-DP versions more computationally efficient than the DSD/SST-SV and DSD/SST-TIP1 versions. Considerations for parallel-computing efficiency also make the DSD/SST-SP and DSD/SST-DP versions more favorable, because increasing the computational cost per time step is better for parallel efficiency than increasing the number of time steps. Furthermore, when higher-order spatial interpolations are used (such as NURBS), it is much more effective to use also higher-order time interpolations, such as what we have in the DSD/SST-SP and DSD/SST-DP versions and what one might have in even higher-order versions in time interpolation.

REMARK 4.13 *The acronym DSD/SST, when written without any of the four option indicators "-DP", "-SP", "-TIP1", "-SV", will imply DSD/SST-DP.*

REMARK 4.14 *For DSD/SST-SP, DSD/SST-TIP1 and DSD/SST-SV, in integration of the incompressibility-constraint term over each space–time slab, as proposed in Takizawa et al. (2010a), we use only one integration point in time, shifted to the upper time level of the slab. All other terms in the space–time finite element formulation are integrated by using Gaussian quadrature points in time, with the number of points set to whatever we intended to have for the overall formulation. With this technique, as pointed in Takizawa et al. (2010a), the incompressibility constraint equation focuses on the velocity field* $(\mathbf{u}^h)_{n+1}^-$.

4.7 Mesh Moving Methods

In interface-tracking (moving-mesh) methods such the ALE and space–time techniques, as the computations proceed, the mesh needs to be updated to accommodate the changes in the spatial domain. It is crucial that this is accomplished as effectively as possible. How the mesh can best be updated depends on several factors, such as the complexity of the interface and overall geometry, how unsteady the interface is, and how the starting mesh was generated. In general, the mesh update could have two components: moving the mesh as long as it is possible and remeshing (i.e., generating fully or partially a new set of nodes and elements) when the element distortion becomes too high.

Most real-world problems require simulations with complex geometries. A complex geometry typically requires an automatic mesh generator to start with, and automatic mesh generators with special features, such as structured layers of elements around solid surfaces, become desirable. The automatic mesh generator technology described in Johnson and Tezduyar (1997), for example, has the capability to build structured layers of elements around solid objects with reasonable geometric complexity, and has been used very effectively in a number of simulations (for early examples see Johnson and Tezduyar, 1996; Tezduyar et al., 1996). With this capability, one can fully control the mesh resolution near solid objects. This feature can be used for more accurate representation of the boundary layers. Sometimes special-purpose mesh generators designed for specific problems can be used. Depending on the complexity of the problem, such mesh generators might involve a high initial design cost, but minimal mesh generation cost. This is the path that was selected for example for the computations reported in Mittal and Tezduyar (1994).

In mesh moving techniques, the only rule the mesh motion needs to follow is that at the interface the normal velocity of the mesh has to match the normal velocity of the fluid. Beyond that, the mesh can be moved in any way desired, with the main objective being to reduce the frequency of remeshing. In 3D simulations, if the remeshing requires calling an automatic mesh generator, the cost of automatic mesh generation becomes a major reason for trying to reduce the frequency of remeshing. If remeshing does not consist of (full or partial) regeneration of just the element connectivities but also involves (full or partial) node regeneration, we need to project the solution from the old mesh to the new one. This involves a search process, which can be carried out in parallel. Still, the computational cost involved in this, and the projection errors introduced by remeshing, add more incentives for reducing the frequency of remeshing.

If the starting mesh is a product of a special-purpose mesh generator and the changes in the shape of the computational domain allow it, the mesh motion could be handled with a special-purpose mesh moving technique. This would be based on moving the nodes according to an explicitly-defined rule and would eliminate remeshing altogether. Simulations can be carried out without calling an automatic mesh generator and without solving any additional equations to determine the motion of the mesh. One of the earliest examples of that, 3D parallel computation of sloshing in a vertically vibrating container, can be found in Tezduyar *et al.* (1993).

In general, however, an automatic mesh moving scheme is needed to move the nodal points. An example of this is the technique introduced in Tezduyar *et al.* (1992b, 1993) and Johnson and Tezduyar (1994), where the motion of the internal nodes is determined by solving the equations of elasticity. In the continuum setting, the fluid-domain displacement may be computed from the following variational formulation (see also Section 2.5.3): find the fluid-domain displacement from its referential configuration, $\hat{\mathbf{y}} \in \mathcal{S}_m$, such that $\forall \, \mathbf{w} \in \mathcal{V}_m$:

$$\int_{\Omega_{\tilde{t}}} \boldsymbol{\epsilon}(\mathbf{w}) \cdot \mathbf{D}\boldsymbol{\epsilon}\left(\hat{\mathbf{y}}(t) - \hat{\mathbf{y}}(\tilde{t})\right) \, d\Omega = 0, \tag{4.129}$$

where $\Omega_{\tilde{t}}$ and $\hat{\mathbf{y}}(\tilde{t})$ are the fluid subdomain and its displacement vector, respectively, at time $\tilde{t} < t$ and considered known, $\mathcal{S}_m$ and $\mathcal{V}_m$ are the sets of trial and test functions for the fluid-domain motion, $\boldsymbol{\epsilon}$ is the strain vector evaluated using the spatial coordinates on $\Omega_{\tilde{t}}$ (see Equation (2.160) for definition), and $\mathbf{D}$ is the elasticity tensor defined in Equation (2.162). The fluid-domain velocity $\hat{\mathbf{u}}$ may be found by differentiating the fluid-domain displacement $\hat{\mathbf{y}}$ with respect to time holding the referential coordinates fixed. The boundary conditions for these mesh motion equations are specified as follows. To prevent the fluid domain and the moving boundary from separating, we require

$$\hat{\mathbf{y}} \cdot \mathbf{n} = \mathbf{y} \cdot \mathbf{n} \quad \text{on} \quad (\Gamma_t)_g \tag{4.130}$$

and

$$\mathbf{w} \cdot \mathbf{n} = 0 \quad \text{on} \quad (\Gamma_t)_g, \tag{4.131}$$

where $(\Gamma_t)_g$ denotes the moving part of the fluid-domain boundary and $\mathbf{y}$ is its displacement.

In the discrete setting, the variational formulation given by Equation (4.129) is solved using the finite-dimensional function sets defined by the fluid-problem mesh, which gives a time-dependent mesh deformation. The mesh deformation is dealt with selectively based on the sizes of the elements by altering the way we account for the Jacobian of the transformation from the element domain to the physical domain (see Tezduyar *et al.*, 1992b, 1993; Johnson and Tezduyar, 1994). The objective is to stiffen the smaller elements, which are typically placed near solid surfaces, more than the larger ones. For this, the elasticity tensor for the mesh motion problem, now denoted by $\mathbf{D}^h$ due to mesh dependence, makes use of the mesh Lamé parameters μ^h and λ^h. These are given by

$$\mu^h = \frac{E_m^h}{2(1 + \nu_m)}, \tag{4.132}$$

$$\lambda^h = \frac{\nu_m E_m^h}{(1 + \nu_m)(1 - 2\nu_m)}, \tag{4.133}$$

where E_m^h is the mesh Young's modulus, defined as

$$E_m^h = E_m \left(\frac{J_{x\xi}}{J_{x\xi}^0}\right)^{-\chi}, \tag{4.134}$$

$J_{x\xi}$ is the Jacobian determinant of the isoparametric element mapping (see Equation (2.47)), $\chi > 0$ is a real parameter, $J_{x\xi}^0$ is an arbitrary global scaling value, and E_m and ν_m are the constant, user-prescribed nominal mesh Young's modulus and Poisson's ratio. As a result of the definition given by Equation (4.134), small elements, which are typically placed near the solid object, become "stiffer" and are less likely to deform as much as the larger elements, which are typically placed in the areas where the solution is not expected to exhibit complex behavior. Note that the magnitude of E_m has no influence on the discrete solution. However, if a direct coupling solution strategy is employed (see Chapter 6), the choice of E_m may influence the condition number of the left-hand-side matrix of the coupled FSI equations. The mesh Poisson's ratio $\nu_m \in [0, 0.5)$ is typically chosen to be 0.3 in our simulations. When this technique was first introduced in Tezduyar et al. (1992b, 1993) and Johnson and Tezduyar (1994), χ was set to 1.0, which resulted in simply dropping the Jacobian from the finite element formulation of the mesh moving (elasticity) equations. This method was augmented in Stein et al. (2003b) to a more extensive kind by introducing a stiffening power that determines the degree to which the smaller elements are rendered stiffer than the larger ones.

There are of course other methods of moving the mesh, including those where the nodal displacement is governed by the Laplace's equation. With the elasticity equations, depending on the relative values of λ and μ, different modes of deformation can be represented. The effect of varying the ratio λ/μ was investigated in Johnson and Tezduyar (1994) by plotting the maximum of the changes in the aspect ratio for all the elements in the domain of the test problem. There are of course also alternatives to the Jacobian-based stiffening technique in dealing with the mesh deformation selectively based on the sizes of the elements, which is related to how the stiffness is distributed among the elements. For example, in the stiffening method proposed in Masud and Hughes (1997) an expression in terms of the element volumes is used. It can be shown that for constant-Jacobian elements, for all the elements with volumes less than 10% of the largest element volume, the difference in the stiffening values reached by the two stiffening techniques is less than 10%. In going down the volume-ranked elements, in most problems of practical interest, one would very quickly reach that 10% point. This is because the range of element volumes in a real-world problem would be rather large, and for elements with comparable dimensions, the 10% in volume would translate to roughly 50% in element length.

There is always room for improving the mesh moving technique used with an interface-tracking technique. For example, in the context of the techniques mentioned above, it was proposed in Tezduyar and Sathe (2007) that the elements are stiffened proportional to an invariant measure of the shear strain associated with the mesh deformation. The specific invariant measure proposed in Tezduyar and Sathe (2007) was the second invariant of the strain deviator tensor. Maintaining the parallel efficiency of the computations is another major incentive for reducing the frequency of remeshing, because parallel efficiency of most automatic mesh generators is substantially lower than that of most flow solvers. For example, reducing the frequency of remeshing to every ten time steps or less Johnson and Tezduyar (1999) would sufficiently reduce the influence of remeshing in terms of its added cost and lack of parallel efficiency. In the collective experience of the authors, because of the advanced mesh moving techniques used, the frequency of remeshing was far less than every ten time steps. This is an important practical reference point when it comes to correctly assessing the advantages and disadvantages of the interface-tracking and interface-capturing techniques.

5

ALE and Space–Time Methods for FSI

This chapter focuses on the ALE and space–time FSI methods. We begin the discussion in the continuous setting and state the weak formulation of the coupled FSI equations with the associated kinematic compatibility conditions at the fluid–structure interface. The ALE formulation of the FSI problem is given next, where, for simplicity, we assume that the fluid and structure meshes are matching at the fluid–structure interface. We then present the space–time formulation of the FSI problem in the context of nonmatching fluid and structure meshes at the interface. We conclude the chapter with a discussion on additional ways of treating fluid and structure interface discretizations.

5.1 FSI Formulation at the Continuous Level

Let $\Omega_0 \subset \mathbb{R}^{n_{sd}}$ represent the combined fluid and structure domain in the initial configuration, which also serves as the reference configuration. Let $\Omega_t \subset \mathbb{R}^{n_{sd}}$ denote the configuration of Ω_0 at the current time t. The domain Ω_0 admits the decomposition

$$\Omega_0 = \overline{(\Omega_1)_0 \cup (\Omega_2)_0}, \tag{5.1}$$

where $(\Omega_1)_0$ and $(\Omega_2)_0$ are the subsets of Ω_0 occupied by the fluid and structure. From this point on in this book, subscripts 1 and 2 will denote the fluid and structure.[1] The decomposition is non-overlapping, that is,

$$(\Omega_1)_0 \cap (\Omega_2)_0 = \emptyset. \tag{5.2}$$

Analogously, for Ω_t, we have

$$\Omega_t = \overline{(\Omega_1)_t \cup (\Omega_2)_t} \tag{5.3}$$

---

[1] We chose the subscripts 1 and 2 to denote the fluid and structure, respectively, because in "fluid–structure interaction" the fluid comes first and the structure comes second. This ordering choice should not be interpreted as the authors' preference for one physical system over the other.

Computational Fluid–Structure Interaction: Methods and Applications, First Edition.
Yuri Bazilevs, Kenji Takizawa and Tayfun E. Tezduyar.
© 2013 John Wiley & Sons, Ltd. Published 2013 by John Wiley & Sons, Ltd.

and

$$(\Omega_1)_t \cap (\Omega_2)_t = \emptyset. \tag{5.4}$$

Furthermore, we adopt a notation that will help us better understand how the fluid–structure interface conditions are handled. In that notation, while subscript "I" refers to the fluid–structure interface, subscript "E" refers to "elsewhere" in the fluid and structure domains or boundaries (see Figure 5.1). We let $(\Gamma_I)_0$ denote the interface between the fluid and structure subdomains in the initial configuration, and we let Γ_I denote its counterpart in the current configuration. The configurations and the relationship between them are shown in Figure 5.1.

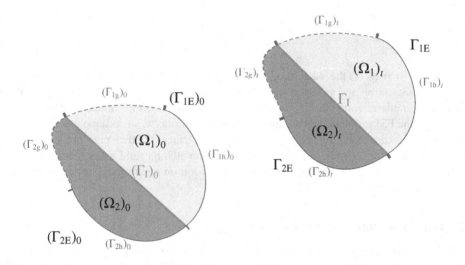

Figure 5.1 The spatial domain for the FSI formulation, including the fluid and structure subdomains, and the interface. The domain is shown in both the reference (left) and current (right) configurations. We note that $\Gamma_{1E} = (\Gamma_{1h})_t \cup (\Gamma_{1g})_t$ and $\Gamma_{2E} = (\Gamma_{2h})_t \cup (\Gamma_{2g})_t$

REMARK 5.1 *The structure subdomain $(\Omega_2)_0$ is typically associated with the material configuration of the structure taken at time $t = 0$. The fluid subdomain $(\Omega_1)_0$ is usually independent of the material configuration occupied by the fluid at $t = 0$. The fluid subdomain $(\Omega_1)_0$ is chosen such that it is convenient for the FSI analysis. The subdomain $(\Omega_2)_t$ is typically associated with the material configuration of the structure at time t, which is the essence of the Largangian approach in structural mechanics. As in the case of $(\Omega_1)_0$, $(\Omega_1)_t$ is also defined from the considerations of convenience and accuracy of the FSI computations. In the moving-mesh approaches, which are the focus of this book, $(\Omega_1)_t$ is obtained as a smooth evolution of $(\Omega_1)_0$ that conforms to the motion of the fluid–structure interface.*

Given the above setup, we state the FSI formulation at the continuous level: find the fluid velocity and pressure, $\mathbf{u} \in \mathcal{S}_u$ and $p \in \mathcal{S}_p$, and the structural displacement, $\mathbf{y} \in \mathcal{S}_y$, such that, $\forall \, \mathbf{w}_1 \in \mathcal{V}_u, q_1 \in \mathcal{V}_p,$ and $\mathbf{w}_2 \in \mathcal{V}_y$:

$$\int_{(\Omega_1)_t} \mathbf{w}_1 \cdot \rho \left(\frac{\partial \mathbf{u}}{\partial t} + \mathbf{u} \cdot \nabla \mathbf{u} - \mathbf{f} \right) d\Omega + \int_{(\Omega_1)_t} \boldsymbol{\varepsilon}(\mathbf{w}_1) : \boldsymbol{\sigma}_1 \, d\Omega$$
$$- \int_{\Gamma_{1E}} \mathbf{w}_1 \cdot \mathbf{h}_{1E} \, d\Gamma + \int_{(\Omega_1)_t} q_1 \nabla \cdot \mathbf{u} \, d\Omega$$
$$+ \int_{(\Omega_2)_t} \mathbf{w}_2 \cdot \rho_2 \left(\frac{d^2 \mathbf{y}}{dt^2} - \mathbf{f}_2 \right) d\Omega + \int_{(\Omega_2)_t} \boldsymbol{\varepsilon}(\mathbf{w}_2) : \boldsymbol{\sigma}_2 \, d\Omega - \int_{\Gamma_{2E}} \mathbf{w}_2 \cdot \mathbf{h}_{2E} \, d\Gamma = 0. \quad (5.5)$$

In Equation (5.5), $\boldsymbol{\sigma}_1$ and $\boldsymbol{\sigma}_2$ are the Cauchy stress tensors for the fluid and structure, respectively, and $\mathbf{h}_{1E}$ and $\mathbf{h}_{2E}$ are the corresponding prescribed traction vectors. Note that for the mass density and body force, for the fluid part we continue using the symbols without the subscript 1, but for the structure part we do use the subscript 2.

The FSI formulation given by Equation (5.5) is subject to the following auxiliary conditions:

$$\mathbf{u} = \frac{d\mathbf{y}}{dt} \quad \text{on } \Gamma_I \quad (5.6)$$

and

$$\mathbf{w}_1 = \mathbf{w}_2 \quad \text{on } \Gamma_I. \quad (5.7)$$

Equation (5.6), the kinematic constraint, equates the fluid velocity with that of the structure at the fluid–structure interface. Equation (5.7) equates the test functions for the fluid and structure linear-momentum equations. To examine the implication of the FSI formulation given by Equation (5.5), we derive the corresponding Euler–Lagrange conditions (see, e.g., Hughes, 2000). For this, we perform an integration-by-parts in Equation (5.5) to obtain, $\forall \, \mathbf{w}_1 \in \mathcal{V}_u$, $q_1 \in \mathcal{V}_p$, and $\mathbf{w}_2 \in \mathcal{V}_y$:

$$\int_{(\Omega_1)_t} \mathbf{w}_1 \cdot \left(\rho \left(\frac{\partial \mathbf{u}}{\partial t} + \mathbf{u} \cdot \nabla \mathbf{u} - \mathbf{f} \right) - \nabla \cdot \boldsymbol{\sigma}_1 \right) d\Omega$$
$$+ \int_{\Gamma_{1E}} \mathbf{w}_1 \cdot (\boldsymbol{\sigma}_1 \mathbf{n}_1 - \mathbf{h}_{1E}) \, d\Gamma$$
$$+ \int_{(\Omega_1)_t} q_1 \nabla \cdot \mathbf{u} \, d\Omega$$
$$+ \int_{(\Omega_2)_t} \mathbf{w}_2 \cdot \left(\rho_2 \left(\frac{d^2 \mathbf{y}}{dt^2} - \mathbf{f}_2 \right) - \nabla \cdot \boldsymbol{\sigma}_2 \right) d\Omega$$
$$+ \int_{\Gamma_{2E}} \mathbf{w}_2 \cdot (\boldsymbol{\sigma}_2 \mathbf{n}_2 - \mathbf{h}_{2E}) \, d\Gamma$$
$$+ \int_{\Gamma_I} (\mathbf{w}_1 \cdot \boldsymbol{\sigma}_1 \mathbf{n}_1 + \mathbf{w}_2 \cdot \boldsymbol{\sigma}_2 \mathbf{n}_2) \, d\Gamma = 0. \quad (5.8)$$

Equation (5.8) implies that the individual fluid and structure equations hold in their respective subdomains, namely,

$$\rho \left(\frac{\partial \mathbf{u}}{\partial t} + \mathbf{u} \cdot \nabla \mathbf{u} - \mathbf{f} \right) - \nabla \cdot \boldsymbol{\sigma}_1 = \mathbf{0} \quad \text{in } (\Omega_1)_t, \quad (5.9)$$

$$\nabla \cdot \mathbf{u} = 0 \quad \text{in} \quad (\Omega_1)_t, \tag{5.10}$$

and

$$\rho_2 \left(\frac{d^2 \mathbf{y}}{dt^2} - \mathbf{f}_2 \right) - \nabla \cdot \boldsymbol{\sigma}_2 = \mathbf{0} \quad \text{in} \quad (\Omega_2)_t. \tag{5.11}$$

Furthermore, the traction boundary conditions are also satisfied, namely,

$$\boldsymbol{\sigma}_1 \mathbf{n}_1 - \mathbf{h}_{1E} = \mathbf{0} \quad \text{on} \quad (\Gamma_{1h})_t \subset \Gamma_{1E} \tag{5.12}$$

and

$$\boldsymbol{\sigma}_2 \mathbf{n}_2 - \mathbf{h}_{2E} = \mathbf{0} \quad \text{on} \quad (\Gamma_{2h})_t \subset \Gamma_{2E}. \tag{5.13}$$

The remaining terms are

$$\int_{\Gamma_I} (\mathbf{w}_1 \cdot \boldsymbol{\sigma}_1 \mathbf{n}_1 + \mathbf{w}_2 \cdot \boldsymbol{\sigma}_2 \mathbf{n}_2) \, d\Gamma = 0. \tag{5.14}$$

Introducing Equation (5.7) into Equation (5.14) gives

$$\int_{\Gamma_I} \mathbf{w}_1 \cdot (\boldsymbol{\sigma}_1 \mathbf{n}_1 + \boldsymbol{\sigma}_2 \mathbf{n}_2) \, d\Gamma = 0. \tag{5.15}$$

Because Equation (5.15) holds $\forall \, \mathbf{w}_1 \in \mathcal{V}_u$, we conclude that

$$\boldsymbol{\sigma}_1 \mathbf{n}_1 + \boldsymbol{\sigma}_2 \mathbf{n}_2 = \mathbf{0} \quad \text{on} \quad \Gamma_I. \tag{5.16}$$

Equation (5.16) is precisely the continuity of the traction vector at the fluid–structure interface.

REMARK 5.2 *Note that the continuity of the traction vector follows from the continuity of the test functions at the fluid–structure interface (see Equation (5.7)). Also note that the continuity of the tractions holds independently of the choice of the constitutive relationship for the structure.*

REMARK 5.3 *The fluid–structure interface conditions given by Equations (5.6) and (5.16) represent the classical assumptions of continuum mechanics, and are most often used in practice. However, the FSI problem might involve other types of interface conditions. For example, in the case of structures made of a porous material (e.g., parachute fabric or vascular wall), the flow rate through the interface is proportional to the local pressure drop across the fluid–structure interface. In the parachute fabric case, Equation (5.6) needs to be modified accordingly. The mathematical formulation and the numerical treatment of such interface conditions will be discussed in the later sections.*

5.2 ALE Formulation of FSI

5.2.1 *Spatially-Discretized ALE FSI Formulation with Matching Fluid and Structure Discretizations*

To formulate the FSI problem using the ALE approach, we begin with the formulation given by Equation (5.5) and write the fluid part of the FSI equations in the ALE frame. This leads to the following ALE FSI formulation: find $\mathbf{u} \in \mathcal{S}_u$, $p \in \mathcal{S}_p$, and $\mathbf{y} \in \mathcal{S}_y$, such that, $\forall\ \mathbf{w}_1 \in \mathcal{V}_u$, $q_1 \in \mathcal{V}_p$, and $\mathbf{w}_2 \in \mathcal{V}_y$:

$$\int_{(\Omega_1)_t} \mathbf{w}_1 \cdot \rho \left(\left.\frac{\partial \mathbf{u}}{\partial t}\right|_{\hat{x}} + (\mathbf{u} - \hat{\mathbf{u}}) \cdot \nabla \mathbf{u} - \mathbf{f} \right) d\Omega + \int_{(\Omega_1)_t} \boldsymbol{\varepsilon}(\mathbf{w}_1) : \boldsymbol{\sigma}_1\, d\Omega$$

$$- \int_{\Gamma_{1E}} \mathbf{w}_1 \cdot \mathbf{h}_{1E}\, d\Gamma + \int_{(\Omega_1)_t} q_1 \nabla \cdot \mathbf{u}\, d\Omega$$

$$+ \int_{(\Omega_2)_t} \mathbf{w}_2 \cdot \rho_2 \left(\frac{d^2\mathbf{y}}{dt^2} - \mathbf{f}_2 \right) d\Omega + \int_{(\Omega_2)_t} \boldsymbol{\varepsilon}(\mathbf{w}_2) : \boldsymbol{\sigma}_2\, d\Omega - \int_{\Gamma_{2E}} \mathbf{w}_2 \cdot \mathbf{h}_{2E}\, d\Gamma = 0. \quad (5.17)$$

We take the continuous formulation given by Equation (5.17) as the starting point, and use Galerkin's method for the structural mechanics part and the ALE-VMS method for the fluid mechanics part (see Equation (4.55)) of the FSI problem. This yields the following semi-discrete ALE formulation of the FSI problem: find $\mathbf{u}^h \in \mathcal{S}_u^h$, $p^h \in \mathcal{S}_p^h$, $\mathbf{y}^h \in \mathcal{S}_y^h$, and $\hat{\mathbf{y}}^h \in \mathcal{S}_m^h$, such that, $\forall\ \mathbf{w}_1^h \in \mathcal{V}_u^h$, $q_1^h \in \mathcal{V}_p^h$, $\mathbf{w}_2^h \in \mathcal{V}_y^h$, and $\mathbf{w}_3^h \in \mathcal{V}_m^h$:

$$\int_{(\Omega_1)_t} \mathbf{w}_1^h \cdot \rho \left(\left.\frac{\partial \mathbf{u}^h}{\partial t}\right|_{\hat{x}} + (\mathbf{u}^h - \hat{\mathbf{u}}^h) \cdot \nabla \mathbf{u}^h - \mathbf{f}^h \right) d\Omega + \int_{(\Omega_1)_t} \boldsymbol{\varepsilon}(\mathbf{w}_1^h) : \boldsymbol{\sigma}_1^h\, d\Omega$$

$$- \int_{\Gamma_{1E}} \mathbf{w}_1^h \cdot \mathbf{h}_{1E}^h\, d\Gamma + \int_{(\Omega_1)_t} q_1^h \nabla \cdot \mathbf{u}^h\, d\Omega$$

$$+ \sum_{e=1}^{n_{el}} \int_{(\Omega_1)_t^e} \tau_{\text{SUPS}} \left((\mathbf{u}^h - \hat{\mathbf{u}}^h) \cdot \nabla \mathbf{w}_1^h + \frac{\nabla q_1^h}{\rho} \right) \cdot \mathbf{r}_{\text{M}}(\mathbf{u}^h, p^h)\, d\Omega$$

$$+ \sum_{e=1}^{n_{el}} \int_{(\Omega_1)_t^e} \rho \nu_{\text{LSIC}} \nabla \cdot \mathbf{w}_1^h r_{\text{C}}(\mathbf{u}^h)\, d\Omega$$

$$- \sum_{e=1}^{n_{el}} \int_{(\Omega_1)_t^e} \tau_{\text{SUPS}} \mathbf{w}_1^h \cdot \left(\mathbf{r}_{\text{M}}(\mathbf{u}^h, p^h) \cdot \nabla \mathbf{u}^h \right) d\Omega$$

$$- \sum_{e=1}^{n_{el}} \int_{(\Omega_1)_t^e} \frac{\nabla \mathbf{w}_1^h}{\rho} : \left(\tau_{\text{SUPS}} \mathbf{r}_{\text{M}}(\mathbf{u}^h, p^h) \right) \otimes \left(\tau_{\text{SUPS}} \mathbf{r}_{\text{M}}(\mathbf{u}^h, p^h) \right) d\Omega$$

$$+ \int_{(\Omega_2)_t} \mathbf{w}_2^h \cdot \rho_2 \left(\frac{d^2\mathbf{y}^h}{dt^2} - \mathbf{f}_2^h \right) d\Omega + \int_{(\Omega_2)_t} \boldsymbol{\varepsilon}(\mathbf{w}_2^h) : \boldsymbol{\sigma}_2^h\, d\Omega$$

$$- \int_{\Gamma_{2E}} \mathbf{w}_2^h \cdot \mathbf{h}_{2E}^h\, d\Gamma$$

$$+ \int_{(\Omega_1)_{\tilde{t}}} \boldsymbol{\epsilon}(\mathbf{w}_3^h) \cdot \mathbf{D}^h \boldsymbol{\epsilon} \left(\hat{\mathbf{y}}^h(t) - \hat{\mathbf{y}}^h(\tilde{t}) \right) d\Omega = 0. \quad (5.18)$$

The above FSI formulation includes the fluid-domain mesh moving terms that appear as the last line of Equation (5.18). The mesh moving terms are governed by the equations of linear elastostatics posed in the fluid-domain configuration at time $\tilde{t} < t$ and subject to the time-dependent displacement of the fluid–structure interface (see Section 4.7, Equation (4.129)).

The fluid-mesh displacement $\hat{\mathbf{y}}^h(t)$ defines the position of the fluid mesh at time t. The mesh velocity $\hat{\mathbf{u}}^h$, which explicitly appears in the ALE FSI formulation, is given as

$$\hat{\mathbf{u}}^h = \left.\frac{\partial \hat{\mathbf{y}}^h}{\partial t}\right|_{\hat{x}}. \tag{5.19}$$

To ensure that the correct fluid–structure coupling is attained in the FSI formulation, the kinematic constraints given by Equations (5.6) and (5.7) must also hold at the discrete level, that is,

$$\mathbf{u}^h = \frac{d\mathbf{y}^h}{dt} \quad \text{on } \Gamma_I \tag{5.20}$$

and

$$\mathbf{w}_1^h = \mathbf{w}_2^h \quad \text{on } \Gamma_I. \tag{5.21}$$

Furthermore, to prevent the fluid and structure meshes from separating at the interface, we require that

$$\hat{\mathbf{y}}^h = \mathbf{y}^h \quad \text{on } \Gamma_I \tag{5.22}$$

and

$$\mathbf{w}_3^h = \mathbf{0} \quad \text{on } \Gamma_I. \tag{5.23}$$

In practice, this means that the fluid and structure meshes must match at the interface. This may be attained by generating a single mesh for the combined fluid/structure domain, which gives a unique set of nodes at the fluid–structure interface. Although the matching fluid–structure interface discretizations may be employed in several FSI applications, we feel this is somewhat limiting. As a result, for greater flexibility, it is desirable to allow for nonmatching fluid–structure interface discretizations. However, in this case, Equations (5.20), (5.21), and (5.22) may not be satisfied in a strong sense. The FSI techniques that overcome this difficulty, and enable the use of nonmatching interface discretizations, will be presented later in this chapter.

REMARK 5.4 *Note that Equation (5.22) specifies all components of the mesh displacement vector as boundary condition at the fluid–structure interface, while Equation (4.130) requires only the normal component to be prescribed. The former condition is necessary to maintain the matching interface discretizations. In the case of the nonmatching interface discretizations, there may be situations where it is beneficial to use the latter condition.*

The terms corresponding to the structural mechanics formulation in Equations (5.17) and (5.18) are in the updated Lagrangian form, as the integrals are evaluated in $(\Omega_2)_t$. It is often convenient to write the structural mechanics formulation in $(\Omega_2)_0$, which amounts to changing variables in the inertial and stress terms:

$$\int_{(\Omega_2)_t} \mathbf{w}_2^h \cdot \rho_2 \frac{d^2 \mathbf{y}^h}{dt^2} \, d\Omega \rightarrow \int_{(\Omega_2)_0} \mathbf{w}_2^h \cdot (\rho_2)_0 \frac{d^2 \mathbf{y}^h}{dt^2} \, d\Omega \tag{5.24}$$

and

$$\int_{(\Omega_2)_t} \boldsymbol{\varepsilon}\left(\mathbf{w}_2^h\right) : \boldsymbol{\sigma}_2^h \, d\Omega \rightarrow \int_{(\Omega_2)_0} \nabla_X \mathbf{w}_2^h : \left(\mathbf{F}^h \mathbf{S}^h\right) d\Omega. \tag{5.25}$$

In Equation (5.25), the superscript h used for the deformation gradient and the second Piola–Kirchhoff stress tensor denotes their dependence on the discrete structural mechanics solution.

To write the vector form of the ALE FSI equations, in addition to the discrete trial and test functions of the fluid mechanics problem (see Equations (4.60)–(4.63)), we define the discrete structure and mesh trial and test functions:

$$\mathbf{y}^h(\mathbf{X},t) = \sum_{\eta^s_{\text{struc}}} \mathbf{y}_A(t) N_A(\mathbf{X}), \tag{5.26}$$

$$\hat{\mathbf{y}}^h(\hat{\mathbf{x}},t) = \sum_{\eta^s_{\text{mesh}}} \hat{\mathbf{y}}_A(t) N_A(\hat{\mathbf{x}}), \tag{5.27}$$

$$\mathbf{w}^h_2(\mathbf{X}) = \sum_{\eta^w_{\text{struc}}} (\mathbf{w}_2)_A N_A(\mathbf{X}), \tag{5.28}$$

$$\mathbf{w}^h_3(\hat{\mathbf{x}}) = \sum_{\eta^w_{\text{mesh}}} (\mathbf{w}_3)_A N_A(\hat{\mathbf{x}}). \tag{5.29}$$

Here, η^s_{struc} and η^w_{struc} are the nodal index sets for the structural equations, and η^s_{mesh} and η^w_{mesh} are the nodal index sets for the mesh moving equations. Note that the above fields are defined on the structure and mesh reference domains. The current-configuration counterparts of Equations (5.26)–(5.29) may be obtained using the push-forward operation given by Equation (4.5). We define four discrete residual vectors corresponding to the fluid mechanics linear-momentum and incompressibility-constraint equations, structural mechanics linear-momentum equations, and mesh moving equations by introducing Equations (4.62), (4.63), (5.28), and (5.29) into Equation (5.18), and assuming that $(\mathbf{w}_1)_A$'s, $(\mathbf{w}_2)_A$'s, $(\mathbf{w}_3)_A$'s, and q_A's are arbitrary constants:

$$\mathbf{N}_{1M} = [(\mathbf{N}_{1M})_{A,i}], \tag{5.30}$$

$$\mathbf{N}_{1C} = [(\mathbf{N}_{1C})_A], \tag{5.31}$$

$$\mathbf{N}_2 = [(\mathbf{N}_2)_{A,i}], \tag{5.32}$$

$$\mathbf{N}_3 = [(\mathbf{N}_3)_{A,i}], \tag{5.33}$$

$$(\mathbf{N}_{1M})_{A,i} = \int_{(\Omega_1)_t} N_A \mathbf{e}_i \cdot \rho \left(\left. \frac{\partial \mathbf{u}^h}{\partial t} \right|_{\hat{\mathbf{x}}} + (\mathbf{u}^h - \hat{\mathbf{u}}^h) \cdot \nabla \mathbf{u}^h - \mathbf{f}^h \right) d\Omega$$

$$+ \int_{(\Omega_1)_t} \boldsymbol{\varepsilon}(N_A \mathbf{e}_i) : \boldsymbol{\sigma}(\mathbf{u}^h, p^h) \, d\Omega - \int_{\Gamma_{1E}} N_A \mathbf{e}_i \cdot \mathbf{h}^h_{1E} \, d\Gamma$$

$$+ \sum_{e=1}^{n_{el}} \int_{(\Omega_1)^e_t} \tau_{\text{SUPS}} \left((\mathbf{u}^h - \hat{\mathbf{u}}^h) \cdot \nabla N_A \mathbf{e}_i \right) \cdot \mathbf{r}_M(\mathbf{u}^h, p^h) \, d\Omega$$

$$+ \sum_{e=1}^{n_{el}} \int_{(\Omega_1)^e_t} \rho \nu_{\text{LSIC}} (\nabla \cdot N_A \mathbf{e}_i) r_C(\mathbf{u}^h) \, d\Omega$$

$$- \sum_{e=1}^{n_{el}} \int_{(\Omega_1)^e_t} \tau_{\text{SUPS}} N_A \mathbf{e}_i \cdot \left(\mathbf{r}_M(\mathbf{u}^h, p^h) \cdot \nabla \mathbf{u}^h \right) d\Omega$$

$$- \sum_{e=1}^{n_{el}} \int_{(\Omega_1)^e_t} \frac{\nabla N_A \mathbf{e}_i}{\rho} : \left(\tau_{\text{SUPS}} \mathbf{r}_M(\mathbf{u}^h, p^h) \right) \otimes \left(\tau_{\text{SUPS}} \mathbf{r}_M(\mathbf{u}^h, p^h) \right) d\Omega, \tag{5.34}$$

$$(N_{1C})_A = \int_{(\Omega_1)_t} N_A \nabla \cdot \mathbf{u}^h \, d\Omega + \sum_{e=1}^{n_{el}} \int_{(\Omega_1)_t^e} \tau_{\text{SUPS}} \frac{\nabla N_A}{\rho} \cdot \mathbf{r}_M\left(\mathbf{u}^h, p^h\right) d\Omega, \tag{5.35}$$

$$(N_2)_{A,i} = \int_{(\Omega_2)_t} N_A \mathbf{e}_i \cdot \rho_2 \left(\frac{d^2 \mathbf{y}^h}{dt^2} - \mathbf{f}_2^h\right) d\Omega + \int_{(\Omega_2)_t} \nabla N_A \mathbf{e}_i : \sigma_2^h \, d\Omega$$
$$- \int_{\Gamma_{2E}} N_A \mathbf{e}_i \cdot \mathbf{h}_{2E}^h \, d\Gamma, \tag{5.36}$$

$$(N_3)_{A,i} = \int_{(\Omega_1)_t} \epsilon(N_A \mathbf{e}_i) \cdot \mathbf{D}^h \epsilon\left(\hat{\mathbf{y}}^h(t) - \hat{\mathbf{y}}^h(\tilde{t})\right) d\Omega. \tag{5.37}$$

We also define the combined fluid mechanics linear-momentum and incompressibility-constraint equation residual as

$$\mathbf{N}_1 = \begin{bmatrix} \mathbf{N}_{1M} \\ \mathbf{N}_{1C} \end{bmatrix}, \tag{5.38}$$

which we will use in Chapter 6.

Let, as before, $\mathbf{U}$, $\dot{\mathbf{U}}$, and $\mathbf{P}$ denote the vectors of nodal degrees-of-freedom of fluid velocity, its time derivative, and pressure, respectively. Also, let $\mathbf{Y}$, $\dot{\mathbf{Y}}$, and $\ddot{\mathbf{Y}}$ denote the vectors of nodal degrees-of-freedom of structure displacement, velocity, and acceleration, respectively. Finally, let $\hat{\mathbf{Y}}$, $\dot{\hat{\mathbf{Y}}}$, and $\ddot{\hat{\mathbf{Y}}}$ denote the vectors of nodal degrees-of-freedom of mesh displacement, velocity, and acceleration, respectively. The vector form of the semi-discrete equations corresponding to the FSI formulation given by Equation (5.18) becomes: find $\mathbf{U}$, $\dot{\mathbf{U}}$, $\mathbf{P}$, $\mathbf{Y}$, $\dot{\mathbf{Y}}$, $\ddot{\mathbf{Y}}$, $\hat{\mathbf{Y}}$, $\dot{\hat{\mathbf{Y}}}$, and $\ddot{\hat{\mathbf{Y}}}$, such that:

$$\mathbf{N}_{1M}(\dot{\mathbf{U}}, \mathbf{U}, \mathbf{P}, \ddot{\mathbf{Y}}, \dot{\mathbf{Y}}, \mathbf{Y}, \ddot{\hat{\mathbf{Y}}}, \dot{\hat{\mathbf{Y}}}, \hat{\mathbf{Y}}) = \mathbf{0}, \tag{5.39}$$

$$\mathbf{N}_{1C}(\dot{\mathbf{U}}, \mathbf{U}, \mathbf{P}, \ddot{\mathbf{Y}}, \dot{\mathbf{Y}}, \mathbf{Y}, \ddot{\hat{\mathbf{Y}}}, \dot{\hat{\mathbf{Y}}}, \hat{\mathbf{Y}}) = \mathbf{0}, \tag{5.40}$$

$$\mathbf{N}_2(\dot{\mathbf{U}}, \mathbf{U}, \mathbf{P}, \ddot{\mathbf{Y}}, \dot{\mathbf{Y}}, \mathbf{Y}, \ddot{\hat{\mathbf{Y}}}, \dot{\hat{\mathbf{Y}}}, \hat{\mathbf{Y}}) = \mathbf{0}, \tag{5.41}$$

$$\mathbf{N}_3(\dot{\mathbf{U}}, \mathbf{U}, \mathbf{P}, \ddot{\mathbf{Y}}, \dot{\mathbf{Y}}, \mathbf{Y}, \ddot{\hat{\mathbf{Y}}}, \dot{\hat{\mathbf{Y}}}, \hat{\mathbf{Y}}) = \mathbf{0}. \tag{5.42}$$

5.2.2 Generalized-α Time Integration of the ALE FSI Equations

The generalized-α method for the time integration of the structural mechanics equations was proposed in Chung and Hulbert (1993). Its extension to the FSI equations was proposed in Bazilevs et al. (2008) and is presented here. We define the nodal solutions at the intermediate time levels as

$$\dot{\mathbf{U}}_{n+\alpha_m} = \dot{\mathbf{U}}_n + \alpha_m(\dot{\mathbf{U}}_{n+1} - \dot{\mathbf{U}}_n), \tag{5.43}$$

$$\mathbf{U}_{n+\alpha_f} = \mathbf{U}_n + \alpha_f(\mathbf{U}_{n+1} - \mathbf{U}_n), \tag{5.44}$$

$$\ddot{\mathbf{Y}}_{n+\alpha_m} = \ddot{\mathbf{Y}}_n + \alpha_m(\ddot{\mathbf{Y}}_{n+1} - \ddot{\mathbf{Y}}_n), \tag{5.45}$$

$$\dot{\mathbf{Y}}_{n+\alpha_f} = \dot{\mathbf{Y}}_n + \alpha_f(\dot{\mathbf{Y}}_{n+1} - \dot{\mathbf{Y}}_n), \tag{5.46}$$

$$\mathbf{Y}_{n+\alpha_f} = \mathbf{Y}_n + \alpha_f(\mathbf{Y}_{n+1} - \mathbf{Y}_n), \tag{5.47}$$

$$\ddot{\hat{\mathbf{Y}}}_{n+\alpha_m} = \ddot{\hat{\mathbf{Y}}}_n + \alpha_m(\ddot{\hat{\mathbf{Y}}}_{n+1} - \ddot{\hat{\mathbf{Y}}}_n), \tag{5.48}$$

$$\dot{\mathbf{Y}}_{n+\alpha_f} = \dot{\mathbf{Y}}_n + \alpha_f(\dot{\mathbf{Y}}_{n+1} - \dot{\mathbf{Y}}_n), \tag{5.49}$$

$$\hat{\mathbf{Y}}_{n+\alpha_f} = \hat{\mathbf{Y}}_n + \alpha_f(\hat{\mathbf{Y}}_{n+1} - \hat{\mathbf{Y}}_n), \tag{5.50}$$

and collocate the fluid, structure, and mesh residuals at these intermediate time levels:

$$\mathbf{N}_{1M}(\dot{\mathbf{U}}_{n+\alpha_m}, \mathbf{U}_{n+\alpha_f}, \mathbf{P}_{n+1}, \ddot{\mathbf{Y}}_{n+\alpha_m}, \dot{\mathbf{Y}}_{n+\alpha_f}, \mathbf{Y}_{n+\alpha_f}, \ddot{\hat{\mathbf{Y}}}_{n+\alpha_m}, \dot{\hat{\mathbf{Y}}}_{n+\alpha_f}, \hat{\mathbf{Y}}_{n+\alpha_f}) = \mathbf{0}, \tag{5.51}$$

$$\mathbf{N}_{1C}(\dot{\mathbf{U}}_{n+\alpha_m}, \mathbf{U}_{n+\alpha_f}, \mathbf{P}_{n+1}, \ddot{\mathbf{Y}}_{n+\alpha_m}, \dot{\mathbf{Y}}_{n+\alpha_f}, \mathbf{Y}_{n+\alpha_f}, \ddot{\hat{\mathbf{Y}}}_{n+\alpha_m}, \dot{\hat{\mathbf{Y}}}_{n+\alpha_f}, \hat{\mathbf{Y}}_{n+\alpha_f}) = \mathbf{0}, \tag{5.52}$$

$$\mathbf{N}_{2}(\dot{\mathbf{U}}_{n+\alpha_m}, \mathbf{U}_{n+\alpha_f}, \mathbf{P}_{n+1}, \ddot{\mathbf{Y}}_{n+\alpha_m}, \dot{\mathbf{Y}}_{n+\alpha_f}, \mathbf{Y}_{n+\alpha_f}, \ddot{\hat{\mathbf{Y}}}_{n+\alpha_m}, \dot{\hat{\mathbf{Y}}}_{n+\alpha_f}, \hat{\mathbf{Y}}_{n+\alpha_f}) = \mathbf{0}, \tag{5.53}$$

$$\mathbf{N}_{3}(\dot{\mathbf{U}}_{n+\alpha_m}, \mathbf{U}_{n+\alpha_f}, \mathbf{P}_{n+1}, \ddot{\mathbf{Y}}_{n+\alpha_m}, \dot{\mathbf{Y}}_{n+\alpha_f}, \mathbf{Y}_{n+\alpha_f}, \ddot{\hat{\mathbf{Y}}}_{n+\alpha_m}, \dot{\hat{\mathbf{Y}}}_{n+\alpha_f}, \hat{\mathbf{Y}}_{n+\alpha_f}) = \mathbf{0}. \tag{5.54}$$

Equations (5.43)–(5.54) are solved for the nodal unknowns at t_{n+1}, assuming that the solution at t_n is given. In addition to Equations (5.43)–(5.54), the relationships between the time derivatives of the nodal degrees-of-freedom is replaced by the discrete Newmark formulas:

$$\mathbf{U}_{n+1} = \mathbf{U}_n + \Delta t_n \left((1-\gamma)\dot{\mathbf{U}}_n + \gamma \dot{\mathbf{U}}_{n+1}\right), \tag{5.55}$$

$$\dot{\mathbf{Y}}_{n+1} = \dot{\mathbf{Y}}_n + \Delta t_n \left((1-\gamma)\ddot{\mathbf{Y}}_n + \gamma \ddot{\mathbf{Y}}_{n+1}\right), \tag{5.56}$$

$$\mathbf{Y}_{n+1} = \mathbf{Y}_n + \Delta t_n \dot{\mathbf{Y}}_n + \frac{\Delta t_n^2}{2}\left((1-2\beta)\ddot{\mathbf{Y}}_n + 2\beta \ddot{\mathbf{Y}}_{n+1}\right), \tag{5.57}$$

$$\dot{\hat{\mathbf{Y}}}_{n+1} = \dot{\hat{\mathbf{Y}}}_n + \Delta t_n \left((1-\gamma)\ddot{\hat{\mathbf{Y}}}_n + \gamma \ddot{\hat{\mathbf{Y}}}_{n+1}\right), \tag{5.58}$$

$$\hat{\mathbf{Y}}_{n+1} = \hat{\mathbf{Y}}_n + \Delta t_n \dot{\hat{\mathbf{Y}}}_n + \frac{\Delta t_n^2}{2}\left((1-2\beta)\ddot{\hat{\mathbf{Y}}}_n + 2\beta \ddot{\hat{\mathbf{Y}}}_{n+1}\right). \tag{5.59}$$

In the above, α_m, α_f, γ, and β are the real-valued parameters that define the time integration method. For a second-order linear ordinary differential equation system with constant coefficients, which is related to the structure and mesh parts of the FSI equations, Chung and Hulbert (1993) showed that second-order accuracy in time is achieved provided that

$$\gamma = \frac{1}{2} + \alpha_m - \alpha_f \tag{5.60}$$

and

$$\beta = \frac{1}{4}(1 + \alpha_m - \alpha_f)^2, \tag{5.61}$$

while unconditional stability is attained provided that

$$\alpha_m \geq \alpha_f \geq 1/2. \tag{5.62}$$

Equations (5.60) and (5.62) also hold true for the first-order linear ordinary differential equation system with constant coefficients, which is related to the fluid mechanics part of the FSI equations. As a result, in principle, the generalized-α method can be applied to the FSI equations in a unified fashion.

To have control over the high-frequency dissipation, α_m and α_f are parameterized by ρ_∞, the spectral radius of the amplification matrix for an infinitely large time step (see Section 4.6.2). Optimal high-frequency damping occurs when all the eigenvalues of the amplification matrix take on the same value, $-\rho_\infty$. For this, according to Jansen et al. (2000), for the fluid mechanics equations

$$(\alpha_m)_1 = \frac{1}{2}\left(\frac{3-(\rho_\infty)_1}{1+(\rho_\infty)_1}\right) \text{ and } (\alpha_f)_1 = \frac{1}{1+(\rho_\infty)_1}, \quad (5.63)$$

while, according to Chung and Hulbert (1993), for the structural mechanics equations

$$(\alpha_m)_2 = \frac{2-(\rho_\infty)_2}{1+(\rho_\infty)_2} \text{ and } (\alpha_f)_2 = \frac{1}{1+(\rho_\infty)_2}. \quad (5.64)$$

In Equations (5.63) and (5.64), and in what follows, the subscripts 1 and 2 distinguish the quantities coming from the two methods. The equations show that for the same values of ρ_∞ (that is, $(\rho_\infty)_1 = (\rho_\infty)_2 = \rho_\infty$) there is a mismatch between $(\alpha_m)_1$ and $(\alpha_m)_2$. This leads to the inconsistent evaluation of the velocity time derivative terms in Equations (5.51)–(5.54), which can result in the loss of second-order accuracy of the generalized-α method applied to the FSI equations. This inconsistency can be eliminated by setting $(\rho_\infty)_1 = (\rho_\infty)_2 = 1$, the case of zero high-frequency damping corresponding to the *midpoint rule*, but this is not sufficiently robust for practical calculations. Instead, in Bazilevs et al. (2008), the authors proposed to adopt expressions given by Equation (5.63) for *both the fluid and structural mechanics parts of the FSI equations*. This choice leads to optimal high-frequency dissipation in the fluid mechanics equations. For the structural mechanics equations, using the expressions derived in Chung and Hulbert (1993) and introducing the parameterizations given by Equation (5.63), the three eigenvalues of the amplification matrix (corresponding to the solution and its two time derivatives) are:

$$\lim_{\Delta t \to \infty} \lambda = \left\{ \frac{-1-3(\rho_\infty)_1}{3+(\rho_\infty)_1}, \frac{-1-3(\rho_\infty)_1}{3+(\rho_\infty)_1}, -(\rho_\infty)_1 \right\}. \quad (5.65)$$

The first two eigenvalues are different from the optimal value of $-(\rho_\infty)_1$, but it is a simple matter to show that they are monotone decreasing functions of $(\rho_\infty)_1$ and are bounded from above and below as

$$\frac{1}{3} \leq \left|\frac{-1-3(\rho_\infty)_1}{3+(\rho_\infty)_1}\right| \leq 1 \quad \forall |(\rho_\infty)_1| \leq 1. \quad (5.66)$$

Thus, the spectral radius of the amplification matrix never exceeds unity in magnitude, and no instabilities are expected to occur for the second-order system. Note that this choice of parameters maintains the second-order accuracy and unconditional stability of the time integration method because the conditions given by Equations (5.60)–(5.62) are satisfied.

5.2.3 Predictor–Multicorrector Algorithm and Linearization of the ALE FSI Equations

We extend the predictor–multicorrector algorithm presented in Section 4.6.2 for the ALE-VMS equations to the ALE FSI equations.

Predictor stage. Given the solution at time level n, we use the same-velocity predictor and set

$$\dot{\mathbf{U}}^0_{n+1} = \frac{(\gamma-1)}{\gamma}\dot{\mathbf{U}}_n, \quad (5.67)$$

$$\mathbf{U}^0_{n+1} = \mathbf{U}_n, \quad (5.68)$$

$$\mathbf{P}^0_{n+1} = \mathbf{P}_n, \quad (5.69)$$

$$\ddot{\mathbf{Y}}_{n+1}^0 = \frac{(\gamma - 1)}{\gamma} \ddot{\mathbf{Y}}_n, \tag{5.70}$$

$$\dot{\mathbf{Y}}_{n+1}^0 = \dot{\mathbf{Y}}_n, \tag{5.71}$$

$$\mathbf{Y}_{n+1}^0 = \mathbf{Y}_n + \Delta t_n \dot{\mathbf{Y}}_n + \frac{\Delta t_n^2}{2} \left((1 - 2\beta) \ddot{\mathbf{Y}}_n + 2\beta \ddot{\mathbf{Y}}_{n+1}^0 \right), \tag{5.72}$$

$$\ddot{\hat{\mathbf{Y}}}_{n+1}^0 = \frac{(\gamma - 1)}{\gamma} \ddot{\hat{\mathbf{Y}}}_n, \tag{5.73}$$

$$\dot{\hat{\mathbf{Y}}}_{n+1}^0 = \dot{\hat{\mathbf{Y}}}_n, \tag{5.74}$$

$$\hat{\mathbf{Y}}_{n+1}^0 = \hat{\mathbf{Y}}_n + \Delta t_n \dot{\hat{\mathbf{Y}}}_n + \frac{\Delta t_n^2}{2} \left((1 - 2\beta) \ddot{\hat{\mathbf{Y}}}_n + 2\beta \ddot{\hat{\mathbf{Y}}}_{n+1}^0 \right), \tag{5.75}$$

where the superscript 0 represents the zeroth value of the iteration counter.

Multicorrector stage. We repeat the following steps for $i = 0, 1, \ldots, (i_{\max} - 1)$, where i is the iteration counter and $i_{\max}$ is the maximum number of nonlinear iterations specified for the current time step.

1. Evaluate the iterates at the intermediate time levels:

$$\dot{\mathbf{U}}_{n+\alpha_m}^i = \dot{\mathbf{U}}_n + \alpha_m (\dot{\mathbf{U}}_{n+1}^i - \dot{\mathbf{U}}_n), \tag{5.76}$$

$$\mathbf{U}_{n+\alpha_f}^i = \mathbf{U}_n + \alpha_f (\mathbf{U}_{n+1}^i - \mathbf{U}_n), \tag{5.77}$$

$$\mathbf{P}_{n+1}^i = \mathbf{P}_{n+1}^i, \tag{5.78}$$

$$\ddot{\mathbf{Y}}_{n+\alpha_m}^i = \ddot{\mathbf{Y}}_n + \alpha_m (\ddot{\mathbf{Y}}_{n+1}^i - \ddot{\mathbf{Y}}_n), \tag{5.79}$$

$$\dot{\mathbf{Y}}_{n+\alpha_f}^i = \dot{\mathbf{Y}}_n + \alpha_f (\dot{\mathbf{Y}}_{n+1}^i - \dot{\mathbf{Y}}_n), \tag{5.80}$$

$$\mathbf{Y}_{n+\alpha_f}^i = \mathbf{Y}_n + \alpha_f (\mathbf{Y}_{n+1}^i - \mathbf{Y}_n), \tag{5.81}$$

$$\ddot{\hat{\mathbf{Y}}}_{n+\alpha_m}^i = \ddot{\hat{\mathbf{Y}}}_n + \alpha_m (\ddot{\hat{\mathbf{Y}}}_{n+1}^i - \ddot{\hat{\mathbf{Y}}}_n), \tag{5.82}$$

$$\dot{\hat{\mathbf{Y}}}_{n+\alpha_f}^i = \dot{\hat{\mathbf{Y}}}_n + \alpha_f (\dot{\hat{\mathbf{Y}}}_{n+1}^i - \dot{\hat{\mathbf{Y}}}_n), \tag{5.83}$$

$$\hat{\mathbf{Y}}_{n+\alpha_f}^i = \hat{\mathbf{Y}}_n + \alpha_f (\hat{\mathbf{Y}}_{n+1}^i - \hat{\mathbf{Y}}_n). \tag{5.84}$$

2. Use the intermediate values to assemble the linear system of equations corresponding to the linearization of Equations (5.51)–(5.54) with respect to the nodal unknowns $\dot{\mathbf{U}}_{n+1}$, $\mathbf{P}_{n+1}$, $\ddot{\mathbf{Y}}_{n+1}$, and $\ddot{\hat{\mathbf{Y}}}_{n+1}$:

$$\left.\frac{\partial \mathbf{N}_{1M}}{\partial \dot{\mathbf{U}}_{n+1}}\right|_i \Delta \dot{\mathbf{U}}_{n+1}^i + \left.\frac{\partial \mathbf{N}_{1M}}{\partial \mathbf{P}_{n+1}}\right|_i \Delta \mathbf{P}_{n+1}^i + \left.\frac{\partial \mathbf{N}_{1M}}{\partial \ddot{\mathbf{Y}}_{n+1}}\right|_i \Delta \ddot{\mathbf{Y}}_{n+1}^i + \left.\frac{\partial \mathbf{N}_{1M}}{\partial \ddot{\hat{\mathbf{Y}}}_{n+1}}\right|_i \Delta \ddot{\hat{\mathbf{Y}}}_{n+1}^i = -\mathbf{N}_{1M}^i, \tag{5.85}$$

$$\left.\frac{\partial \mathbf{N}_{1C}}{\partial \dot{\mathbf{U}}_{n+1}}\right|_i \Delta \dot{\mathbf{U}}_{n+1}^i + \left.\frac{\partial \mathbf{N}_{1C}}{\partial \mathbf{P}_{n+1}}\right|_i \Delta \mathbf{P}_{n+1}^i + \left.\frac{\partial \mathbf{N}_{1C}}{\partial \ddot{\mathbf{Y}}_{n+1}}\right|_i \Delta \ddot{\mathbf{Y}}_{n+1}^i + \left.\frac{\partial \mathbf{N}_{1C}}{\partial \ddot{\hat{\mathbf{Y}}}_{n+1}}\right|_i \Delta \ddot{\hat{\mathbf{Y}}}_{n+1}^i = -\mathbf{N}_{1C}^i, \tag{5.86}$$

$$\left.\frac{\partial \mathbf{N}_2}{\partial \dot{\mathbf{U}}_{n+1}}\right|_i \Delta \dot{\mathbf{U}}_{n+1}^i + \left.\frac{\partial \mathbf{N}_2}{\partial \mathbf{P}_{n+1}}\right|_i \Delta \mathbf{P}_{n+1}^i + \left.\frac{\partial \mathbf{N}_2}{\partial \ddot{\mathbf{Y}}_{n+1}}\right|_i \Delta \ddot{\mathbf{Y}}_{n+1}^i + \left.\frac{\partial \mathbf{N}_2}{\partial \ddot{\hat{\mathbf{Y}}}_{n+1}}\right|_i \Delta \ddot{\hat{\mathbf{Y}}}_{n+1}^i = -\mathbf{N}_2^i, \tag{5.87}$$

$$\left.\frac{\partial \mathbf{N}_3}{\partial \dot{\mathbf{U}}_{n+1}}\right|_i \Delta \dot{\mathbf{U}}_{n+1}^i + \left.\frac{\partial \mathbf{N}_3}{\partial \mathbf{P}_{n+1}}\right|_i \Delta \mathbf{P}_{n+1}^i + \left.\frac{\partial \mathbf{N}_3}{\partial \ddot{\mathbf{Y}}_{n+1}}\right|_i \Delta \ddot{\mathbf{Y}}_{n+1}^i + \left.\frac{\partial \mathbf{N}_3}{\partial \ddot{\hat{\mathbf{Y}}}_{n+1}}\right|_i \Delta \ddot{\hat{\mathbf{Y}}}_{n+1}^i = -\mathbf{N}_3^i. \quad (5.88)$$

The linear equation system is solved for the increments of the fluid mechanics, structural mechanics, and mesh moving unknowns. We postpone the discussion of the different options that may be employed to solve the above linear equation system until the next chapter.

3. Update the solution:

$$\dot{\mathbf{U}}_{n+1}^{i+1} = \dot{\mathbf{U}}_{n+1}^i + \Delta \dot{\mathbf{U}}_{n+1}^i, \quad (5.89)$$

$$\mathbf{U}_{n+1}^{i+1} = \mathbf{U}_{n+1}^i + \gamma \Delta t_n \Delta \dot{\mathbf{U}}_{n+1}^i, \quad (5.90)$$

$$\mathbf{P}_{n+1}^{i+1} = \mathbf{P}_{n+1}^i + \Delta \mathbf{P}_{n+1}^i, \quad (5.91)$$

$$\ddot{\mathbf{Y}}_{n+1}^{i+1} = \ddot{\mathbf{Y}}_{n+1}^i + \Delta \ddot{\mathbf{Y}}_{n+1}^i, \quad (5.92)$$

$$\dot{\mathbf{Y}}_{n+1}^{i+1} = \dot{\mathbf{Y}}_{n+1}^i + \gamma \Delta t_n \Delta \ddot{\mathbf{Y}}_{n+1}^i, \quad (5.93)$$

$$\mathbf{Y}_{n+1}^{i+1} = \mathbf{Y}_{n+1}^i + \beta \Delta t_n^2 \Delta \ddot{\mathbf{Y}}_{n+1}^i, \quad (5.94)$$

$$\ddot{\hat{\mathbf{Y}}}_{n+1}^{i+1} = \ddot{\hat{\mathbf{Y}}}_{n+1}^i + \Delta \ddot{\hat{\mathbf{Y}}}_{n+1}^i, \quad (5.95)$$

$$\dot{\hat{\mathbf{Y}}}_{n+1}^{i+1} = \dot{\hat{\mathbf{Y}}}_{n+1}^i + \gamma \Delta t_n \Delta \ddot{\hat{\mathbf{Y}}}_{n+1}^i, \quad (5.96)$$

$$\hat{\mathbf{Y}}_{n+1}^{i+1} = \hat{\mathbf{Y}}_{n+1}^i + \beta \Delta t_n^2 \Delta \ddot{\hat{\mathbf{Y}}}_{n+1}^i. \quad (5.97)$$

We make use of the following left-hand-side matrices (i.e., tangent matrices) in Step 2 of the multicorrector stage:

- The left-hand-side matrices for the fluid mechanics equations, $\frac{\partial \mathbf{N}_{1M}}{\partial \dot{\mathbf{U}}_{n+1}}$, $\frac{\partial \mathbf{N}_{1M}}{\partial \mathbf{P}_{n+1}}$, $\frac{\partial \mathbf{N}_{1C}}{\partial \dot{\mathbf{U}}_{n+1}}$, and $\frac{\partial \mathbf{N}_{1C}}{\partial \mathbf{P}_{n+1}}$, are given by Equations (4.92)–(4.98).

- The left-hand-side matrix for the structure may be obtained directly from the linearized structural mechanics formulation (see Equation (1.155)):

$$\frac{\partial \mathbf{N}_2}{\partial \ddot{\mathbf{Y}}_{n+1}} = \left[K_{AB}^{ik} \right], \quad (5.98)$$

$$K_{AB}^{ik} = \alpha_m \int_{(\Omega_2)_0} N_A (\rho_2)_0 N_B \, d\Omega \, \delta_{ik} + \alpha_f \beta \Delta t_n^2 \int_{(\Omega_2)_0} \frac{\partial N_A}{\partial X_J} D_{iJkL}^X \frac{\partial N_B}{\partial X_L} \, d\Omega, \quad (5.99)$$

where D_{iJkL}^X's are the components of the tangent stiffness tensor in the reference configuration (see also Equation (1.152)):

$$D_{iJkL}^X = F_{iI} \mathbb{C}_{IJKL} F_{kK} + \delta_{ik} S_{JL}. \quad (5.100)$$

The left-hand-side matrix in Equation (5.98) is written with respect to the reference-configuration variables. It is also possible to evaluate it in the current configuration. For this, we change variables to the current configuration and obtain

$$\frac{\partial \mathbf{N}_2}{\partial \ddot{\mathbf{Y}}_{n+1}} = \left[K_{AB}^{ik} \right], \quad (5.101)$$

$$K_{AB}^{ik} = \alpha_m \int_{(\Omega_2)_t} N_A \rho_2 N_B \, d\Omega \, \delta_{ik} + \alpha_f \beta \Delta t_n^2 \int_{(\Omega_2)_t} \frac{\partial N_A}{\partial x_j} D_{ijkl}^x \frac{\partial N_B}{\partial x_l} \, d\Omega. \quad (5.102)$$

Here D_{ijkl}^x's are the components of the tangent stiffness tensor in the current configuration:

$$D_{ijkl}^x = J^{-1} F_{iI} F_{jJ} \mathbb{C}_{IJKL} F_{kK} F_{lL} + \delta_{ik} J^{-1} F_{jJ} S_{JL} F_{lL} \qquad (5.103)$$

$$= J^{-1} F_{iI} F_{jJ} \mathbb{C}_{IJKL} F_{kK} F_{lL} + \delta_{ik} (\sigma_2)_{jl}, \qquad (5.104)$$

where the second equality follows from the definition of the Cauchy stress tensor. Both the reference- and current-configuration definitions of the structural left-hand-side matrix are equivalent.

- The left-hand-side matrix for the mesh moving equations, $\frac{\partial \mathbf{N}_3}{\partial \hat{\mathbf{Y}}_{n+1}}$, corresponds to that of the linear elastostatics formulation (see Equations (2.167) and (2.168)), scaled by the factor $\alpha_f \beta \Delta t_n^2$.

- The derivatives of the discrete residuals of the fluid mechanics equations with respect to the mesh moving nodal unknowns, $\frac{\partial \mathbf{N}_{1M}}{\partial \hat{\mathbf{Y}}_{n+1}}$ and $\frac{\partial \mathbf{N}_{1C}}{\partial \hat{\mathbf{Y}}_{n+1}}$, are called the *shape derivatives*. These are required for the consistent linearization of the discrete FSI equations, and are discussed in the recent work of Fernandez and Moubachir (2005), Dettmer and Peric (2008) and Bazilevs et al. (2008). The detailed derivation of the shape derivatives is given in Bazilevs et al. (2008). It was also shown in Bazilevs et al. (2008) that omitting the shape derivatives from the linearization of the FSI equations has little effect on the convergence of the Newton–Raphson iterations. Furthermore, omitting the shape derivatives decouples the mesh solve from the rest of the linear system in Step 2 of the multicorrector stage, which, in turn, leads to significant computational savings. This approach was called "quasi-direct coupling" in Tezduyar et al. (2004, 2006a,b) (see Section 6.1.2). In the "direct coupling" method proposed in Tezduyar et al. (2004, 2006a,b) and Tezduyar (2007a), the matrix-vector products associated with the shape derivatives are evaluated with the numerical element-vector-based (NEVB) computation technique. All other matrix-vector products involved in iterative solution of the coupled equation blocks are evaluated with the analytical EVB (AEVB) technique Tezduyar (2001b, 2004a, 2007a), resulting in a mixed AEVB/NEVB method Tezduyar (2001b, 2004a, 2007a) (see Section 6.1.3).

5.3 Space–Time Formulation of FSI

The space–time FSI formulation covered in this section is the Stabilized Space–Time FSI (SSTFSI) technique, which was introduced in Tezduyar and Sathe (2007) and improved in Takizawa and Tezduyar (2011, 2012b). The SSTFSI technique is based on the DSD/SST method (Tezduyar, 1992, 2003a; Tezduyar et al., 1992a,c; Takizawa and Tezduyar, 2011, 2012b; see Sections 4.4 and 4.6.3).

5.3.1 Core Formulation

The SSTFSI formulation is in the context of nonmatching fluid and structure meshes at the interface. We will include in the formulation the two additional stabilization terms of the DSD/SST-VMST (ST-VMS) method (see Equation (4.113)). With the subscripts "I" and "E" defined earlier, we will use a function-space notation that is slightly expanded compared to the earlier sections (see Figure 5.2). With that, the convective form of the SSTFSI-VMST

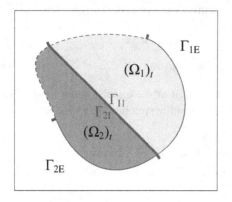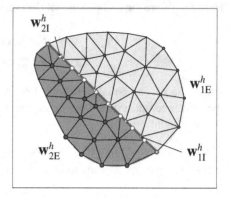

Figure 5.2 The spatial domain for the SSTFSI formulation, including the fluid and structure subdomains, and the interfaces at time t (left). We note that $(\Omega_1)_t$ and Γ_{1E} are subsets of Q_n and P_n, respectively, corresponding to instant t. Subscript t can be replaced with 0, n, or $n+1$. The test functions are divided into those associated with the fluid–structure interface ($\mathbf{w}_{1I}^h$ and $\mathbf{w}_{2I}^h$) and elsewhere ($\mathbf{w}_{1E}^h$ and $\mathbf{w}_{2E}^h$) (right)

formulation can be written as follows:

$$\int_{Q_n} \mathbf{w}_{1E}^h \cdot \rho \left(\frac{\partial \mathbf{u}^h}{\partial t} + \mathbf{u}^h \cdot \nabla \mathbf{u}^h - \mathbf{f}^h \right) dQ + \int_{Q_n} \boldsymbol{\varepsilon}(\mathbf{w}_{1E}^h) : \boldsymbol{\sigma}_1(\mathbf{u}^h, p^h) dQ$$

$$- \int_{P_n} \mathbf{w}_{1E}^h \cdot \mathbf{h}_{1E}^h dP + \int_{Q_n} q_{1E}^h \nabla \cdot \mathbf{u}^h dQ + \int_{(\Omega_1)_n} (\mathbf{w}_{1E}^h)_n^+ \cdot \rho \left((\mathbf{u}^h)_n^+ - (\mathbf{u}^h)_n^- \right) d\Omega$$

$$+ \sum_{e=1}^{(n_{\mathrm{el}})_n} \int_{Q_n^e} \frac{\tau_{\mathrm{SUPS}}}{\rho} \left[\rho \left(\frac{\partial \mathbf{w}_{1E}^h}{\partial t} + \mathbf{u}^h \cdot \nabla \mathbf{w}_{1E}^h \right) + \nabla q_{1E}^h \right] \cdot \mathbf{r}_{\mathrm{M}}(\mathbf{u}^h, p^h) dQ$$

$$+ \sum_{e=1}^{(n_{\mathrm{el}})_n} \int_{Q_n^e} \rho \nu_{\mathrm{LSIC}} \nabla \cdot \mathbf{w}_{1E}^h r_{\mathrm{C}}(\mathbf{u}^h) dQ$$

$$- \sum_{e=1}^{(n_{\mathrm{el}})_n} \int_{Q_n^e} \tau_{\mathrm{SUPS}} \mathbf{w}_{1E}^h \otimes \mathbf{r}_{\mathrm{M}}(\mathbf{u}^h, p^h) : \left(\nabla \mathbf{u}^h \right) dQ$$

$$- \sum_{e=1}^{(n_{\mathrm{el}})_n} \int_{Q_n^e} \frac{\tau_{\mathrm{SUPS}}^2}{\rho} \left(\nabla \mathbf{w}_{1E}^h \right) : \mathbf{r}_{\mathrm{M}}(\mathbf{u}^h, p^h) \otimes \mathbf{r}_{\mathrm{M}}(\mathbf{u}^h, p^h) dQ = 0, \quad (5.105)$$

$$\int_{Q_n} q_{1I}^h \nabla \cdot \mathbf{u}^h dQ + \sum_{e=1}^{(n_{\mathrm{el}})_n} \int_{Q_n^e} \frac{\tau_{\mathrm{SUPS}}}{\rho} \nabla q_{1I}^h \cdot \mathbf{r}_{\mathrm{M}}(\mathbf{u}^h, p^h) dQ = 0, \quad (5.106)$$

$$\int_{(\Gamma_{1I})_{n+1}} \left(\mathbf{w}_{1I}^h \right)_{n+1}^- \cdot \left(\left(\mathbf{u}_{1I}^h \right)_{n+1}^- - \mathbf{u}_{2I}^h \right) d\Gamma = 0, \quad (5.107)$$

$$\int_{Q_n} \left(\mathbf{w}_{1I}^h\right)_{n+1}^{-} \cdot \rho \left(\frac{\partial \mathbf{u}^h}{\partial t} + \mathbf{u}^h \cdot \nabla \mathbf{u}^h - \mathbf{f}^h\right) dQ + \int_{Q_n} \varepsilon\left(\left(\mathbf{w}_{1I}^h\right)_{n+1}^{-}\right) : \sigma_1(\mathbf{u}^h, p^h) dQ$$

$$- \int_{P_n} \left(\mathbf{w}_{1I}^h\right)_{n+1}^{-} \cdot \mathbf{h}_{1I}^h dP$$

$$+ \sum_{e=1}^{(n_{el})_n} \int_{Q_n^e} \tau_{\text{SUPS}} \rho \left(\frac{\partial \left(\mathbf{w}_{1I}^h\right)_{n+1}^{-}}{\partial t} + \mathbf{u}^h \cdot \nabla \left(\mathbf{w}_{1I}^h\right)_{n+1}^{-}\right) \cdot \mathbf{r}_M(\mathbf{u}^h, p^h) dQ$$

$$+ \sum_{e=1}^{(n_{el})_n} \int_{Q_n^e} \rho \nu_{\text{LSIC}} \nabla \cdot \left(\mathbf{w}_{1I}^h\right)_{n+1}^{-} r_C(\mathbf{u}^h) dQ$$

$$- \sum_{e=1}^{(n_{el})_n} \int_{Q_n^e} \tau_{\text{SUPS}} \left(\mathbf{w}_{1I}^h\right)_{n+1}^{-} \otimes \mathbf{r}_M(\mathbf{u}^h, p^h) : \left(\nabla \mathbf{u}^h\right) dQ$$

$$- \sum_{e=1}^{(n_{el})_n} \int_{Q_n^e} \frac{\tau_{\text{SUPS}}^2}{\rho} \left(\nabla \left(\mathbf{w}_{1I}^h\right)_{n+1}^{-}\right) : \mathbf{r}_M(\mathbf{u}^h, p^h) \otimes \mathbf{r}_M(\mathbf{u}^h, p^h) dQ = 0, \tag{5.108}$$

$$\int_{\Gamma_{2I}} \mathbf{w}_{2I}^h \cdot \left(\mathbf{h}_{2I}^h + \mathbf{h}_{1I}^h\right) d\Gamma = 0, \tag{5.109}$$

$$\int_{(\Omega_2)_0} \mathbf{w}_2^h \cdot (\rho_2)_0 \left(\frac{d^2 \mathbf{y}^h}{dt^2} - \mathbf{f}_2\right) d\Omega + \int_{(\Omega_2)_0} \nabla_X \mathbf{w}_2^h : \left(\mathbf{F}^h \mathbf{S}^h\right) d\Omega$$
$$- \int_{\Gamma_{2E}} \mathbf{w}_{2E}^h \cdot \mathbf{h}_{2E}^h d\Gamma - \int_{\Gamma_{2I}} \mathbf{w}_{2I}^h \cdot \mathbf{h}_{2I}^h d\Gamma = 0, \tag{5.110}$$

$$\int_{(\Omega_1)_{\tilde{t}}} \epsilon\left(\mathbf{w}_{3E}^h\right) \cdot \mathbf{D}^h \epsilon\left(\hat{\mathbf{y}}_{n+1}^h - \hat{\mathbf{y}}^h(\tilde{t})\right) d\Omega = 0, \tag{5.111}$$

$$\int_{(\Gamma_{1I})_{\tilde{t}}} \mathbf{w}_{3I}^h \cdot \left(\hat{\mathbf{y}}_{n+1}^h - \mathbf{y}_{n+1}^h\right) d\Gamma = 0. \tag{5.112}$$

In this formulation, $(\mathbf{u}_{1I}^h)_{n+1}^{-}$, $\mathbf{h}_{1I}^h$, and $\mathbf{h}_{2I}^h$ (the fluid velocity, fluid stress, and structural stress at the fluid–structure interface) are treated as separate unknowns, and Equations (5.107), (5.108) and (5.109) can be seen as corresponding to these three unknowns, respectively. The structural displacement rate at the interface, $\mathbf{u}_2^h$, is derived from $\mathbf{y}^h$ (see Equation (5.6)).

REMARK 5.5 *Since we use the space–time formulation only for the fluid mechanics part, the subscript "1" is omitted from Q and P.*

REMARK 5.6 *We do not have the equation generated by $\left(\mathbf{w}_{1I}^h\right)_n^{+}$ because we set $\left(\mathbf{u}_{1I}^h\right)_n^{+}$ to $\left(\mathbf{u}_{1I}^h\right)_n^{-}$.*

As in Tezduyar and Sathe (2007), we can replace Equation (5.108) with the following equation:

$$\int_{P_n} \left(\mathbf{w}_{II}^h\right)_{n+1}^- \cdot \mathbf{h}_{II}^h \, dP = -\int_{P_n} \left(\mathbf{w}_{II}^h\right)_{n+1}^- \cdot p^h \mathbf{n} \, dP + \int_{Q_n} 2\mu\varepsilon\left(\left(\mathbf{w}_{II}^h\right)_{n+1}^-\right) : \varepsilon(\mathbf{u}^h) \, dQ$$
$$+ \sum_{e=1}^{(n_{el})_n} \int_{Q_n^e} \left(\mathbf{w}_{II}^h\right)_{n+1}^- \cdot \nabla \cdot \left(2\mu\varepsilon(\mathbf{u}^h)\right) \, dQ. \quad (5.113)$$

We note that Equation (5.113) has been derived by assuming that viscous-flux jump terms across inter-element borders are negligible. Alternatively, one can leave that projection equation in its form prior to the integration-by-parts:

$$\int_{P_n} \left(\mathbf{w}_{II}^h\right)_{n+1}^- \cdot \mathbf{h}_{II}^h \, dP = \int_{P_n} \left(\mathbf{w}_{II}^h\right)_{n+1}^- \cdot \left(-p^h \mathbf{I} + 2\mu\varepsilon(\mathbf{u}^h)\mathbf{n}\right) \, dP, \quad (5.114)$$

and this would require also the projection of $\varepsilon(\mathbf{u}^h)$ from the element interiors to the nodes.

REMARK 5.7 *The conservative form of the SSTFSI-VMST method can be obtained by replacing Equations (5.105) and (5.108) with the equations corresponding to the conservative form of the DSD/SST-VMST formulation, which is represented by Equation (4.109).*

REMARK 5.8 *The original SSTFSI formulation, which we will call SSTFSI-SUPS, can be obtained by dropping the last two terms in Equations (5.105) and (5.108). For backward compatibility, the acronym SSTFSI, when written without any of the two option indicators "-SUPS" and "-VMST", will imply SSTFSI-SUPS.*

REMARK 5.9 *In Tezduyar and Sathe (2007), the versions of the SSTFSI technique corresponding to the DSD/SST-DP, DSD/SST-SP, DSD/SST-TIP1, and DSD/SST-SV formulations (see Remarks 4.8–4.11) were called "SSTFSI-DP," "SSTFSI-SP," "SSTFSI-TIP1," and "SSTFSI-SV," respectively.*

REMARK 5.10 *The acronym SSTFSI, when written without any of the four option indicators "-DP", "-SP", "-TIP1", "-SV", will imply SSTFSI-DP.*

REMARK 5.11 *The statements made in Remark 4.12 about the versions of the DSD/SST formulation also apply to the corresponding versions of the SSTFSI technique.*

In computations where we account for the porosity of the membrane fabric, as formulated in Tezduyar and Sathe (2007), Equation (5.107) is replaced with

$$\int_{(\Gamma_{II})_{n+1}} \left(\mathbf{w}_{II}^h\right)_{n+1}^- \cdot \left((\mathbf{u}_{II}^h)_{n+1}^- - \mathbf{u}_{2I}^h + k_{\text{PORO}} \left(\mathbf{n} \cdot \mathbf{h}_{II}^h\right) \mathbf{n}\right) d\Gamma = 0, \quad (5.115)$$

where k_{PORO} is the porosity coefficient. This coefficient is typically given in units of "CFM." When a fabric with a porosity coefficient of 1 CFM is subjected to a pressure differential of 1/2 in of water, the amount of flow crossing is 1 ft^3/min across a sample size of 1 ft^2, which translates to a normal velocity of 1 ft/min. In our current implementation, in Equation (5.115) we take into account only the pressure component of $\mathbf{h}_{1\mathrm{I}}^h$.

REMARK 5.12 *In FSI computations with membranes and shells, the pressure at the interface has split nodal values corresponding to the fluid surfaces above and below the membrane or shell structure. It was proposed in Tezduyar and Sathe (2007) to use such split nodal values for pressure also at the boundaries (i.e., edges) of a membrane structure submerged in the fluid. As pointed out in Tezduyar and Sathe (2007), our computations show that this provides additional numerical stability for the edges of the membrane.*

5.3.2 Interface Projection Techniques for Nonmatching Fluid and Structure Interface Discretizations

The formulation given by Equations (5.105)–(5.112) is based on allowing for cases when the fluid and structure meshes at the interface are not identical. If they are identical, as pointed out in Tezduyar and Sathe (2007), the same formulation can still be used and, with proper interface projection algorithms, the formulation becomes equivalent to a monolithic method. Here we describe some additional interface projection techniques.

5.3.2.1 Least-Squares Projection

Solving a projection equation, such as Equations (5.107), (5.109), or (5.112), involves a search for quadrature points. This process is time consuming and may not be efficient in parallel computing. To avoid a search at every time step, one can formulate the projections over a single reference configuration as follows:

$$\int_{(\Gamma_{1\mathrm{I}})_{\mathrm{REF}}} (\mathbf{w}_{1\mathrm{I}})_{n+1}^{-} \cdot \left(\left(\mathbf{u}_{1\mathrm{I}}^h \right)_{n+1}^{-} - \mathbf{u}_{2\mathrm{I}}^h \right) \, \mathrm{d}\Gamma = 0, \tag{5.116}$$

$$\int_{(\Gamma_{2\mathrm{I}})_{\mathrm{REF}}} \mathbf{w}_{2\mathrm{I}}^h \cdot \left(\mathbf{h}_{2\mathrm{I}}^h + \mathbf{h}_{1\mathrm{I}}^h \right) \, \mathrm{d}\Gamma = 0, \tag{5.117}$$

$$\int_{(\Gamma_{1\mathrm{I}})_{\mathrm{REF}}} \mathbf{w}_{3\mathrm{I}}^h \cdot \left(\hat{\mathbf{y}}_{n+1}^h - \mathbf{y}_{n+1}^h \right) \, \mathrm{d}\Gamma = 0, \tag{5.118}$$

where $(\Gamma_{1\mathrm{I}})_{\mathrm{REF}}$ and $(\Gamma_{2\mathrm{I}})_{\mathrm{REF}}$ represent some reference configurations of $\Gamma_{1\mathrm{I}}$ and $\Gamma_{2\mathrm{I}}$, respectively.

5.3.2.2 Numerical Substitution and Direct Substitution

The linear equation systems involved in each nonlinear iteration are solved iteratively. Therefore how Equations (5.116) and (5.117) are handled determines how well the fluid and structure systems are coupled. As mentioned in Takizawa *et al.* (2010b), these two projection equations are solved by "numerical substitution," which essentially consists of sublevel GMRES iterations. This technique is applicable to Equations (5.107)–(5.109), (5.112)–(5.114), and (5.116)–(5.118). If the fluid and structure meshes at the interface are identical, then the projections given by Equations (5.107), (5.109), and (5.112) simplify to "direct substitution." With that, the formulation becomes monolithic.

5.3.2.3 Separated Stress Projection (SSP)

In the SSP option proposed in Tezduyar *et al.* (2008a), the pressure and viscous parts of the stress at the fluid interface are projected to the structure interface separately, pressure as a scalar and viscous stress as a vector. The projected parts are then combined while integrating the interface stresses in the structural mechanics equations. In the SSP option, the projections given by Equations (5.113) and (5.109) are replaced with the following projections:

$$\int_{P_n} \left(\mathbf{w}_{1\mathrm{I}}^h\right)_{n+1}^- \cdot \left(\mathbf{h}_v^h\right)_{1\mathrm{I}} \, \mathrm{d}P = \int_{Q_n} 2\mu\varepsilon\left(\left(\mathbf{w}_{1\mathrm{I}}^h\right)_{n+1}^-\right) : \varepsilon(\mathbf{u}^h) \, \mathrm{d}Q$$

$$+ \sum_{e=1}^{(n_{\mathrm{el}})_n} \int_{Q_n^e} \left(\mathbf{w}_{1\mathrm{I}}^h\right)_{n+1}^- \cdot \nabla \cdot \left(2\mu\varepsilon(\mathbf{u}^h)\right) \, \mathrm{d}Q, \quad (5.119)$$

$$\int_{\Gamma_{2\mathrm{I}}} q_{2\mathrm{I}}^h \left(p_{2\mathrm{I}}^h - p_{1\mathrm{I}}^h\right) \mathrm{d}\Gamma = 0, \quad (5.120)$$

$$\int_{\Gamma_{2\mathrm{I}}} \mathbf{w}_{2\mathrm{I}}^h \cdot \left(\left(\mathbf{h}_v^h\right)_{2\mathrm{I}} + \left(\mathbf{h}_v^h\right)_{1\mathrm{I}}\right) \mathrm{d}\Gamma = 0, \quad (5.121)$$

$$\mathbf{h}_{2\mathrm{I}}^h = -p_{2\mathrm{I}}^h \mathbf{n}_{2\mathrm{I}} + \left(\mathbf{h}_v^h\right)_{2\mathrm{I}}, \quad (5.122)$$

where $\mathbf{h}_v^h$ is the viscous part of the stress vector, $p_{1\mathrm{I}}^h$ is the pressure at the fluid interface, $p_{2\mathrm{I}}^h$ and $q_{2\mathrm{I}}^h$ are the projection of that pressure to the structure interface and its corresponding test function, and $\mathbf{n}_{2\mathrm{I}}$ is the unit normal vector at the structure interface. The stress vector at the structure interface, given by Equation (5.122), is evaluated while integrating the interface stresses in the structural mechanics equations. Therefore, in the way Equation (5.122) is used, $\mathbf{n}_{2\mathrm{I}}$ is evaluated at the integration point, and $p_{2\mathrm{I}}^h$ and $\left(\mathbf{h}_v^h\right)_{2\mathrm{I}}$ are the interpolated values at the integration point.

As an alternative to the projection given by Equation (5.119), one can leave that projection equation in its form prior to the integration-by-parts:

$$\int_{P_n} \left(\mathbf{w}_{1\mathrm{I}}^h\right)_{n+1}^- \cdot \mathbf{h}_{1\mathrm{I}}^h \, \mathrm{d}P = \int_{P_n} \left(\mathbf{w}_{1\mathrm{I}}^h\right)_{n+1}^- \cdot 2\mu\varepsilon(\mathbf{u}^h)\mathbf{n}_{1\mathrm{I}} \, \mathrm{d}P, \quad (5.123)$$

which would correspond to Equation (5.114) and would again require the projection of $\varepsilon(\mathbf{u}^h)$ from the element interiors to the nodes.

REMARK 5.13 *It was proposed in Takizawa et al. (2010b) that the "mass" matrix associated with the first term in Equation (5.119) be lumped, pointing out that with this lumping the projection would become equivalent to a direct substitution, which would make the computations more efficient. This mass lumping would also be applicable to the third term in Equation (5.108) and first term in Equations (5.113), (5.114), and (5.123).*

REMARK 5.14 *As pointed out in Takizawa et al. (2010b), a smoother stress distribution is observed with a lumped mass matrix than with a consistent mass matrix.*

REMARK 5.15 *For the computations reported in Takizawa et al. (2010b, 2011c), the fluid interface stress vector is reconstructed from its pressure and viscous parts before projecting it to the structure. In that sense, the approach used in Takizawa et al. (2010b) and Takizawa et al. (2011c) has only some of the ingredients of the SSP technique, but those ingredients are still helpful in increasing the accuracy and efficiency.*

REMARK 5.16 *As proposed in Takizawa et al. (2011c), in general, we want to use the SSP option fully, in which we not only calculate the pressure and viscous parts of the fluid interface stress separately but also project them separately and re-combine them at the structure interface after the projection. In fact, the SSP option was used fully in the computations reported in Takizawa et al. (2012b). As pointed out in Takizawa et al. (2011c) and Tezduyar et al. (2011), although the amount of data projected increases by a factor 4/3, the results are more accurate when the structure mesh at the interface is finer than the fluid mesh.*

5.4 Advanced Mesh Update Techniques

A number of advanced mesh update techniques based on the Jacobian-based stiffening have been developed since the inception of the technique and we outline them here.

5.4.1 Solid-Extension Mesh Moving Technique (SEMMT)

In dealing with fluid–solid interfaces, in the mesh moving technique introduced in Tezduyar *et al.* (1992b, 1993) and Johnson and Tezduyar (1994), the structured layers of elements generated around the solid objects to fully control the mesh resolution move as "glued" to the solid objects, undergoing a rigid-body motion. No equations are solved for the motion of the nodes in these layers, because these nodal motions are not governed by the equations of elasticity. This results in some cost reduction. But more importantly, the user retains the full control of the mesh resolution in these layers. For early examples of automatic mesh moving combined with structured layers of elements undergoing rigid-body motion with solid objects, see (Tezduyar *et al.*, 1993; Johnson and Tezduyar, 1994). Earlier examples of element layers undergoing rigid-body motion, in combination with deforming structured meshes, can be found in Tezduyar (1992).

In computation of flows with fluid–solid interfaces where the solid is deforming, the motion of the fluid mesh near the interface cannot be represented by a rigid-body motion. Depending on the deformation mode of the solid, the mesh moving technique described in Section 4.7 may need to be used. In such cases, the thin fluid elements near the solid surface become a challenge for the mesh moving technique. In the SEMMT Tezduyar (2001a, 2003b), it was proposed to treat those thin fluid elements almost like an extension of the solid elements. In the SEMMT, in solving the equations of elasticity governing the motion of the fluid nodes, higher rigidity is assigned to these thin elements compared to the other fluid elements. Two ways of accomplishing this were proposed in Tezduyar (2001a, 2003b): solving the elasticity equations for the nodes connected to the thin elements separate from the elasticity equations for the other nodes, or together. If they are solved separately, for the thin elements, as boundary conditions at the interface with the other elements, traction-free conditions would be used. The separate solution option is referred to as "SEMMT – Multiple Domain (SEMMT–MD)" and the unified solution option as "SEMMT – Single Domain (SEMMT–SD)." In Stein and Tezduyar (2002) and Stein et al. (2004), test computations were presented to demonstrate how the SEMMT functions as part of the elasticity-equation-based mesh moving technique with Jacobian-based stiffening. Both SEMMT options described above were employed. The test computations included mesh deformation tests (Stein and Tezduyar, 2002; Stein et al., 2004) and a 2D FSI model problem Stein et al. (2004). For completeness, we include here some of those test computations.

5.4.1.1 General Test Conditions and Mesh Quality Measures

The tests were carried out with the standard technique (where $\chi = 1.0$ for all the elements, and all the nodes are moved together), SEMMT–SD (with $\chi = 2.0$ for the inner elements and $\chi = 1.0$ for the outer elements), and SEMMT–MD (with $\chi = 1.0$ for all elements in both domains). The mesh over which the elasticity equations are solved is updated at each increment. To evaluate the effectiveness of different mesh moving techniques, two measures of mesh quality are defined, similar to those in Johnson and Tezduyar (1996). They are *element area change* (f_A^e) and *element shape change* (f_{AR}^e):

$$f_A^e = \left|\log\left(\frac{A^e}{A_0^e}\right)\right|, \tag{5.124}$$

$$f_{AR}^e = \left|\log\left(\frac{AR^e}{AR_0^e}\right)\right|. \tag{5.125}$$

Here subscript "0" refers to the undeformed mesh (i.e., the mesh obtained after the last remesh), and AR^e is the element aspect ratio, defined as

$$AR^e = \frac{(\ell^e_{max})^2}{A^e}, \tag{5.126}$$

where ℓ^e_{max} is the maximum edge length for element e. We define array norms for the set of element mesh quality measures as

$$\|f_A\|_p = \left\{\sum_e (f_A^e)^p\right\}^{1/p}, \tag{5.127}$$

$$\|f_{AR}\|_p = \left\{\sum_e (f_{AR}^e)^p\right\}^{1/p}, \tag{5.128}$$

where f_A and f_{AR} are the arrays of mesh quality values f_A and f_{AR} for all elements of interest, and p is the norm indicator. In the following examples, we use $p = \infty$, and

$$\|f_A\|_\infty = \max_e (f_A^e), \tag{5.129}$$

$$\|f_{AR}\|_\infty = \max_e (f_{AR}^e). \tag{5.130}$$

Thus, for a given set of elements, global area and shape changes are defined to be the maximum values of the element area and shape changes, respectively.

5.4.1.2 Mesh Deformation Tests

In this set of tests we use a 2D unstructured mesh consisting of triangular elements and an embedded structure with zero thickness. The mesh spans a region of $|x| \leq 1.0$ and $|y| \leq 1.0$. The structure spans $y = 0.0$ and $|x| \leq 0.5$. Three layers of elements (with $\ell_y = 0.01$) are placed along each side of the structure, with 50 element edges along the structure (i.e., $\ell_x = 0.02$). Figure 5.3 shows the mesh and its close up view near the structure. The test cases involve three different types of prescribed motion or deformation for the structure: rigid-body translation in the y-direction, rigid-body rotation about the origin, and prescribed bending. In the case of prescribed bending, the structure deforms from a line to a circular arc, with no stretch in the structure and no net vertical or horizontal displacement. For each test case the maximum displacement or deformation is reached over 50 increments. Those maximum values are $\Delta y = 0.5$ for the translation test, a rotation of $\Delta \theta = \pi/4$ for the rotation test, and bending to a half circle ($\theta = \pi$) for the bending test.

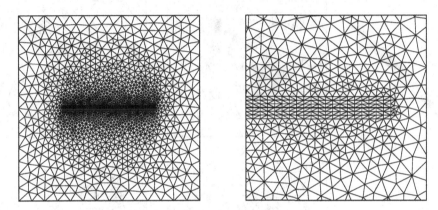

Figure 5.3 2D test mesh for SEMMT

Figure 5.4 shows, for SEMMT–MD, the deformed mesh for the translation, rotation, and bending tests. It is clear that the SEMMT approach reduces distortion in the inner elements. Figure 5.5 shows, for the bending test, the two mesh quality measures (defined based on all the elements) for the standard technique, SEMMT–SD, and SEMMT–MD. The curves display the mesh quality as a function of the bending magnitude and show an improvement in overall mesh quality for the SEMMT over the standard technique. When we compare performance of the SEMMT with the standard technique for the inner elements (i.e., layers), we see a more dramatic improvement than we do for the overall mesh motion. Figure 5.6 shows, for the translation test, the two mesh quality measures (defined based on the inner elements) plotted as functions of the translation magnitude for the standard technique, SEMMT–SD, and SEMMT–MD. The SEMMT yields significant reductions in the mesh distortion compared to the standard technique. Figure 5.7 shows, for the rotation test, the two mesh quality measures (defined based on the inner elements) plotted as functions of the rotation magnitude for the standard technique, SEMMT–SD, and SEMMT–MD. Again, the SEMMT yields significant reductions in the mesh distortion. Figure 5.8 shows, for the bending test, the two mesh quality measures (defined based on the inner elements) plotted as functions of the bending magnitude for the standard technique, SEMMT–SD, and SEMMT–MD. Once again, the SEMMT

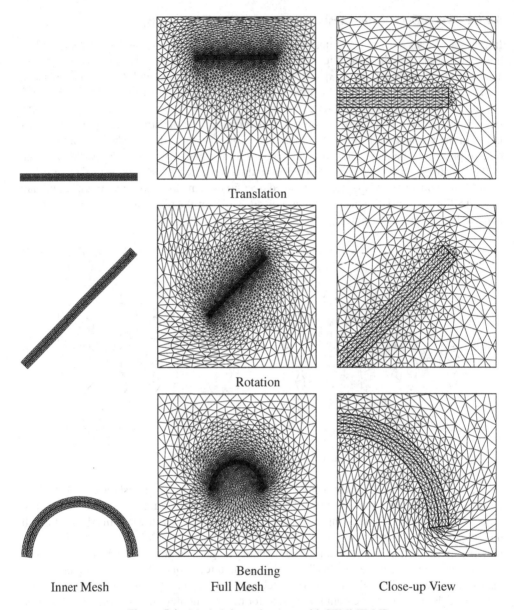

Figure 5.4 Mesh deformation tests with SEMMT–MD

approach yields much less mesh distortion. For more on this set of tests (see Stein *et al.*, 2004).

5.4.2 *Move-Reconnect-Renode Mesh Update Method (MRRMUM)*

The MRRMUM was proposed in Tezduyar *et al.* (2005). In the MRRMUM (see Tezduyar *et al.*, 2005, 2006c), two remeshing options were defined, with each proposed for use when it

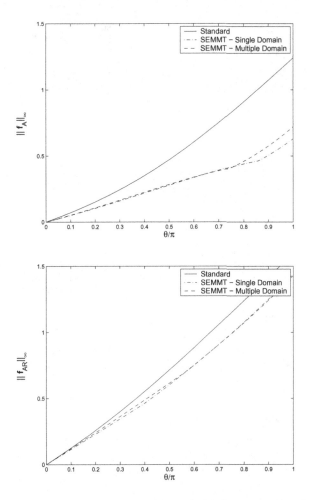

Figure 5.5 Bending test mesh quality (defined based on all the elements) as function of bending magnitude, for the standard mesh moving technique, SEMMT–SD, and SEMMT–MD

is most effective to do so. In the "reconnect" option, only the way the nodes are connected is changed and thus only the elements are replaced (fully or partially) with a new set of elements. The mesh generator developed in Fujisawa et al. (2003), for example, provides the reconnect option. In the "renode" option, the existing nodes are replaced (fully or partially) with a new set of nodes. This, of course, results in also replacing the existing elements with a new set of elements. Because the reconnect option is simpler and involves less projection errors, it is preferable to the renode option. In the MRRMUM, we move the mesh for as many time steps as we can, reconnect only as frequently as we need to, and renode only when doing so is the only remaining option.

In Tezduyar et al. (2005), for the prescribed rigid-body rotation of a parachute, the performances of the two remeshing options described above were compared. By examining the aerodynamical forces acting on the parachute in all three directions, performances of

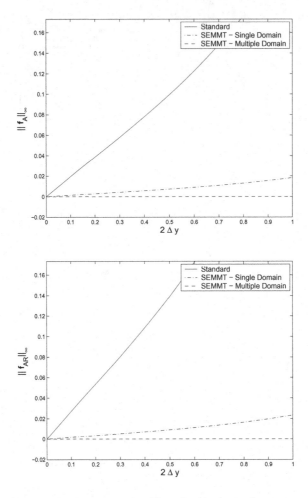

Figure 5.6 Translation test mesh quality (defined based on the inner elements) as function of translation magnitude, for the standard mesh moving technique, SEMMT–SD, and SEMMT–MD

remeshing with the reconnect and renode options were evaluated. The evaluations showed that the force oscillations seen immediately after the remeshing are reduced substantially with the reconnect option.

5.4.3 Pressure Clipping

Pressure clipping was introduced in Johnson and Tezduyar (1996) for the purpose of reducing the pressure spikes typically encountered after the mesh-to-mesh projection following a remeshing. After such a projection, the incompressibility constraint is slightly violated, but it is recovered at the next nonlinear iteration. However, at that nonlinear iteration, the pressure, as it performs its duty of enforcing the constraint, changes more than it should.

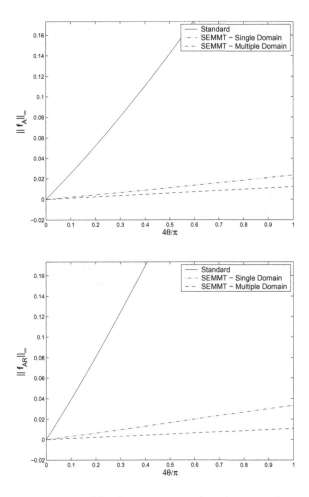

Figure 5.7 Rotation test mesh quality (defined based on the inner elements) as function of rotation magnitude, for the standard mesh moving technique, SEMMT–SD, and SEMMT–MD

Therefore, in the time step following a remeshing, in calculating the fluid mechanics forces at the fluid–solid interface, clipped values of the pressure are used. The clipped values are obtained by the least-squares projection from the pressure values prior to remeshing. It was also proposed in Tezduyar and Sathe (2007) to use those clipped values as the initial guess for the nonlinear iterations of the subsequent time step. Using the pressure clipping in conjunction with the MRRMUM should further improve the quality of the solution already improved by using, whenever we can, the reconnect option of remeshing. The pressure clipping technique was also used successfully in Takizawa and Tezduyar (2011) with the sequentially-coupled FSI technique. The purpose in that case was to build a good starting condition for the fine-mesh computation of the sequence, using data obtained from the coarse-mesh computations already carried out.

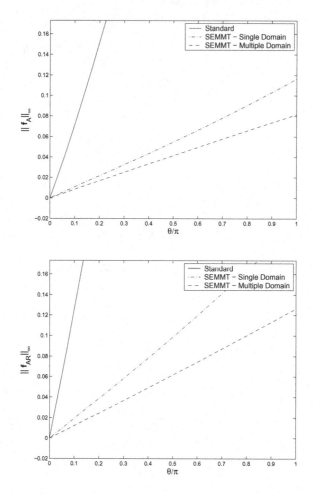

Figure 5.8 Bending test mesh quality (defined based on the inner elements) as function of bending magnitude, for the standard mesh moving technique, SEMMT–SD, and SEMMT–MD

5.5 FSI Geometric Smoothing Technique (FSI-GST)

The FSI-GST was introduced in Tezduyar and Sathe (2007) for computations where the geometric complexity of the structure would require a fluid mechanics mesh that is not affordable or not desirable or just not manageable in mesh moving. In this technique, the structural mesh and displacement rates at the interface are projected to the fluid mesh after a geometric smoothing. In the geometric smoothing, a value (mesh coordinate or displacement rate) at a node is replaced by a weighted average of the values at that node and a limited set of nearby nodes. When projecting the stress values from the smoothened interface to the structure, one of the simplest ways is to just transfer those values to the corresponding nodes of the structure. In some computations, one may need not an isotropic geometric smoothing but a directional smoothing along some preferred direction. The FSI Directional Geometric Smoothing Technique (FSI-DGST), also introduced in Tezduyar and Sathe (2007), was motivated by such computations. In the FSI-DGST, whenever possible, the interface mesh is generated in such a fashion that the preferred smoothing directions can approximately be represented by the gridlines of the interface mesh. Then the weighted averaging for a node on such a gridline would involve a limited set of nearby nodes only along that gridline. The directional smoothing concept is similar to the directional "upwind" concept of the SUPG formulation, where the residual-based numerical dissipation is active only in the streamline direction.

6

Advanced FSI and Space–Time Techniques

In this Chapter we discuss some of the advanced FSI and space–time techniques developed by the authors. We begin with a presentation of the solution strategies for the coupled FSI equations, discuss how certain features of the space–time approach and DSD/SST method can lead to advanced computational techniques, and describe a contact algorithm developed for FSI computations. These advanced FSI and space–time procedures enable accurate and efficient solution of challenging problems of contemporary engineering interest and significance.

6.1 Solution of the Fully-Discretized Coupled FSI Equations

Full discretization of the FSI formulation described in the previous sections leads to coupled, nonlinear equation systems that need to be solved at every time step. In a conceptual form that is partitioned with respect to the models represented, such nonlinear equation systems can be written as follows:

$$\mathbf{N}_1(\mathbf{d}_1, \mathbf{d}_2, \mathbf{d}_3) = \mathbf{0}, \tag{6.1}$$

$$\mathbf{N}_2(\mathbf{d}_1, \mathbf{d}_2, \mathbf{d}_3) = \mathbf{0}, \tag{6.2}$$

$$\mathbf{N}_3(\mathbf{d}_1, \mathbf{d}_2, \mathbf{d}_3) = \mathbf{0}, \tag{6.3}$$

where $\mathbf{d}_1$, $\mathbf{d}_2$, and $\mathbf{d}_3$ are the vectors of nodal unknowns corresponding to generic unknown functions $\mathbf{u}_1$, $\mathbf{u}_2$, and $\mathbf{u}_3$, respectively. In the context of an FSI problem, the generic functions $\mathbf{u}_1$, $\mathbf{u}_2$, and $\mathbf{u}_3$ represent the fluid, structure and mesh unknowns, respectively. For the space–time formulation of the fluid mechanics problem, $\mathbf{d}_1$ represents unknowns associated with the finite element formulation written for the space–time slab between the time levels n to $n + 1$ (see Tezduyar, 1992, 2003a; Tezduyar *et al.*, 1992c; Tezduyar and Sathe, 2007; Takizawa and Tezduyar, 2011, 2012b). Solution of these equations with the Newton–Raphson method would necessitate at every Newton–Raphson step solution of the following linear equation system:

$$\mathbf{A}_{11}\mathbf{x}_1 + \mathbf{A}_{12}\mathbf{x}_2 + \mathbf{A}_{13}\mathbf{x}_3 = \mathbf{b}_1, \tag{6.4}$$

$$\mathbf{A}_{21}\mathbf{x}_1 + \mathbf{A}_{22}\mathbf{x}_2 + \mathbf{A}_{23}\mathbf{x}_3 = \mathbf{b}_2, \tag{6.5}$$

$$\mathbf{A}_{31}\mathbf{x}_1 + \mathbf{A}_{32}\mathbf{x}_2 + \mathbf{A}_{33}\mathbf{x}_3 = \mathbf{b}_3, \tag{6.6}$$

Computational Fluid–Structure Interaction: Methods and Applications, First Edition.
Yuri Bazilevs, Kenji Takizawa and Tayfun E. Tezduyar.
© 2013 John Wiley & Sons, Ltd. Published 2013 by John Wiley & Sons, Ltd.

where $\mathbf{b}_1 = -\mathbf{N}_1$, $\mathbf{b}_2 = -\mathbf{N}_2$, and $\mathbf{b}_3 = -\mathbf{N}_3$ are the residuals of the nonlinear equations, $\mathbf{x}_1$, $\mathbf{x}_2$, and $\mathbf{x}_3$ are the correction increments for $\mathbf{d}_1$, $\mathbf{d}_2$, and $\mathbf{d}_3$, and $\mathbf{A}_{\beta\gamma} = \partial \mathbf{N}_\beta / \partial \mathbf{d}_\gamma$ are the FSI problem left-hand-side matrices discussed in the previous chapter.

REMARK 6.1 *In FSI computations with a fluctuating traction boundary condition at the outflow, to improve the convergence of the nonlinear iterations, it was proposed in Tezduyar and Sathe (2007) to calculate the initial guess for p_{n+1} with the expression $p_{n+1}^0 = p_n + (\Delta p_{\text{OUTF}})_n$. In this expression, $(\Delta p_{\text{OUTF}})_n$ is a measure of the change in the outflow traction from time level n to $n + 1$.*

Methods for solving the fully-discretized coupled FSI equations are of two varieties: loosely- and strongly-coupled, also referred to as staggered and monolithic. The strongly-coupled solution methods can further be categorized as block-iterative, direct, and quasi-direct coupling techniques. In loosely-coupled approaches, the equations for fluid, solid, and mesh movement are solved sequentially, in an uncoupled fashion. Typically, within each time step, the increment in the fluid solution is computed on a fixed spatial domain, the fluid forces on the structure are collected, and the structural-solution increment is computed, which is followed by an update of the mesh position. This enables the use of existing fluid and structural solvers, a significant motivation for adopting this approach. However, convergence problems could be encountered, typically when the structure is light and the fluid is heavy, and when an incompressible fluid is fully enclosed by the structure. In strongly-coupled approaches, the equations for fluid, solid, and mesh deformation are solved simultaneously in a fully coupled fashion. The main advantage is that monolithic solvers are more robust. Many of the convergence problems encountered with the staggered approaches are completely avoided. The monolithic approach necessitates writing a coupled fluid–structure solver, thus precluding the use of existing fluid and structure solvers.

6.1.1 Block-Iterative Coupling

In the block-iterative coupling (Tezduyar, 2003c,d, 2004a,b, 2007a; Tezduyar *et al.*, 2004, 2006a,b; Tezduyar and Sathe, 2007), the fluid, structure, and mesh systems are treated as separate blocks, and the nonlinear iterations are carried out one block at a time. In solving a block of equations for the block of unknowns it is associated with, we use the most current values of the other blocks of unknowns. Assuming a cyclic order of $1 \to 2 \to 3$, in an iteration step taking us from iterative solution i to $i + 1$, the following three blocks of equations are solved:

$$\left. \frac{\partial \mathbf{N}_1}{\partial \mathbf{d}_1} \right|_{(\mathbf{d}_1^i, \mathbf{d}_2^i, \mathbf{d}_3^i)} \Delta \mathbf{d}_1^i = -\mathbf{N}_1 \left(\mathbf{d}_1^i, \mathbf{d}_2^i, \mathbf{d}_3^i \right), \qquad (6.7)$$

$$\mathbf{d}_1^{i+1} = \mathbf{d}_1^i + \Delta \mathbf{d}_1^i, \qquad (6.8)$$

$$\left. \frac{\partial \mathbf{N}_2}{\partial \mathbf{d}_2} \right|_{(\mathbf{d}_1^{i+1}, \mathbf{d}_2^i, \mathbf{d}_3^i)} \Delta \mathbf{d}_2^i = -\mathbf{N}_2 \left(\mathbf{d}_1^{i+1}, \mathbf{d}_2^i, \mathbf{d}_3^i \right), \qquad (6.9)$$

$$\mathbf{d}_2^{i+1} = \mathbf{d}_2^i + \Delta \mathbf{d}_2^i, \qquad (6.10)$$

$$\left. \frac{\partial \mathbf{N}_3}{\partial \mathbf{d}_3} \right|_{(\mathbf{d}_1^{i+1}, \mathbf{d}_2^{i+1}, \mathbf{d}_3^i)} \Delta \mathbf{d}_3^i = -\mathbf{N}_3 \left(\mathbf{d}_1^{i+1}, \mathbf{d}_2^{i+1}, \mathbf{d}_3^i \right), \qquad (6.11)$$

$$\mathbf{d}_3^{i+1} = \mathbf{d}_3^i + \Delta \mathbf{d}_3^i. \qquad (6.12)$$

Each of the three blocks of linear equations systems given by Equations (6.7), (6.9), and (6.11) is also solved iteratively, using the GMRES search technique (Saad and Schultz, 1986).

In FSI computations where the structure is light, structural response becomes very sensitive to small changes in the fluid mechanics forces. In such cases, when the coupling between the three blocks of equations given by Equations (6.4)–(6.6) is handled with a block-iterative coupling technique rather than a direct coupling technique, convergence becomes difficult to achieve. In Subsections 6.1.2 and 6.1.3 we describe "more direct" techniques for handling the coupling. A shortcut approach was proposed in Tezduyar (2003c,d, 2004a) (and was also described in Tezduyar, 2004b, 2007a; Tezduyar et al., 2004, 2006a,b) for improving the convergence of the block-iterative coupling technique. In this approach, to reduce "over-correcting" (i.e., "over-incrementing") the structural displacements during the block iterations, the mass matrix contribution to $\mathbf{A}_{22}$ is increased. This is achieved without altering $\mathbf{b}_1$, $\mathbf{b}_2$ or $\mathbf{b}_3$ (i.e., without altering $\mathbf{N}_1(\mathbf{d}_1, \mathbf{d}_2, \mathbf{d}_3)$, $\mathbf{N}_2(\mathbf{d}_1, \mathbf{d}_2, \mathbf{d}_3)$ or $\mathbf{N}_3(\mathbf{d}_1, \mathbf{d}_2, \mathbf{d}_3)$), and therefore when the block iterations converge, they converge to the solution of the problem with the correct structural mass.

6.1.2 Quasi-Direct Coupling

In quasi-direct coupling (Tezduyar et al., 2004, 2006a,b; Tezduyar and Sathe, 2007), the fluid+structure and mesh systems are treated as two separate blocks, and the nonlinear iterations are carried out one block at a time. In solving a block of equations for the block of unknowns it is associated with, we use the most current values of the other block of unknowns. In an iteration step taking us from iterative solution i to $i+1$, the following two blocks of equations are solved:

$$\left.\frac{\partial \mathbf{N}_1}{\partial \mathbf{d}_1}\right|_{(\mathbf{d}_1^i, \mathbf{d}_2^i, \mathbf{d}_3^i)} \Delta \mathbf{d}_1^i + \left.\frac{\partial \mathbf{N}_1}{\partial \mathbf{d}_2}\right|_{(\mathbf{d}_1^i, \mathbf{d}_2^i, \mathbf{d}_3^i)} \Delta \mathbf{d}_2^i = -\mathbf{N}_1\left(\mathbf{d}_1^i, \mathbf{d}_2^i, \mathbf{d}_3^i\right), \qquad (6.13)$$

$$\left.\frac{\partial \mathbf{N}_2}{\partial \mathbf{d}_1}\right|_{(\mathbf{d}_1^i, \mathbf{d}_2^i, \mathbf{d}_3^i)} \Delta \mathbf{d}_1^i + \left.\frac{\partial \mathbf{N}_2}{\partial \mathbf{d}_2}\right|_{(\mathbf{d}_1^i, \mathbf{d}_2^i, \mathbf{d}_3^i)} \Delta \mathbf{d}_2^i = -\mathbf{N}_2\left(\mathbf{d}_1^i, \mathbf{d}_2^i, \mathbf{d}_3^i\right), \qquad (6.14)$$

$$\mathbf{d}_1^{i+1} = \mathbf{d}_1^i + \Delta \mathbf{d}_1^i, \qquad (6.15)$$

$$\mathbf{d}_2^{i+1} = \mathbf{d}_2^i + \Delta \mathbf{d}_2^i, \qquad (6.16)$$

$$\left.\frac{\partial \mathbf{N}_3}{\partial \mathbf{d}_3}\right|_{(\mathbf{d}_1^{i+1}, \mathbf{d}_2^{i+1}, \mathbf{d}_3^i)} \Delta \mathbf{d}_3^i = -\mathbf{N}_3\left(\mathbf{d}_1^{i+1}, \mathbf{d}_2^{i+1}, \mathbf{d}_3^i\right), \qquad (6.17)$$

$$\mathbf{d}_3^{i+1} = \mathbf{d}_3^i + \Delta \mathbf{d}_3^i. \qquad (6.18)$$

Each of the two blocks of linear equations systems given by Equations (6.13), (6.14) and (6.17) is also solved iteratively, using the GMRES search technique.

REMARK 6.2 *In the iterative solution of the combined fluid+structure (i.e., 1+2) block with the GMRES search technique and a diagonal preconditioner, depending on the nature of the problem, one of these two parts might pose a greater convergence challenge than the other one. The convergence challenges might be created by an incompressibility constraint, having thin or shallow computational domains, or some other factor. The scaling provided by diagonal preconditioning is unlikely to remedy such disparities in the convergence challenges* offered by the two parts, which are typically exhibited as disparities in the residual-decay rates for the two parts rather than disparities in the residual magnitudes. In some cases, the scaling provided by diagonal preconditioning might not even be able to properly account for the disparities in the residual magnitudes corresponding to the fluid and structure parts. "Selective Scaling" was proposed in Tezduyar and Sathe (2007) to place, in GMRES iterations, greater emphasis on the part posing greater convergence challenge. With this additional scaling (beyond diagonal preconditioning), in constructing the Krylov vectors of the GMRES search technique, the relative weights given to the residual vectors associated with the fluid and structure parts are determined based on the relative convergence challenges posed by those two parts. It was proposed in Tezduyar and Sathe (2007) to determine those relative weights on a case-by-case basis as well as on a more automated basis, where the weights increase with decreasing residual-decay rates.

REMARK 6.3 *In Takizawa et al. (2011c) Selective Scaling was extended to also shifting the emphasis between the parts of the fluid mechanics equations corresponding to the momentum conservation and incompressibility constraint.*

6.1.3 Direct Coupling

In direct coupling (Tezduyar *et al.*, 2004, 2006a,b; Tezduyar and Sathe, 2007), the fluid+structure+mesh system is treated as a single block, and the linear equation system given by Equations (6.4)–(6.6) is solved iteratively:

$$\mathbf{P}_{11} \mathbf{z}_1 + \mathbf{P}_{12} \mathbf{z}_2 + \mathbf{P}_{13} \mathbf{z}_3 = \mathbf{b}_1 - (\mathbf{A}_{11}\mathbf{x}_1 + \mathbf{A}_{12}\mathbf{x}_2 + \mathbf{A}_{13}\mathbf{x}_3), \quad (6.19)$$

$$\mathbf{P}_{21} \mathbf{z}_1 + \mathbf{P}_{22} \mathbf{z}_2 + \mathbf{P}_{23} \mathbf{z}_3 = \mathbf{b}_2 - (\mathbf{A}_{21}\mathbf{x}_1 + \mathbf{A}_{22}\mathbf{x}_2 + \mathbf{A}_{23}\mathbf{x}_3), \quad (6.20)$$

$$\mathbf{P}_{31} \mathbf{z}_1 + \mathbf{P}_{32} \mathbf{z}_2 + \mathbf{P}_{33} \mathbf{z}_3 = \mathbf{b}_3 - (\mathbf{A}_{31}\mathbf{x}_1 + \mathbf{A}_{32}\mathbf{x}_2 + \mathbf{A}_{33}\mathbf{x}_3), \quad (6.21)$$

where $\mathbf{P}_{\beta\gamma}$'s represent the blocks of the preconditioning matrix $\mathbf{P}$. We note that the quasi-direct and block-iterative coupling techniques also involve solution of such linear systems. The linear system is separated into a 2×2 and a 1×1 block in quasi-direct coupling and three 1×1 blocks in block-iterative coupling. The most computing-intensive part of Equations (6.19)–(6.21) is the evaluation of the matrix–vector products of the form $\mathbf{A}_{\beta\gamma}\mathbf{x}_\gamma$ (for $\beta, \gamma = 1, 2, \ldots, N$ and no sum). In the FSI computations carried out by the authors and their research groups, those evaluations are performed with the sparse-matrix-based computation technique (see Kalro and Tezduyar, 1998, 2000) as well as element-vector-based (EVB) computation techniques (see Johan *et al.*, 1991, 1995; Tezduyar, 2001b, 2004a, 2007a; Tezduyar and Sathe, 2007). The EVB techniques do not require computation of any matrices, not even at the element level. The EVB computations can be carried out in two ways: Numerical EVB (NEVB) and analytical EVB (AEVB) computations.

6.1.3.1 NEVB Computations

In the NEVB computation technique, which is also called the matrix-free computation technique (see Johan *et al.*, 1991, 1995), a matrix-vector product of the form $\mathbf{Ax}$, which is the directional derivative of $\mathbf{N}$ in $\mathbf{x}$ direction, is evaluated by the expression:

$$\mathbf{Ax} = \mathop{\mathbf{A}}_{e=1}^{n_{\text{el}}} \left[\frac{\mathbf{N}^e(\mathbf{d} + \epsilon \mathbf{x}) - \mathbf{N}^e(\mathbf{d})}{\epsilon} \right], \quad (6.22)$$

where $\mathbf{N}^e$ is the element-level vector representing the contribution of element e to $\mathbf{N}$, and ϵ is a small parameter used in the numerical calculation of the limit representing the directional derivative. This concept was extended in Tezduyar (2001b, 2004a, 2007a) and Tezduyar and Sathe (2007) to FSI computations, where we need to evaluate the matrix–vector products of the form $\mathbf{A}_{\beta\gamma}\mathbf{x}_\gamma$:

$$\mathbf{A}_{\beta\gamma}\mathbf{x}_\gamma = \underset{e=1}{\overset{n_{\text{el}}}{\mathbf{A}}} \left[\frac{\mathbf{N}^e_\beta(\ldots, \mathbf{d}_\gamma + \epsilon_{\beta\gamma}\mathbf{x}_\gamma, \ldots) - \mathbf{N}^e_\beta(\ldots, \mathbf{d}_\gamma, \ldots)}{\epsilon_{\beta\gamma}} \right], \tag{6.23}$$

where $\mathbf{N}^e_\beta$ is the element-level vector representing the contribution of element e to $\mathbf{N}_\beta$, and $\epsilon_{\beta\gamma}$ is the limit-evaluation parameter selected for the unknown set γ in the equation set β. If we decide to use a single limit-evaluation parameter ϵ_β for all the unknown sets in the equation set β, then the computations can be carried out as

$$\sum_{\gamma=1}^{N} \mathbf{A}_{\beta\gamma}\mathbf{x}_\gamma = \underset{e=1}{\overset{n_{\text{el}}}{\mathbf{A}}} \left[\frac{\mathbf{N}^e_\beta(\mathbf{d} + \epsilon_\beta\mathbf{x}) - \mathbf{N}^e_\beta(\mathbf{d})}{\epsilon_\beta} \right]. \tag{6.24}$$

REMARK 6.4 *Using a single limit-evaluation parameter for all the unknown sets in the equation set β would be computationally more economical. On the other hand, using a different limit-evaluation parameter for each unknown set would give us the option of taking separately into account the dependence of $\mathbf{N}_\beta$ on each unknown $\mathbf{d}_\gamma$, including how $\frac{\partial \mathbf{N}_\beta}{\partial \mathbf{d}_\gamma}$ varies with $\mathbf{d}_\gamma$. This is an important consideration because of the multi-physics and multiscale nature of FSI computations.*

6.1.3.2 AEVB Computations

The AEVB computation technique (see Tezduyar, 2001b, 2004a, 2007a; Tezduyar and Sathe, 2007) can be used for evaluating the matrix-vector products of the form $\mathbf{A}_{\beta\gamma}\mathbf{x}_\gamma$ if deriving expressions for such matrix-vector products is not more involved than we would like and we prefer not to deal with limit-evaluation parameters and numerical evaluation of directional derivatives.

Let us suppose that the nonlinear vector function $\mathbf{N}_\beta$ corresponds to a finite element integral form $\mathbf{B}_\beta(\mathbf{W}_\beta, \mathbf{u}_1, \ldots, \mathbf{u}_N)$. Here $\mathbf{W}_\beta$ represents the vector of nodal values associated with the weighting function $\mathbf{w}_\beta$, which generates the nonlinear equation block β. Let us also suppose that we are able to, without major difficulty, derive the expressions for the first-order terms in the expansion of $\mathbf{B}_\beta(\mathbf{W}_\beta, \mathbf{u}_1, \ldots, \mathbf{u}_N)$ in $\mathbf{u}_\gamma$. Those first-order terms in $\Delta\mathbf{u}_\gamma$ will be represented by the finite element integral form $\mathbf{G}_{\beta\gamma}(\mathbf{W}_\beta, \mathbf{u}_1, \ldots, \mathbf{u}_N, \Delta\mathbf{u}_\gamma)$. For example, $\mathbf{G}_{11}(\mathbf{W}_1, \mathbf{u}_1, \ldots, \mathbf{u}_N, \Delta\mathbf{u}_1)$ will represent the first-order terms obtained by expanding the finite element formulation of the fluid mechanics equations (i.e., momentum equation and incompressibility constraint) in fluid mechanics unknowns (i.e., fluid velocity and pressure). We note that the integral form $\mathbf{G}_{\beta\gamma}$ will generate $\frac{\partial \mathbf{N}_\beta}{\partial \mathbf{d}_\gamma}$. Consequently, as it was pointed out in Tezduyar (2001b, 2004a, 2007a) and Tezduyar and Sathe (2007), the product $\mathbf{A}_{\beta\gamma}\mathbf{x}_\gamma$ can be evaluated as follows:

$$\mathbf{A}_{\beta\gamma}\mathbf{x}_\gamma = \frac{\partial \mathbf{N}_\beta}{\partial \mathbf{d}_\gamma}\mathbf{x}_\gamma = \underset{e=1}{\overset{n_{\text{el}}}{\mathbf{A}}} \mathbf{G}_{\beta\gamma}(\mathbf{W}_\beta, \mathbf{u}_1, \ldots, \mathbf{u}_N, \mathbf{v}_\gamma), \tag{6.25}$$

where $\mathbf{v}_\gamma$ is a function interpolated from $\mathbf{x}_\gamma$ in the same way $\mathbf{u}_\gamma$ is interpolated from $\mathbf{d}_\gamma$.

In the mixed AEVB/NEVB computation technique (Tezduyar, 2001b, 2004a, 2007a; Tezduyar and Sathe, 2007), in evaluation of $\mathbf{A}_{\beta\gamma}\mathbf{x}_\gamma$ for each combination of β and γ, depending on the nature of what is involved in that particular evaluation, one can select between the AEVB and NEVB computation techniques.

In FSI computations with the direct coupling, the evaluation of the $\mathbf{A}_{13}\mathbf{x}_3$ matrix-vector product (i.e., the action of the shape derivatives on the vector $\mathbf{x}_3$) with the sparse-matrix-based or AEVB techniques is rather involved. In the direct coupling (Tezduyar et al., 2004, 2006a,b; Tezduyar, 2007a; Tezduyar and Sathe, 2007), the matrix-vector product $\mathbf{A}_{13}\mathbf{x}_3$ is computed with the NEVB technique:

$$\mathbf{A}_{13}\mathbf{x}_3 = \underset{e=1}{\overset{n_{el}}{\mathbf{A}}} \left[\frac{\mathbf{N}_1^e(\mathbf{d}_1, \mathbf{d}_2, \mathbf{d}_3 + \epsilon_{13}\mathbf{x}_3) - \mathbf{N}_1^e(\mathbf{d}_1, \mathbf{d}_2, \mathbf{d}_3)}{\epsilon_{13}} \right]. \tag{6.26}$$

REMARK 6.5 *Remark 6.2, with the wording expanded to the combined fluid+structure+ mesh (i.e., 1+2+3) system, becomes applicable to the direct coupling.*

6.2 Segregated Equation Solvers and Preconditioners

To simplify the context for the fundamental concepts of this section, let us first consider only the Navier–Stokes equations of incompressible flows, without any structure. The stabilized formulation of the fluid mechanics problem leads to a nonlinear equation system that needs to be solved at every time step. In a form that is partitioned (segregated) with respect to velocity and pressure, that nonlinear equation system can be written as follows:

$$\mathbf{N}_U(\mathbf{d}_U, \mathbf{d}_P) = \mathbf{0}, \tag{6.27}$$

$$\mathbf{N}_P(\mathbf{d}_U, \mathbf{d}_P) = \mathbf{0}, \tag{6.28}$$

where $\mathbf{d}_U$ and $\mathbf{d}_P$ are the vectors of nodal unknowns corresponding to velocity and pressure, respectively. Solution of this nonlinear equation system with the Newton–Raphson method would necessitate at every Newton–Raphson step solution of the following linear equation system:

$$\mathbf{A}_{UU}\mathbf{x}_U + \mathbf{A}_{UP}\mathbf{x}_P = \mathbf{b}_U, \tag{6.29}$$

$$\mathbf{A}_{PU}\mathbf{x}_U + \mathbf{A}_{PP}\mathbf{x}_P = \mathbf{b}_P, \tag{6.30}$$

where $\mathbf{b}_U$ and $\mathbf{b}_P$ are the residuals of the nonlinear equations, $\mathbf{x}_U$ and $\mathbf{x}_P$ are the correction increments for $\mathbf{d}_U$ and $\mathbf{d}_P$, and $\mathbf{A}_{\beta\gamma} = \partial \mathbf{N}_\beta / \partial \mathbf{d}_\gamma$, with $\beta, \gamma = U, P$.

6.2.1 Segregated Equation Solver for Nonlinear Systems (SESNS)

The SESNS technique originates from the segregated solvers reported in Brooks and Hughes (1982), Tezduyar et al. (1990, 1991, 1992d) and Tezduyar (1992). In Brooks and Hughes (1982), a segregated solver was first used with the SUPG formulation based on elements with bilinear velocity and constant pressure. The overly dissipative nature of the one-step SUPG formulation with constant pressure motivated the introduction of the multi-step SUPG formulations reported in Tezduyar et al. (1990), where the segregated solution approach was extended to multi-step methods. In Tezduyar et al. (1991), segregated solvers were provided for the SUPG formulation based on elements with higher-order interpolations for velocity and pressure, in the context of both one-step and multi-step formulations. In Tezduyar (1992) and Tezduyar et al. (1992d), a segregated solver was provided for the SUPG/PSPG formulation based on elements with equal-order interpolations for velocity and pressure, where, because of the PSPG stabilization, the submatrix $\mathbf{A}_{PP}$ was no longer zero.

In SESNS, instead of solving the equation system given by Equations (6.29) and (6.30) in its given form, we solve its approximate version where $\mathbf{A}_{UU}$ is approximated by a diagonal matrix $\mathbf{D}_{UU}$:

$$\mathbf{D}_{UU}\mathbf{x}_U + \mathbf{A}_{UP}\mathbf{x}_P = \mathbf{b}_U, \tag{6.31}$$

$$\mathbf{A}_{PU}\mathbf{x}_U + \mathbf{A}_{PP}\mathbf{x}_P = \mathbf{b}_P, \tag{6.32}$$

where $\mathbf{D}_{UU} = \mathrm{DIAG}(\mathbf{A}_{UU})$. From Equations (6.31) and (6.32), we obtain the following sets of equations:

$$\mathbf{x}_U + \mathbf{D}_{UU}^{-1}\mathbf{A}_{UP}\mathbf{x}_P = \mathbf{D}_{UU}^{-1}\mathbf{b}_U, \tag{6.33}$$

$$\left(\mathbf{A}_{PU}\mathbf{D}_{UU}^{-1}\mathbf{A}_{UP} - \mathbf{A}_{PP}\right)\mathbf{x}_P = \mathbf{A}_{PU}\mathbf{D}_{UU}^{-1}\mathbf{b}_U - \mathbf{b}_P. \tag{6.34}$$

Equation (6.34) can be solved iteratively with the GMRES method. After solving Equation (6.34) for $\mathbf{x}_P$, we substitute that solution into Equation (6.33) and compute $\mathbf{x}_U$. Doing this twice would be similar to using a predictor–multicorrector algorithm with two passes. With two passes, we can get second-order accuracy in time, but, because of stability considerations, we still would have a limit on the time-step size.

6.2.2 Segregated Equation Solver for Linear Systems (SESLS)

In SESLS, we do not replace the equation system given by Equations (6.29) and (6.30) with its approximate version. We solve it with preconditioned iterations. By using the concepts we used in Equations (6.19)–(6.21) and a comparable notation, we write the iterative solution of Equations (6.29) and (6.30) as follows:

$$\mathbf{P}_{UU}\mathbf{z}_U + \mathbf{P}_{UP}\mathbf{z}_P = \mathbf{b}_U - (\mathbf{A}_{UU}\mathbf{x}_U + \mathbf{A}_{UP}\mathbf{x}_P), \tag{6.35}$$

$$\mathbf{P}_{PU}\mathbf{z}_U + \mathbf{P}_{PP}\mathbf{z}_P = \mathbf{b}_P - (\mathbf{A}_{PU}\mathbf{x}_U + \mathbf{A}_{PP}\mathbf{x}_P). \tag{6.36}$$

We define:

$$\mathbf{r}_U = \mathbf{b}_U - (\mathbf{A}_{UU}\mathbf{x}_U + \mathbf{A}_{UP}\mathbf{x}_P), \tag{6.37}$$

$$\mathbf{r}_P = \mathbf{b}_P - (\mathbf{A}_{PU}\mathbf{x}_U + \mathbf{A}_{PP}\mathbf{x}_P), \tag{6.38}$$

select the preconditioning matrix blocks as

$$\mathbf{P}_{UU} = \mathbf{D}_{UU}, \qquad \mathbf{P}_{UP} = \mathbf{A}_{UP}, \tag{6.39}$$

$$\mathbf{P}_{PU} = \mathbf{A}_{PU}, \qquad \mathbf{P}_{PP} = \mathbf{A}_{PP}, \tag{6.40}$$

and rewrite Equations (6.35) and (6.36) as follows:

$$\mathbf{D}_{UU}\mathbf{z}_U + \mathbf{A}_{UP}\mathbf{z}_P = \mathbf{r}_U, \qquad (6.41)$$

$$\mathbf{A}_{PU}\mathbf{z}_U + \mathbf{A}_{PP}\mathbf{z}_P = \mathbf{r}_P. \qquad (6.42)$$

Now, we do to Equations (6.41) and (6.42) exactly what we did to Equations (6.31) and (6.32) and obtain:

$$\mathbf{z}_U + \mathbf{D}_{UU}^{-1}\mathbf{A}_{UP}\mathbf{z}_P = \mathbf{D}_{UU}^{-1}\mathbf{r}_U, \qquad (6.43)$$

$$\left(\mathbf{A}_{PU}\mathbf{D}_{UU}^{-1}\mathbf{A}_{UP} - \mathbf{A}_{PP}\right)\mathbf{z}_P = \mathbf{A}_{PU}\mathbf{D}_{UU}^{-1}\mathbf{r}_U - \mathbf{r}_P. \qquad (6.44)$$

Equation (6.44) can be solved iteratively with the GMRES method. We will call these iterations "sublevel" (or "lower") iterations. After solving Equation (6.44) for $\mathbf{z}_P$, we substitute that solution into Equation (6.43) and compute $\mathbf{z}_U$. This completes the solution of Equations (6.41) and (6.42), which is equivalent to applying the preconditioner defined by Equations (6.39) and (6.40) in the iterative solution of Equations (6.29) and (6.30).

The SESLS is simply an extension of the SESNS concept to the linear equation systems that need to be solved at every Newton–Raphson step. For more sophisticated iteration techniques with sublevels (see Sameh and Sarin, 1999, 2002; Manguoglu *et al.*, 2008, 2009, 2010, 2011a,b).

6.2.3 Segregated Equation Solver for Fluid–Structure Interactions (SESFSI)

The SESFSI is based on extending the SESLS concept to FSI computations. In describing this solver, for the identification of the equation and unknown blocks, we will use an expanded notation that will remain local to this subsection. The vectors **x**, **b**, **z**, and **r** are defined as

$$\mathbf{x} = \begin{pmatrix} \mathbf{x}_E \\ \mathbf{x}_I \\ \mathbf{x}_Y \\ \mathbf{x}_S \\ \mathbf{x}_F \\ \mathbf{x}_H \\ \mathbf{x}_P \end{pmatrix}, \quad \mathbf{b} = \begin{pmatrix} \mathbf{b}_E \\ \mathbf{b}_I \\ \mathbf{b}_Y \\ \mathbf{b}_S \\ \mathbf{b}_F \\ \mathbf{b}_H \\ \mathbf{b}_P \end{pmatrix}, \quad \mathbf{z} = \begin{pmatrix} \mathbf{z}_E \\ \mathbf{z}_I \\ \mathbf{z}_Y \\ \mathbf{z}_S \\ \mathbf{z}_F \\ \mathbf{z}_H \\ \mathbf{z}_P \end{pmatrix}, \quad \mathbf{r} = \begin{pmatrix} \mathbf{r}_E \\ \mathbf{r}_I \\ \mathbf{r}_Y \\ \mathbf{r}_S \\ \mathbf{r}_F \\ \mathbf{r}_H \\ \mathbf{r}_P \end{pmatrix}, \qquad (6.45)$$

where the index translation is given in Table 6.1.

Table 6.1 SESFSI index translation

E	all other fluid velocities
I	fluid velocity $(\mathbf{u}^h)_{n+1}^-$ at the interface ($\mathbf{u}_{n+1}^h$ for ALE)
Y	structural displacements at the interface
S	all other structural displacements
F	interface stresses acting on the structure
H	interface stresses acting on the fluid
P	fluid pressure

The matrix $\mathbf{A}$ is expressed as follows:

$$\mathbf{A} = \begin{pmatrix} \mathbf{A}_{EE} & \mathbf{A}_{EI} & & & & & & \mathbf{A}_{EP} \\ & \mathbf{A}_{II} & \mathbf{A}_{IY} & & & & & \mathbf{A}_{IP} \\ & & \mathbf{A}_{YY} & \mathbf{A}_{YS} & \mathbf{A}_{YF} & & & \\ & & \mathbf{A}_{SY} & \mathbf{A}_{SS} & & & & \\ & & & & \mathbf{A}_{FF} & \mathbf{A}_{FH} & & \\ \mathbf{A}_{HE} & \mathbf{A}_{HI} & & & & \mathbf{A}_{HH} & \mathbf{A}_{HP} \\ \mathbf{A}_{PE} & \mathbf{A}_{PI} & & & & & \mathbf{A}_{PP} \end{pmatrix}. \quad (6.46)$$

We note that the matrix $\mathbf{A}_{IP}$ is generated by the porosity term, when in Equation (5.115) we take into account only the pressure component of $\mathbf{h}_{II}^h$. If we take into account $\mathbf{h}_{II}^h$ fully, then the coupling matrix generated would be $\mathbf{A}_{IH}$ instead of $\mathbf{A}_{IP}$.

Now, similar to what we did by means of Equations (6.39) and (6.40), we define various preconditioner options for the SESFSI. The first one is a very simple preconditioner, and we will call it $\mathbf{P}_{SIMP}$:

$$\mathbf{P}_{SIMP} = \begin{pmatrix} \mathbf{D}_{EE} & \mathbf{0} & & & & & & \mathbf{A}_{EP} \\ & \mathbf{L}_{II} & \mathbf{0} & & & & & \mathbf{0} \\ & & \mathbf{D}_{YY} & \mathbf{0} & \mathbf{0} & & & \\ & & \mathbf{0} & \mathbf{D}_{SS} & & & & \\ & & & & \mathbf{L}_{FF} & \mathbf{0} & & \\ \mathbf{0} & \mathbf{0} & & & & \mathbf{L}_{HH} & \mathbf{0} \\ \mathbf{A}_{PE} & \mathbf{A}_{PI} & & & & & \mathbf{A}_{PP} \end{pmatrix}, \quad (6.47)$$

where $\mathbf{D}_{EE} = \mathrm{DIAG}(\mathbf{A}_{EE})$, $\mathbf{L}_{II} = \mathrm{LUMP}(\mathbf{A}_{II})$, $\mathbf{D}_{YY} = \mathrm{DIAG}(\mathbf{A}_{YY})$, $\mathbf{D}_{SS} = \mathrm{DIAG}(\mathbf{A}_{SS})$, $\mathbf{L}_{FF} = \mathrm{LUMP}(\mathbf{A}_{FF})$, $\mathbf{L}_{HH} = \mathrm{LUMP}(\mathbf{A}_{HH})$, and the operator "LUMP" represents the matrix-lumping operation. We note that the replacement of $\mathbf{A}_{IP}$ by $\mathbf{0}$ constitutes an approximation only if the structure has porosity.

The second preconditioner is an augmented version, in the sense that, compared to $\mathbf{P}_{SIMP}$, it includes more of the coupling matrices, and we will call it $\mathbf{P}_{AUGM}$:

$$\mathbf{P}_{AUGM} = \begin{pmatrix} \mathbf{D}_{EE} & \mathbf{0} & & & & & & \mathbf{A}_{EP} \\ & \mathbf{L}_{II} & \mathbf{0} & & & & & \mathbf{A}_{IP} \\ & & \mathbf{D}_{YY} & \mathbf{0} & \mathbf{A}_{YF} & & & \\ & & \mathbf{0} & \mathbf{D}_{SS} & & & & \\ & & & & \mathbf{L}_{FF} & \mathbf{0} & & \\ \mathbf{A}_{HE} & \mathbf{A}_{HI} & & & & \mathbf{L}_{HH} & \mathbf{0} \\ \mathbf{A}_{PE} & \mathbf{A}_{PI} & & & & & \mathbf{A}_{PP} \end{pmatrix}. \quad (6.48)$$

The third preconditioner, compared to $\mathbf{P}_{SIMP}$, includes the coupling matrices that lead to a more direct treatment of the projections, and we will call it $\mathbf{P}_{DPRO}$:

$$\mathbf{P}_{DPRO} = \begin{pmatrix} \mathbf{D}_{EE} & \mathbf{0} & & & & & & \mathbf{A}_{EP} \\ & \mathbf{L}_{II} & \mathbf{B}_{IY} & & & & & \mathbf{0} \\ & & \mathbf{D}_{YY} & \mathbf{0} & \mathbf{0} & & & \\ & & \mathbf{0} & \mathbf{D}_{SS} & & & & \\ & & & & \mathbf{L}_{FF} & \mathbf{B}_{FH} & & \\ \mathbf{0} & \mathbf{0} & & & & \mathbf{L}_{HH} & \mathbf{B}_{HP} \\ \mathbf{A}_{PE} & \mathbf{A}_{PI} & & & & & \mathbf{A}_{PP} \end{pmatrix}. \quad (6.49)$$

To define $\mathbf{B}_{\text{IY}}$ and $\mathbf{B}_{\text{FH}}$, we first consider the number of nodal points in blocks E, I, Y, S, F, H, and P, and denote them by $(n_n)_\text{E}$, $(n_n)_\text{I}$, $(n_n)_\text{Y}$, $(n_n)_\text{S}$, $(n_n)_\text{F}$, $(n_n)_\text{H}$, and $(n_n)_\text{P}$, respectively.

If $(n_n)_\text{I} < (n_n)_\text{Y}$, for each node in Block I, we select the nearest node in Block Y, with the condition that a node in Block Y can be selected only once as the nearest node. We use the index YI to denote the part of Block Y associated with that set of nodes. We use the index YO to denote the part of Block Y associated with the other (i.e., remaining) nodes in Block Y. Based on this, we partition $\mathbf{D}_{\text{YY}}$ into $\mathbf{D}_{\text{YIYI}}$ and $\mathbf{D}_{\text{YOYO}}$. With that, we define $\mathbf{B}_{\text{IY}}$ as follows:

$$\begin{pmatrix} \mathbf{L}_{\text{II}} & \mathbf{B}_{\text{IY}} \\ & \mathbf{D}_{\text{YY}} \end{pmatrix} = \begin{pmatrix} \mathbf{L}_{\text{II}} & -\mathbf{L}_{\text{II}} & 0 \\ & \mathbf{D}_{\text{YIYI}} & \\ & & \mathbf{D}_{\text{YOYO}} \end{pmatrix} \quad \text{for } (n_n)_\text{I} < (n_n)_\text{Y}. \tag{6.50}$$

If $(n_n)_\text{I} > (n_n)_\text{Y}$, for each node in Block Y, we select the nearest node in Block I. Index IY denotes the part of Block I associated with that set of nodes, and index IO denotes the part of Block I associated with the other nodes in Block I. Based on this, we partition $\mathbf{I}_{\text{II}}$ into $\mathbf{I}_{\text{IYIY}}$ and $\mathbf{L}_{\text{IOIO}}$. With that, we define $\mathbf{B}_{\text{IY}}$ as follows:

$$\begin{pmatrix} \mathbf{L}_{\text{II}} & \mathbf{B}_{\text{IY}} \\ & \mathbf{D}_{\text{YY}} \end{pmatrix} = \begin{pmatrix} \mathbf{L}_{\text{IOIO}} & & 0 \\ & \mathbf{L}_{\text{IYIY}} & -\mathbf{L}_{\text{IYIY}} \\ & & \mathbf{D}_{\text{YY}} \end{pmatrix} \quad \text{for } (n_n)_\text{I} > (n_n)_\text{Y}. \tag{6.51}$$

Using the nearest-node concept, if $(n_n)_\text{F} < (n_n)_\text{H}$, we partition Block H into blocks HF and HO, partition $\mathbf{L}_{\text{HH}}$ into $\mathbf{L}_{\text{HFHF}}$ and $\mathbf{L}_{\text{HOHO}}$, and define $\mathbf{B}_{\text{FH}}$ as follows:

$$\begin{pmatrix} \mathbf{L}_{\text{FF}} & \mathbf{B}_{\text{FH}} \\ & \mathbf{L}_{\text{HH}} \end{pmatrix} = \begin{pmatrix} \mathbf{L}_{\text{FF}} & -\mathbf{L}_{\text{FF}} & 0 \\ & \mathbf{L}_{\text{HFHF}} & \\ & & \mathbf{L}_{\text{HOHO}} \end{pmatrix} \quad \text{for } (n_n)_\text{F} < (n_n)_\text{H}. \tag{6.52}$$

If $(n_n)_\text{F} > (n_n)_\text{H}$, we partition Block F into blocks FH and FO, partition $\mathbf{L}_{\text{FF}}$ into $\mathbf{L}_{\text{FHFH}}$ and $\mathbf{L}_{\text{FOFO}}$, and define $\mathbf{B}_{\text{FH}}$ as follows:

$$\begin{pmatrix} \mathbf{L}_{\text{FF}} & \mathbf{B}_{\text{FH}} \\ & \mathbf{L}_{\text{HH}} \end{pmatrix} = \begin{pmatrix} \mathbf{L}_{\text{FOFO}} & & 0 \\ & \mathbf{L}_{\text{FHFH}} & -\mathbf{L}_{\text{FHFH}} \\ & & \mathbf{L}_{\text{HH}} \end{pmatrix} \quad \text{for } (n_n)_\text{F} > (n_n)_\text{H}. \tag{6.53}$$

To define $\mathbf{B}_{\text{HP}}$, we first partition Block H into blocks H1, H2, and H3 (corresponding to the three spatial directions) and partition $\mathbf{L}_{\text{HH}}$ into $\mathbf{L}_{\text{H1H1}}$, $\mathbf{L}_{\text{H2H2}}$ and $\mathbf{L}_{\text{H3H3}}$. We note that $\mathbf{L}_{\text{H2H2}} = \mathbf{L}_{\text{H1H1}}$ and $\mathbf{L}_{\text{H3H3}} = \mathbf{L}_{\text{H1H1}}$. We also partition Block P into blocks PH and PD (corresponding to the nodal values of the pressure at the interface and elsewhere in the fluid domain) and partition $\mathbf{A}_{\text{PP}}$ into $\mathbf{A}_{\text{PHPH}}$, $\mathbf{A}_{\text{PHPD}}$, $\mathbf{A}_{\text{PDPH}}$, and $\mathbf{A}_{\text{PDPD}}$. In addition, we define three diagonal matrices $\mathbf{D}^1$, $\mathbf{D}^2$, and $\mathbf{D}^3$, representing the three spatial components of the unit normal vectors at the interface nodes. These diagonal matrices are defined as

$$\mathbf{D}^j = \left[(\mathbf{n}_A)^j \delta_{AB} \right] \quad \text{(no sum)} \quad j = 1, 2, 3, \tag{6.54}$$

where δ_{AB} are the components of the identity tensor, and $(\mathbf{n}_A)^j$ is the jth component of the unit normal vector at the interface node A. With that, we define $\mathbf{B}_{HP}$ as follows:

$$\begin{pmatrix} \mathbf{L}_{HH} & \mathbf{B}_{HP} \\ & \mathbf{A}_{PP} \end{pmatrix} = \begin{pmatrix} \mathbf{L}_{H1H1} & & & -\mathbf{L}_{H1H1}\,\mathbf{D}^1 & \\ & \mathbf{L}_{H2H2} & & -\mathbf{L}_{H2H2}\,\mathbf{D}^2 & \\ & & \mathbf{L}_{H3H3} & -\mathbf{L}_{H3H3}\,\mathbf{D}^3 & \\ & & & \mathbf{A}_{PHPH} & \mathbf{A}_{PHPD} \\ & & & \mathbf{A}_{PDPH} & \mathbf{A}_{PDPD} \end{pmatrix}. \tag{6.55}$$

6.3 New-Generation Space–Time Formulations

In this and next four sections, we further capitalize on some of the desirable features of the DSD/SST method. As pointed out in Takizawa and Tezduyar (2011), different components (i.e., unknowns), and the corresponding test functions, can be discretized with different sets of temporal basis functions. This was shown in Takizawa and Tezduyar (2011) by introducing a secondary mapping $\Theta_\zeta(\theta) \in [-1, 1]$, where $\Theta_\zeta(\theta)$ is a strictly-increasing function, and rewriting the generalized space–time basis function for the element indices (a, α) as

$$(N_a^\alpha)_\zeta = T^\alpha \left(\Theta_\zeta(\theta) \right) N_a(\boldsymbol{\xi}). \tag{6.56}$$

Here ζ indicates the component, which can also be "t." We can also use different functions T^α for different components, which is for example what we have in DSD/SST-SP formulation, and this could be in combination with the secondary mapping. Prescribed and unknown variables can be represented over different space–time slabs, because we only need to supply the prescribed values at the integration points of the space–time slab used in representing the unknown variables. Most of the time higher-order basis functions can represent complex functions with fewer control points. This is very helpful in decreasing the I/O intensity, such as in computations with the multiscale Sequentially-Coupled FSI techniques (see Tezduyar et al., 2009a,b, 2010a; Takizawa and Tezduyar, 2011 and Section 8.7).

The earlier set of new-generation space–time formulations (SP, TIP1, and SV) were introduced essentially to reduce the number of equations solved simultaneously (see Remark 4.12). Consider the linear temporal basis functions depicted in Figure 6.1. The SP option reduces the pressure to a single unknown by using a piecewise constant basis function ($n_{ent} = 1$). The SV option reduces the velocity to a single unknown vector by using $n_{ent} = 2$ with the

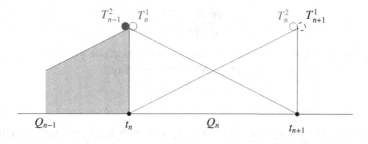

Figure 6.1 Linear temporal basis functions

"frozen" option. Here, the "frozen" option means removing the test function corresponding to $\alpha = 1$ and setting $\phi_n^1 = \phi_{n-1}^2$ (see Figure 6.2). The frozen option uses only a portion of the function space. As shown by the stability and accuracy analysis for the DSD/SST formulation of the advection equation, which was reported in Takizawa et al. (2011f) briefly and was given in Takizawa and Tezduyar (2011) fully, $n_{\text{ent}} = 2$ with the frozen option is more accurate than using a piecewise constant function, which would be first-order accurate. It was pointed out in Takizawa and Tezduyar (2011) that for $n_{\text{ent}} > 2$ there is no good reason to use the frozen option, because, for example, instead of using $n_{\text{ent}} = 3$ with the frozen option, we can just use $n_{\text{ent}} = 2$, which is more straightforward.

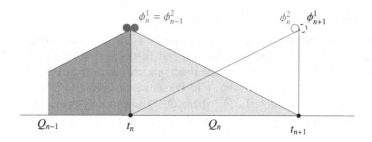

Figure 6.2 "Frozen" option

Three different basis functions were specified in Takizawa and Tezduyar (2011), to be used with the mesh representation, momentum equation, and incompressibility constraint, which were denoted by the subscripts **x**, **u**, and p, respectively.

6.3.1 Mesh Representation

In general, for the moving-mesh and FSI cases, the temporal basis functions for the mesh represent the mesh path. As pointed out in Takizawa and Tezduyar (2011), we need the flexibility to specify the speed along the path. Consider the mesh position vector given by

$$\mathbf{x}(\theta) = T^\alpha \left(\Theta_\mathbf{x}(\theta) \right) N_a \mathbf{x}_a^\alpha, \qquad (6.57)$$

where the repeated indices imply summation over the applicable range. It was proposed in Takizawa and Tezduyar (2011) to use $\Theta_\mathbf{x}(\theta)$ to specify the velocity. We seek a function $\Theta_\mathbf{x}$ such that

$$\hat{\mathbf{u}} = \frac{d\theta}{dt} \frac{d\Theta_\mathbf{x}}{d\theta} \frac{dT^\alpha}{d\Theta_\mathbf{x}} N_a \mathbf{x}_a^\alpha \qquad (6.58)$$

represents the desired mesh velocity along the path. This approach gives us the option to form a space–time parametric space beyond the NURBS capability.

6.3.2 Momentum Equation

The lowest order is the SV option, where $(n_{\text{ent}})_\mathbf{u} = 2$ with the frozen option. This has 2nd order accuracy (see Takizawa and Tezduyar, 2011). To increase the accuracy, we can employ $(n_{\text{ent}})_\mathbf{u} = 2$ or $(n_{\text{ent}})_\mathbf{u} = 3$ and higher order with the NURBS basis functions.

6.3.3 Incompressibility Constraint

In the SP option, we employ a piecewise constant function for the pressure p^h and its test function q^h, that is, $(n_{\text{ent}})_p = 1$. The original SP option sometimes generates small oscillations in the velocity because of the way it deals with the incompressibility constraint. This problem was resolved in Takizawa et al. (2010a) by moving the time integration point for the incompressibility constraint to $\theta = 1$ (see Remark 4.14). Because there has to be a balance between the number of velocity and pressure equations, changing the number of equations for the incompressibility constraint has certain restrictions. The SP option is one of the successful options.

6.4 Time Representation

Let us represent time $t \in (0, T)$ with p^{th} order NURBS basis functions, R^β ($\beta = 0, \ldots, n_{\text{ct}} - 1$). The basis functions are defined on the parametric space described by the open knot vector $\{\vartheta_1, \ldots, \vartheta_{n_{\text{kt}}}\}$, where n_{ct} and n_{kt} are the number of control points and knots. Then, time t can be written as

$$t = \sum_{\beta=0}^{n_{\text{ct}}-1} t_c^\beta R^\beta(\vartheta), \tag{6.59}$$

where t_c^β represents the temporal-control point. In the case of the space–time formulation there is a mesh corresponding to each temporal-control point t_c^β. As proposed in Takizawa et al. (2012c) and Takizawa and Tezduyar (2012b), we use a strictly-increasing mapping function $\Theta_t(\theta)$ to relate the element coordinate for a knot span and the NURBS parametric space:

$$\vartheta = \frac{(1 - \Theta_t(\theta))\vartheta_{e+p+1} + (1 + \Theta_t(\theta))\vartheta_{e+p+2}}{2}, \tag{6.60}$$

where e represents the element index ($e = 0, \ldots, n_{\text{elt}} - 1$, where n_{elt} is the number of elements). We have $n_{\text{ct}} = n_{\text{kt}} - p - 1$, and assuming no knot multiplicity inside the knot vector, $n_{\text{elt}} = n_{\text{kt}} - 2p - 1$. The element shape functions are defined as

$$T_e^\alpha(\Theta_t(\theta)) = R^{e+\alpha-1}(\vartheta). \tag{6.61}$$

In the time interval of element e, we represent t with the local shape functions:

$$t(\Theta_t(\theta)) = \sum_{\alpha=1}^{p+1} t_c^{e+\alpha-1} T_e^\alpha(\Theta_t(\theta)). \tag{6.62}$$

REMARK 6.6 *Similar to how it was in Section 6.3 with $\Theta_x(\theta)$, the re-parametrization with the mapping function $\Theta_t(\theta)$ adds flexibility to temporal representation, which would be attractive in some cases. For example, an arc can be represented by NURBS, however we cannot represent a constant speed on the arc. The re-parametrization allows us to have a constant speed on the arc (see Section 6.4.4).*

6.4.1 Time Marching Problem

Consider a set of NURBS basis functions that is being used in representing the data that we will work with. Figure 6.3 shows an example. Starting from this, as proposed in Takizawa

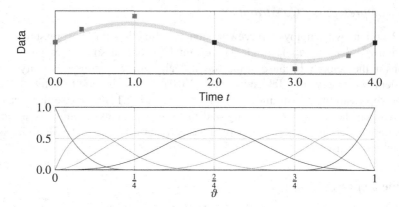

Figure 6.3 Data represented with NURBS. The data and control variables (top). The basis functions corresponding to each control variables (bottom)

et al. (2012c) and Takizawa and Tezduyar (2012b), we can form a new basis set by knot insertion with the objective that all elements become patches as shown in Figure 6.4. After that, we can use this basis set in our space–time computation, where the knot spans would be the time intervals of the space–time slabs, and represent the data exactly. An alternative process, which was also proposed in Takizawa *et al.* (2012c) and Takizawa and Tezduyar (2012b), has the same functionality, but without the need to explicitly represent the data we are working with using the new basis set. In that process, we simply form a new basis set where each element is a patch, and the data is represented in the formulation in terms of its own basis set. In general, the basis set that we simply form does not need to be the same as the one that could be obtained by the process described earlier, but if it is, then the two processes result in equivalent solution methods. Figure 6.5 shows the new basis functions that we simply form.

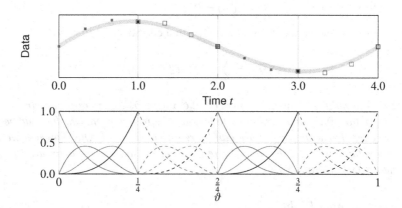

Figure 6.4 The data represented with basis functions after the knot insertion. The data and control variables (top). The basis functions corresponding to each control variables (bottom)

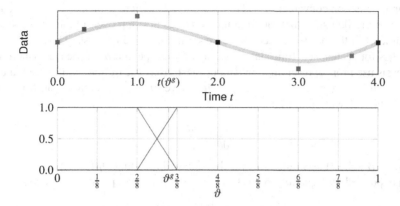

Figure 6.5 The data and control variables (top). The basis functions that we simply form for a given interval for the space–time computation (bottom). To integrate over the interval, in the NURBS representation of the data we need to search for the corresponding element and parametric coordinate for the time $t(\vartheta^g)$ of each quadrature point ϑ^g, and interpolate the value from the data

Since we need to work with different basis sets, we map one parametric space to the other through physical time t. With the function defined by Equation (6.59), time t can be obtained from the parametric space ϑ. Here we consider the inverse functionality; that is, $t \rightarrow \vartheta$. Finding the parametric space coordinate was proposed in Takizawa et al. (2012d) and Takizawa and Tezduyar (2012b) as follows:

1. Find the element e that is represented by the knot span $(\vartheta_{e+p+1}, \vartheta_{e+p+2})$. The process requires only time values at each element boundary, and we can quickly obtain the element index e by using a binary search technique.

2. Calculate θ for a given t by using Newton–Raphson iterations as follows:

$$\theta^{i+1} = \theta^i - \left(t - t\left(\Theta_t\left(\theta^i\right)\right)\right)\left(\left.\frac{dt}{d\theta}\right|^i\right)^{-1}, \tag{6.63}$$

where superscript "i" is the iteration counter, $t\left(\Theta_t\left(\theta^i\right)\right)$ can be calculated from Equation (6.62), and

$$\left.\frac{dt}{d\theta}\right|^i = \sum_{\alpha=1}^{p+1} t_c^{e+\alpha-1} \left.\frac{dT_e^\alpha}{d\Theta_t}\right|_{\Theta_t(\theta^i)} \left.\frac{d\Theta_t}{d\theta}\right|_{\theta^i}. \tag{6.64}$$

We use as the initial guess $\theta^0 = 0$.

3. Compute ϑ from Equation (6.60).

6.4.2 Design of Temporal NURBS Basis Functions

In the previous section we described how to find the parametric space value corresponding to physical time. Here we describe from Takizawa et al. (2012c) and Takizawa and Tezduyar (2012b) some specific temporal representations.

For implementation convenience and computational efficiency, we restrict the time interval of the space–time slab such that the time interval does not step over a time corresponding to a temporal knot in the basis set used for representing the data or mesh. Thus the supporting set of meshes for each space–time slabs consists of only specific $p + 1$ meshes, where p is the order of basis used for representing the data or mesh. Because of that requirement, a uniform element size, that is, $t(\vartheta_{e+p+2}) - t(\vartheta_{e+p+1}) = \Delta t$, where $\Delta t = \frac{T}{n_{\text{elt}}}$, is convenient. Moreover, we might have the following requirement:

$$\frac{dt}{d\theta} = \frac{\Delta t}{2}. \tag{6.65}$$

In the case of B-spline basis functions, for the identical mapping $\Theta_t(\theta) = \theta$, we can satisfy the condition expressed by Equation (6.65) by selecting the control points as follows:

$$t_c^\beta = t_c^{\beta-1} + \frac{\vartheta_{\beta+p+1} - \vartheta_{\beta+1}}{p(\vartheta_{n_{\text{kt}}} - \vartheta_1)} T, \tag{6.66}$$

for $\beta = 1, \ldots, n_{\text{ct}} - 1$ and $t_c^0 = 0$.

6.4.3 Approximation in Time

Let $\mathbf{x}_A^s$ be the sampling values of a time-varying spatial position vector $\mathbf{x}_A$ at sampling times t^s ($s = 0, \ldots, n_{\text{sp}} - 1$, where n_{sp} is the number of sampling points). For example, $\mathbf{x}_A$ could be the position vector for spatial node A, or it could be the position vector for a point on a surface geometry extracted from video data. For each A, as proposed in Takizawa et al. (2012c) and Takizawa and Tezduyar (2012b), we represent the path corresponding to the sampling points with NURBS. This serves two purposes: bringing smoothness to the temporal representation, and better representation accuracy for less control points. First, we form a linear finite element mesh in time, consisting of two-node elements. Then, we use least-squares projection to translate that to NURBS representation:

$$\int_0^T \mathbf{R}_A^h \cdot \left(\mathbf{x}_A^h - \boldsymbol{\chi}_A^h\right) dt = 0, \tag{6.67}$$

where $\mathbf{R}_A^h$ and $\mathbf{x}_A^h$ are the test function and NURBS representation in time, and $\boldsymbol{\chi}_A^h$ is the linear representation in time. Thus, we obtain the control point $\mathbf{x}_A^\beta$ corresponding to each time control point t_c^β. Figure 6.6 is an example.

REMARK 6.7 *This is a simple projection. However, the concept is applicable to more complicated formulations to obtain smoother motion.*

6.4.4 An Example: Circular-Arc Motion

We describe this example from Takizawa et al. (2012c) and Takizawa and Tezduyar (2012b).

6.4.4.1 Path Representation

NURBS temporal basis functions can be used in representing a particle path on a circular arc. Let us select as the origin the center of the circle to which the arc belongs. The particle travels

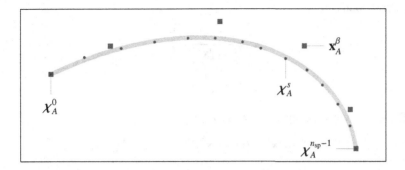

Figure 6.6 NURBS representation for a time-varying spatial position vector. The circles are the spatial position vector at each sampling time. The squares are the temporal-control points and the smooth curve is represented by them

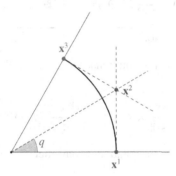

Figure 6.7 A circular arc represented by quadratic NURBS

from $\mathbf{x}^1$ to $\mathbf{x}^3$, $\|\mathbf{x}^1\| = \|\mathbf{x}^3\|$, as shown in Figure 6.7. It is known that a circular arc can be represented exactly by three control points with quadratic NURBS basis functions (it is valid only for $q < \frac{\pi}{2}$). The weights are $w_1 = w_3 = 1$, and $w_2 = \cos q$, where

$$\cos 2q = \frac{\mathbf{x}^1 \cdot \mathbf{x}^3}{r^2}, \qquad r = \|\mathbf{x}^1\| = \|\mathbf{x}^3\|. \tag{6.68}$$

This results in the following temporal basis functions:

$$T^1(\Theta) = \frac{(1-\Theta)^2}{2\left((1+\Theta^2) + w_2\left(1-\Theta^2\right)\right)}, \tag{6.69}$$

$$T^2(\Theta) = \frac{w_2(1-\Theta^2)}{(1+\Theta^2) + w_2\left(1-\Theta^2\right)}, \tag{6.70}$$

$$T^3(\Theta) = \frac{(1+\Theta)^2}{2\left((1+\Theta^2) + w_2\left(1-\Theta^2\right)\right)}. \tag{6.71}$$

and the control points are $\mathbf{x}^1$,

$$\mathbf{x}^2 = \frac{r}{w_2} \frac{\mathbf{x}^1 + \mathbf{x}^3}{\|\mathbf{x}^1 + \mathbf{x}^3\|} \tag{6.72}$$

$$= \frac{1}{2w_2^2} \left(\mathbf{x}^1 + \mathbf{x}^3 \right), \tag{6.73}$$

and $\mathbf{x}^3$. Thus, the arc can be represented as follows:

$$\mathbf{x}(\Theta_x) = \mathbf{x}^1 T^1(\Theta_x) + \mathbf{x}^2 T^2(\Theta_x) + \mathbf{x}^3 T^3(\Theta_x). \tag{6.74}$$

6.4.4.2 Constant Angular Velocity

First by using Equation (6.73), we rearrange Equation (6.74) as follows:

$$\mathbf{x}(\Theta_x) = \underbrace{\left[T^1(\Theta_x) + \frac{1}{2w_2^2} T^2(\Theta_x) \right]}_{Q^1(\Theta_x)} \mathbf{x}^1 + \underbrace{\left[T^3(\Theta_x) + \frac{1}{2w_2^2} T^2(\Theta_x) \right]}_{Q^3(\Theta_x)} \mathbf{x}^3, \tag{6.75}$$

where Q^1 and Q^3 are introduced for notational convenience. Taking cross product with the unit vector along $\mathbf{x}^2$, we get

$$\frac{\mathbf{x}^1 + \mathbf{x}^3}{\|\mathbf{x}^1 + \mathbf{x}^3\|} \times \mathbf{x}(\Theta_x) = \frac{\mathbf{x}^1 \times \mathbf{x}^3}{r^2 \sin(2q)} r \sin(\omega t), \tag{6.76}$$

where $-\frac{\Delta t}{2} \leq t \leq \frac{\Delta t}{2}$ and $\omega \Delta t = 2q$ for notational convenience. From Equation (6.76), we get

$$\frac{\mathbf{x}^1 + \mathbf{x}^3}{2r \cos q} \times \mathbf{x}(\Theta_x) = \frac{\mathbf{x}^1 \times \mathbf{x}^3}{2r \sin q \cos q} \sin(\omega t). \tag{6.77}$$

From Equations (6.75) and (6.77), we get

$$\left(Q^3 - Q^1 \right) = \frac{\sin(\omega t)}{\sin q}. \tag{6.78}$$

From Equation (6.78), it can be shown that

$$\Theta_x = \frac{\sin q}{1 - \cos q} \frac{\sin(\omega t)}{1 + \cos(\omega t)}. \tag{6.79}$$

We assume that time is represented with the same basis functions, and therefore

$$t(\Theta_t) = \frac{\Delta t}{2} \left(T^3(\Theta_t) - T^1(\Theta_t) \right) = \frac{\Delta t \Theta_t}{1 + \Theta_t^2 + (1 - \Theta_t^2) \cos q}, \tag{6.80}$$

and

$$\frac{dt}{d\theta} = \Delta t \frac{1 - \Theta_t^2 + (1 + \Theta_t^2) \cos q}{\left(1 + \Theta_t^2 + (1 - \Theta_t^2) \cos q\right)^2} \frac{d\Theta_t}{d\theta}. \tag{6.81}$$

- If we select $\Theta_t = \theta$:
  Equation (6.79) can be evaluated by substituting for t from Equation (6.80), and the derivative $\frac{dt}{d\theta}$ becomes

$$\frac{dt}{d\theta} = \Delta t \frac{1 - \theta^2 + (1 + \theta^2)\cos q}{(1 + \theta^2 + (1 - \theta^2)\cos q)^2}, \qquad (6.82)$$

and

$$\frac{d\Theta_x}{d\theta} = \frac{2q \sin q}{1 - \cos q} \frac{1 + \cos(\omega t)}{(1 + \cos(\omega t))^2} \frac{1 - \theta^2 + (1 + \theta^2)\cos q}{(1 + \theta^2 + (1 - \theta^2)\cos q)^2}. \qquad (6.83)$$

- If we select $\frac{dt}{d\theta} = \frac{\Delta t}{2}$:
  From Equation (6.81):

$$\frac{d\Theta_t}{d\theta} = \frac{1}{2} \frac{\left(1 + \Theta_t^2 + (1 - \Theta_t^2)\cos q\right)^2}{1 - \Theta_t^2 + (1 + \Theta_t^2)\cos q}. \qquad (6.84)$$

The mapping for the particle path becomes:

$$\Theta_x = \frac{\sin q}{1 - \cos q} \frac{\sin(q\theta)}{1 + \cos(q\theta)}, \qquad (6.85)$$

and

$$\frac{d\Theta_x}{d\theta} = \frac{q \sin q}{1 - \cos q} \frac{1}{1 + \cos(q\theta)}. \qquad (6.86)$$

6.5 Simple-Shape Deformation Model (SSDM)

Suppose we want to track the motion/deformation of an object with surface shape that is too complex to track in full detail, giving us just the option of tracking only a finite number of points belonging to this complex shape. In the simple-shape deformation model (SSDM), which was proposed in Takizawa *et al.* (2012c), we assume that those tracked points are associated with a simple shape (SS) instead of the actual, complex shape. NURBS is used for the spatial representation of the SS. We note that the SS is larger than the complex shape.

Starting with the reference configuration, the SS, the complex shape and the tracked points all can be seen in a common parametric space (Figure 6.8). The complex shape can be

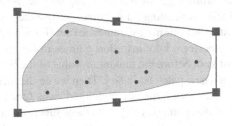

Figure 6.8 SSDM. Complex shape is shaded. Circles are tracked points. SS is represented by squares (control points).

represented by finite elements or NURBS. Control points of the SS at different times during tracking are determined by a least-squares fit. The fit minimizes the difference between the positions on the SS (with respect to the reference configuration) of the tracked points and the positions of the actual tracked points. The complex shape at a given temporal-control point is determined by interpolation from the parametric space in the case of the finite element representation, and by least-square projection in the case of NURBS representation. The least-squares integration is over the parametric space of the complex shape, and we minimize, with respect to the control points of the complex shape, the difference between the complex-shape and simple shape representations.

In the full space–time representation, the method we described above is applied to temporal-control values that are determined as described in Section 6.4.3, instead of the actual physical locations.

6.6 Mesh Update Techniques in the Space–Time Framework

6.6.1 Mesh Computation and Representation

Given the surface mesh, we compute the volume mesh using the mesh moving technique explained in Section 4.7. Here, as proposed in Takizawa *et al.* (2012c), we apply this technique to computing the meshes that will serve as temporal-control points. This allows us to do mesh computations with longer time in between, but get the mesh-related information, such as the coordinates and their time derivatives, from the temporal representation whenever we need. Obviously this also reduces the storage amount and access associated with the meshes. However, because of the longer time between the control meshes, linear interpolation of the surfaces between control points in time might be needed in computing those meshes with the mesh moving technique mentioned.

REMARK 6.8 *We note that getting the meshes used in the computations from the temporal representation can be done independently of which time direction was used in computing the control meshes.*

6.6.2 Remeshing Technique

In many computations remeshing becomes unavoidable. Two choices were proposed in Takizawa *et al.* (2012c). To explain those two choices, let us assume that when we try to move from control mesh M_c^β to $M_c^{\beta+1}$, we find the quality of $M_c^{\beta+1}$ to be less than desirable. In the first choice, which is called "trimming," we remesh going back to $M_c^{\beta-p+1}$. Then whenever our solution process needs a mesh, depending on the time, we use the control meshes belonging to either only the un-remeshed set or only the remeshed set (Figure 6.9).

In the second choice, we perform knot insertion p times in the temporal representation of the surface at the right-most knot before the maximum value of the basis function corresponding to $t_c^{\beta+1}$, making that knot a new patch boundary. Then we do the mesh moving computation for the control meshes associated with the newly-defined basis functions, not only the one at the new patch boundary, but also going back $(p-1)$ basis functions (Figure 6.10).

We use the second choice in computations, because we believe that in many cases the need for remeshing is generated by a topological change, which we can avoid going over with a large step if we use the knot insertion process.

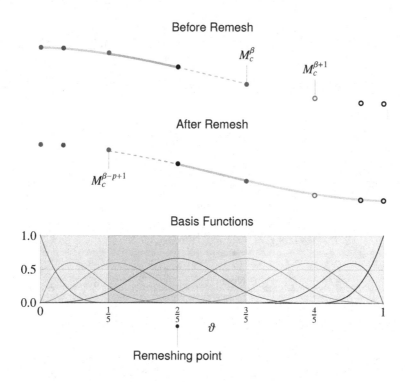

Figure 6.9 Remeshing and trimming NURBS. A set of un-remeshed meshes (top). A set of remeshed meshes (middle). Common basis functions (bottom)

6.7 Fluid Mechanics Computation with Temporal NURBS Mesh

We solve the fluid dynamics equations with the DSD/SST formulation. Here we describe, from Takizawa *et al.* (2012c,d), two techniques related to the moving-mesh problem.

6.7.1 No-Slip Condition on a Prescribed Boundary

Suppose we have a prescribed mesh motion, and no-slip conditions on part of the boundary of that mesh. Those Dirichlet conditions can be obtained from the mesh boundary motion.

Prior to solving the equations using a space–time slab Q_n, we use a least-squares projection for each prescribed node A as follows:

$$\int_{t_n}^{t_{n+1}} \boldsymbol{R}_A^h \cdot \left(\mathbf{u}_A^h - \frac{\mathrm{d}\mathbf{x}_A^h}{\mathrm{d}t} \right) \mathrm{d}t = 0, \tag{6.87}$$

where $\boldsymbol{R}_A^h$ represents the test function, $\mathbf{u}_A^h$ is represented by temporal-control velocities (unknown) and the corresponding basis functions in time, and the mesh velocity is obtained by the derivative of the mesh displacement, which is also represented by temporal-control positions and their basis functions. We note that $\mathbf{u}_A^h$ at time t_n approaching from below and above might be different.

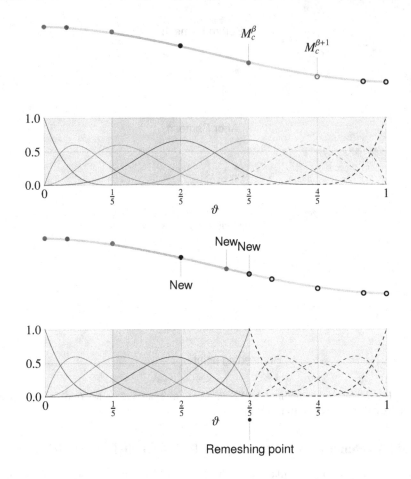

Figure 6.10 Remeshing with knot insertion. For the set of un-remeshed meshes, there are p newly-defined basis functions and the corresponding control points are marked "New." We carry out the mesh moving computations for those meshes

6.7.2 Starting Condition

Starting a fluid dynamics computation is not always easy, especially in the presence of moving boundaries. For example, pre-FSI computation techniques were developed in Tezduyar *et al.* (2007b, 2009a, 2010b) and Takizawa *et al.* (2011d) to build a starting condition for FSI computations. Here we describe the technique proposed in Takizawa *et al.* (2012c) as a pre-computation sequence for flow computations with prescribed boundary motion.

Suppose we want to compute with a mesh temporally represented with NURBS $(M_c^0, M_c^1, \cdots)$. We define the temporal-control value $\mathbf{x}_c^k$ to be used interchangeably with the kth temporal-control mesh M_c^k.

Advanced FSI and Space–Time Techniques

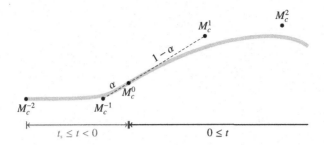

Figure 6.11 Mesh representation for the starting condition with quadratic NURBS in time

6.7.2.1 With Quadratic NURBS Representation

Figure 6.11 is an example of mesh representation with quadratic NURBS in time. We generate two additional meshes as follows:

$$M_c^{-1} = \frac{M_c^0 - \alpha M_c^1}{1 - \alpha}, \tag{6.88}$$

$$M_c^{-2} = M_c^{-1}, \tag{6.89}$$

where $0 < \alpha < 1$ is an extrapolation parameter. The mesh M_c^{-1} is an extrapolation. The corresponding temporal-control point for M_c^{-1} is

$$t_c^{-1} = \frac{t_c^0 - \alpha t_c^1}{1 - \alpha}, \tag{6.90}$$

with the only requirement being $t_c^{-2} \equiv t_s < t_c^{-1}$, which determines the length of the precomputation. For the computations reported in this book, in temporal representation of the mesh, as NURBS basis functions, we use quadratic B-spline functions defined by the knot vector $\{0, 0, 0, 1, 1, 1\}$. Based on the preliminary computations, using even higher-order NURBS basis functions was proposed in Takizawa *et al.* (2012c), so that the acceleration is continuous. This was explained in Takizawa *et al.* (2012d), and is described in the next subsection.

6.7.2.2 With Cubic NURBS Representation

Figure 6.12 is an example of mesh representation with cubic NURBS in time. Here we construct a starting-condition temporal patch with C^2-continuity with the next patch:

$$\lim_{t \to 0^-} \mathbf{x} = \lim_{t \to 0^+} \mathbf{x}, \tag{6.91}$$

$$\lim_{t \to 0^-} \frac{d\mathbf{x}}{dt} = \lim_{t \to 0^+} \frac{d\mathbf{x}}{dt}, \tag{6.92}$$

$$\lim_{t \to 0^-} \frac{d^2\mathbf{x}}{dt^2} = \lim_{t \to 0^+} \frac{d^2\mathbf{x}}{dt^2}, \tag{6.93}$$

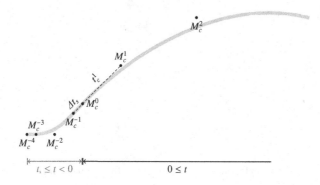

Figure 6.12 Mesh representation for the starting condition with cubic NURBS in time

where **x** represents the mesh position defined with temporal-control meshes, and we want the mesh velocity and acceleration to be zero at the beginning of that temporal patch:

$$\frac{d\mathbf{x}}{dt}\bigg|_{t=t_s} = \mathbf{0}, \quad (6.94)$$

$$\frac{d^2\mathbf{x}}{dt^2}\bigg|_{t=t_s} = \mathbf{0}. \quad (6.95)$$

REMARK 6.9 *Because this condition provides zero velocity and acceleration, we can compute with a static mesh prior to the computation in this temporal patch to develop the flow field.*

To achieve above conditions we use cubic B-spline functions defined by the knot vector $\{0, 0, 0, 0, \frac{1}{2}, 1, 1, 1, 1\}$. Because the last temporal-control mesh for the starting-condition patch is M_c^0, the condition given by Equation (6.91) is automatically satisfied. The conditions given by Equations (6.94) and (6.95) can be satisfied by setting $M_c^{-4} = M_c^{-3} = M_c^{-2}$, with any set of corresponding control points for time: $t_c^{-4} \equiv t^s < t_c^{-3} < t_c^{-2}$. However, satisfying the conditions given by Equations (6.92) and (6.93) is not trivial.

The following is an example using cubic B-splines. The derivatives of the mesh position at $t \to 0^+$ are as follows:

$$\frac{d\mathbf{x}}{d\vartheta}\bigg|_{\vartheta=\vartheta_1} = \frac{3}{\vartheta_5 - \vartheta_2}\left(\mathbf{x}_c^1 - \mathbf{x}_c^0\right), \quad (6.96)$$

$$\frac{d^2\mathbf{x}}{d\vartheta^2}\bigg|_{\vartheta=\vartheta_1} = \frac{6}{\vartheta_5 - \vartheta_3}\left(\frac{\mathbf{x}_c^2 - \mathbf{x}_c^1}{\vartheta_6 - \vartheta_3} - \frac{\mathbf{x}_c^1 - \mathbf{x}_c^0}{\vartheta_5 - \vartheta_2}\right), \quad (6.97)$$

where ϑ is parametric coordinate[1] defined on the first patch. Similarly, the derivatives of the time are as follows:

$$\frac{dt}{d\vartheta}\bigg|_{\vartheta=\vartheta_1} = \frac{3}{\vartheta_5 - \vartheta_2} t_c^1, \quad (6.98)$$

---

[1] We note that the index of the knot vector starts from 1.

… Advanced FSI and Space–Time Techniques … 163

$$\left.\frac{d^2 t}{d\vartheta^2}\right|_{\vartheta=\vartheta_1} = \frac{6}{\vartheta_5 - \vartheta_3}\left(\frac{t_c^2 - t_c^1}{\vartheta_6 - \vartheta_3} - \frac{t_c^1}{\vartheta_5 - \vartheta_2}\right). \quad (6.99)$$

Thus, the mesh velocity and acceleration can be determined as follows:

$$\lim_{t \to 0^+} \frac{d\mathbf{x}}{dt} = \frac{\mathbf{x}_c^1 - \mathbf{x}_c^0}{t_c^1}, \quad (6.100)$$

$$\lim_{t \to 0^+} \frac{d^2\mathbf{x}}{dt^2} = \left.\left(\frac{dt}{d\vartheta}\right)^{-2}\left(\frac{d^2\mathbf{x}}{d\vartheta^2} - \frac{d\mathbf{x}}{dt}\frac{d^2 t}{d\vartheta^2}\right)\right|_{\vartheta=\vartheta_1}, \quad (6.101)$$

with Equations (6.97)–(6.99). We note that if the temporal-control points are given by Equation (6.66), the second term of Equation (6.101) is zero.

Similarly, the derivatives at $t \to 0^-$ are as follows:

$$\lim_{t \to 0^-} \frac{d\mathbf{x}}{dt} = -\frac{\mathbf{x}_c^0 - \mathbf{x}_c^{-1}}{t_c^{-1}}, \quad (6.102)$$

$$\lim_{t \to 0^-} \frac{d^2\mathbf{x}}{dt^2} = \frac{1}{3(t_c^{-1})^2}\left(\mathbf{x}_c^{-2} - 3\mathbf{x}_c^{-1} + 2\mathbf{x}_c^0 + \frac{t_c^{-2} - 3t_c^{-1}}{t_c^{-1}}\left(\mathbf{x}_c^0 - \mathbf{x}_c^{-1}\right)\right). \quad (6.103)$$

With that, there are many ways of satisfying the conditions given by Equations (6.92) and (6.93). Here we set $t_c^{-1} = -\Delta t_s$. We also define the control points for the time in such a way that the part of Equation (6.103) related to $\frac{d^2 t}{d\vartheta^2}$ at $t \to 0^-$ is zero; that is, $t_c^{-2} = 3t_c^{-1}$. Thus, we can set $\mathbf{x}_c^{-1}$ and $\mathbf{x}_c^{-2}$ as

$$\mathbf{x}_c^{-1} = \mathbf{x}_c^0 - \frac{\Delta t_s}{t_c^1}\left(\mathbf{x}_c^1 - \mathbf{x}_c^0\right), \quad (6.104)$$

$$\mathbf{x}_c^{-2} = \mathbf{x}_c^0 - 3\left(\frac{\Delta t_s}{t_c^1}\left(\mathbf{x}_c^1 - \mathbf{x}_c^0\right) - \Delta t_s^2 \lim_{t \to 0^+} \frac{d^2\mathbf{x}}{dt^2}\right). \quad (6.105)$$

Furthermore we set $t_c^{-3} = -5\Delta t_s$ and $t_s = -6\Delta t_s$ to have an overall linear relationship between the time and the parametric space. We leave Δt_s as a free parameter that we can adjust for various purposes, such as having smaller acceleration and smaller displacement from the M_c^0.

REMARK 6.10 *In addition to using NURBS basis functions for the temporal representation of the motion and deformation of the solid surfaces and volume meshes computed, we can use NURBS as temporal basis functions in our space–time computations. With that, we would have temporal advantages similar to the spatial advantages one would have by using NURBS as spatial basis functions.*

6.8 The Surface-Edge-Node Contact Tracking (SENCT-FC) Technique

The SENCT-FC technique was motivated by the numerical challenges involved in parachute FSI computations with contact between the parachutes when they are used in clusters. In a contact algorithm we need for such or comparable FSI computations, the objective is to prevent the structural surfaces from coming closer than a predetermined minimum distance we would like to maintain to protect the quality of the fluid mechanics mesh between the structural surfaces. The SENCT technique was introduced in Tezduyar and Sathe (2007) for this purpose. Two versions of the SENCT technique were proposed in Tezduyar and Sathe (2007). In the SENCT-Force (SENCT-F) technique, the contacted node is subjected to penalty forces that are inversely proportional to the projection distances to the contacting surfaces, edges, and nodes. In the SENCT-Displacement (SENCT-D) technique, the displacement of the contacted node is adjusted to correlate with the motion of the contacting surfaces, edge, and nodes. The SENCT technique is described in more detail in Sathe and Tezduyar (2008), which includes a number of test computations. For FSI problems with nonmatching fluid and structure meshes at the interface, it was proposed in Remark 1 of Tezduyar et al. (2008a) to formulate the contact model based on the fluid mechanics mesh at the interface (see also Remark 9.1 in Section 9.2). This version of the SENCT was denoted with the option key "-M1."

The SENCT-FC technique, which was used in the parachute cluster computations reported in Takizawa et al. (2011f), was introduced and described in detail in Takizawa et al. (2011e). It has some features in common with the SENCT-F technique but is more robust. Also, compared to the SENCT-F technique, the forces are applied in a conservative fashion, and the letter "C" in "FC" stands for "conservative." Later in this section, we will comment more on this aspect of the differences between SENCT-FC and SENCT-F (see Remark 6.12). The new technique is used as SENCT-FC-M1 in the computations. It can be seen as having three parts: contact detection, force representation, and solving the contact force equations. In the following sections we describe those from Takizawa et al. (2011e).

6.8.1 Contact Detection and Node Sets

To detect contact, we calculate the distance between a point and the closest point to it. The closest point is searched on surfaces, which can somehow be kept to a limited number. For example, we may want to exclude self-contact. Here, we use the nodes on the fluid surface for contact detection, however this technique is applicable for any other kind of points (e.g., integration points). The technique starts by finding the closest point in the same way as in the earlier SENCT techniques described in Tezduyar and Sathe (2007) and Sathe and Tezduyar (2008); that is, the closet point on one of nodes, edges or surface elements. The node, edge, or surface element containing the closest point is referred to in this book as a "segment." Here we define $\mathbf{d}_A = \mathbf{x}_A^C - \mathbf{x}_A$ as the distance vector for each node and the closest point, which is represented by $\mathbf{x}^C$. In the case of $\|\mathbf{d}_A\| < \epsilon_A$, this node is in contact. The predetermined minimum distance is defined as follows:

$$\epsilon_A \equiv \epsilon_A^S + \epsilon_A^C, \qquad (6.106)$$

where $\epsilon_A^S \geq 0$ and $\epsilon_A^C \geq 0$ are the length parameters for node A and the closest point, respectively. Figure 6.13 shows an example. The set of contacted nodes is defined as follows:

$$\eta_D = \{A \mid \|\mathbf{d}_A\| < \epsilon_A, \ \forall A \in \eta\}, \qquad (6.107)$$

where η represents all possible contacted nodes on the fluid surface. Each contacted node A has some contacting nodes, which are denoted by γ_A. All nodes in contact are represented as follows:

$$\eta_C = \bigcup_{A \in \eta_D} \{A + \gamma_A\}. \qquad (6.108)$$

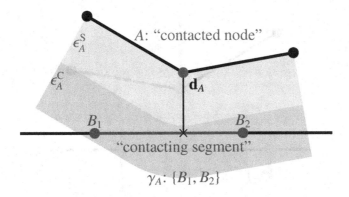

Figure 6.13 Contact detection and definitions of "contacted node" and "contacting-segment nodes"

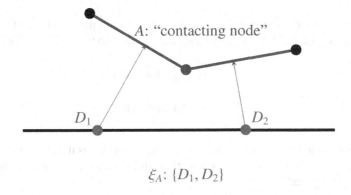

Figure 6.14 The set ξ_A contains the nodes contacted by node A

In general $\eta_D \subseteq \eta_C$. As a reverse relationship, we define ξ_A as the nodes contacted by node A. These set definitions are shown in Figures 6.13 and 6.14.

6.8.2 Contact Force and Reaction Force

We introduce a virtual contact force $\boldsymbol{\varphi}_A$ for each contacted node A and a reaction force $\boldsymbol{\varphi}_B^R$ for node $B \in \gamma_A$. First, we model the force and the reaction forces as follows:

$$\boldsymbol{\varphi}_A = -\varphi_A \mathbf{n}_A, \tag{6.109}$$

$$\mathbf{n}_A = \frac{\mathbf{d}_A}{\|\mathbf{d}_A\|}, \tag{6.110}$$

and

$$\boldsymbol{\varphi}_B^R = E_{BA} \varphi_A \mathbf{n}_A, \tag{6.111}$$

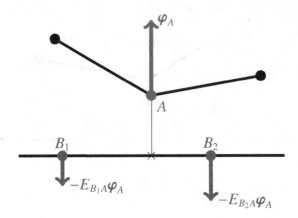

Figure 6.15 Contact force for the contacted node A and the reaction forces for the contacting nodes γ_A

where E_{BA} is a scalar for each node $B \in \gamma_A$. Figure 6.15 shows the definition of the force and the reaction forces. The scalar values can be solved by using the following equations:

$$\varphi_A + \sum_{B \in \gamma_A} \varphi_B^R = 0, \qquad (6.112)$$

$$\sum_{B \in \gamma_A} \left(\mathbf{x}_B - \mathbf{x}_A^C\right) \times \varphi_B^R = 0. \qquad (6.113)$$

The first equation is the balance of forces and the second equation is the no-moment condition. There is a unique set of solutions E_{BA} for Equations (6.112) and (6.113) in the cases of node, edge, and triangle segments.

REMARK 6.11 *We note that in the case of a node, edge, and triangle segment, the scalar factor E_{BA} is the same as the shape function value $N_B\left(\mathbf{x}_A^C\right)$.*

The total force for each contacted node A is

$$\mathbf{f}_A = -\varphi_A \mathbf{n}_A + \sum_{D \in \xi_A} E_{AD} \varphi_D \mathbf{n}_D \qquad \forall A \in \eta_D, \qquad (6.114)$$

and for the other nodes it is

$$\mathbf{f}_A = \sum_{D \in \xi_A} E_{AD} \varphi_D \mathbf{n}_D \qquad \forall A \in \eta_C - \eta_D. \qquad (6.115)$$

Figure 6.16 shows all the forces acting on node A.

REMARK 6.12 *The contact forces are applied in SENCT-FC in a conservative fashion, because while we calculate the forces acting on a contacted node, we also calculate the corresponding reaction forces acting on the contacting nodes, as represented by Equations (6.109) and (6.111) and eventually by Equations (6.114) and (6.115). In SENCT-F on the other hand, only the forces acting on the contacted nodes are calculated, assuming that all the*

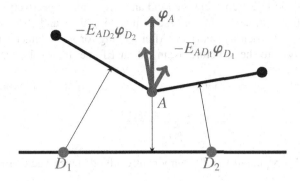

Figure 6.16 All the forces acting on node A. Node A is both "contacted" and "contacting"

contacting nodes would also become contacted nodes during the search process, and therefore the contact forces are not applied in a conservative fashion.

We rewrite the ith component of total force for node $A \in \eta_C$ as follows:

$$f_{Ai} = \left(E_{AC}\delta_{ij} - \delta_{AC}\delta_{ij}\right)\delta_{CD}n_{Dj}\varphi_D, \quad (6.116)$$

where we use the summation convention ($C, D \in \eta_D$ and $j = 1, \ldots, n_{sd}$). With the matrix-vector notation, the above equation becomes

$$\mathbf{F} = \mathbf{QV\Phi}, \quad (6.117)$$

where

$$\mathbf{F} = [f_{Ai}], \quad (6.118)$$

$$\mathbf{Q} = \left[E_{AC}\delta_{ij} - \delta_{AC}\delta_{ij}\right], \quad (6.119)$$

$$\mathbf{V} = \left[\delta_{CD}n_{Dj}\right], \quad (6.120)$$

$$\mathbf{\Phi} = [\varphi_D]. \quad (6.121)$$

6.8.3 Solving for the Contact Force

We use the following equation for the contacted node A:

$$\mathbf{d}_A \cdot \mathbf{d}_A = \epsilon_A^2. \quad (6.122)$$

For implementation convenience, we use block-iterative coupling between Equation (6.122) and the fluid+structure block of the FSI system. In that framework, the block corresponding Equation (6.122) becomes:

$$\mathbf{n}_A^i \cdot \left(\left(\frac{\partial \mathbf{d}_A}{\partial \mathbf{x}_1}\right)^i \frac{\partial \mathbf{x}_1}{\partial \mathbf{x}_2} \left(\frac{\partial \mathbf{x}_2}{\partial \mathbf{F}_2}\right)^i \frac{\partial \mathbf{F}_2}{\partial \mathbf{F}_1}(\Delta \mathbf{F})^i\right) = \frac{\epsilon_A^2 - \|\mathbf{d}_A^i\|^2}{2\|\mathbf{d}_A^i\|}, \quad (6.123)$$

where subscripts "1" and "2" represent the fluid and structure, respectively, and superscript i denotes the ith nonlinear iteration. In the case of matching fluid and structure meshes at the interface, $\frac{\partial \mathbf{x}_1}{\partial \mathbf{x}_2}$ and $\frac{\partial \mathbf{F}_2}{\partial \mathbf{F}_1}$ are identity matrices. After solving for this contact force, we form the total force and apply it to the structure as an external force, then solve the fluid+structure block.

We describe each term in more detail below. The closest point can be expressed as follows:

$$\mathbf{x}_A^C = \sum_{B \in \gamma_A} N_B\left(\mathbf{x}_A^C\right) \mathbf{x}_B. \tag{6.124}$$

We define $H_{AB} \equiv N_B\left(\mathbf{x}_A^C\right)$, and the ith component of the distance vector for $A \in \eta_D$ can be written as

$$d_{Ai} = \left(H_{AB}\delta_{ij} - \delta_{AB}\delta_{ij}\right) x_{Bj}, \tag{6.125}$$

where $B \in \eta_C$. With the matrix-vector notation, it becomes

$$\mathbf{D} = \mathbf{S}\mathbf{X}, \tag{6.126}$$

where

$$\mathbf{D} = [d_{Ai}], \tag{6.127}$$
$$\mathbf{S} = \left[\left(H_{AB}\delta_{ij} - \delta_{AB}\delta_{ij}\right)\right], \tag{6.128}$$
$$\mathbf{X} = \left[x_{Bj}\right]. \tag{6.129}$$

Thus,

$$\left(\frac{\partial \mathbf{D}}{\partial \mathbf{X}}\right) = \mathbf{S}. \tag{6.130}$$

Because of Remark 6.11, $\mathbf{S}$ is the transpose of $\mathbf{Q}$. Here we define three more matrices:

$$\mathbf{C} = \frac{\partial \mathbf{x}_1}{\partial \mathbf{x}_2}, \tag{6.131}$$

$$\mathbf{Z} = \frac{\partial \mathbf{x}_2}{\partial \mathbf{F}_2}, \tag{6.132}$$

$$\mathbf{B} = \frac{\partial \mathbf{F}_2}{\partial \mathbf{F}_1}. \tag{6.133}$$

Thus, we obtain the following equation system:

$$\left[\left(\mathbf{Q}^i \mathbf{V}^i\right)^T \left(\mathbf{C}\mathbf{Z}^i \mathbf{B}\right)\left(\mathbf{Q}^i \mathbf{V}^i\right)\right] \Delta \Phi^i = \Psi^i, \tag{6.134}$$

where

$$\Psi^i = \left[\frac{\epsilon_A^2 - \|\mathbf{d}_A^i\|^2}{2\|\mathbf{d}_A^i\|}\right]. \tag{6.135}$$

We approximate $\mathbf{Z}^i$ by $\beta \Delta t^2 \mathbf{M}^{-1}$, where $\mathbf{M}$ is the structure mass matrix and β is part of the Hilber–Hughes–Taylor Hilber *et al.* (1977) scheme. In the case of uncoupled multiple contacts, because of the nodal ordering we selected, the coefficient matrix multiplying $\Delta \mathbf{\Phi}^i$ is in block-diagonal form. We split it into the block-diagonal matrices and solve each block directly by using LAPACK. Then, we apply the force $\mathbf{B}\mathbf{Q}^i \mathbf{V}^i \Delta \mathbf{\Phi}^i$ to the structure just like an external force.

REMARK 6.13 *In the case of node-to-node contact, the forces of the two contacted nodes are not linearly independent. Therefore, we exclude one of the equations.*

7

General Applications and Examples of FSI Modeling

In this chapter we present general computational examples of FSI and flow problems with moving interfaces. The examples illustrate how the theory and methods presented in the previous chapters work. The first example is 2D flow past an elastic beam. This is a popular FSI benchmark problem, which includes some aspects of an FSI computation with moving meshes: coupled, unsteady fluid and structural mechanics, and large deformation of the fluid mechanics mesh. The second example is 2D flow past a rigid airfoil attached to a torsion spring. The use of a special mesh moving technique, in which the mesh near the airfoil rotates with the airfoil, is illustrated. This maintains the mesh quality in the boundary layer during the computations. The computation also serves as a test to compare the numerical performances of the SUPS and VMST versions of the DSD/SST formulation. The third example is 3D computation of the inflation of a balloon with an incompressible fluid. The fluid incompressibility imposes a constraint on the structural deformation and creates difficulties for the convergence of staggered FSI coupling methods. On the other hand, the quasi-direct FSI coupling method performs very well for this class of problems. The fourth example is 3D computation of flow through and around a windsock. A special geometric smoothing technique is used in the projection between the structure and fluid meshes at the interface, sheltering the fluid mechanics mesh from excessive distortion and making the problem computable without any remeshing. The last example is 3D computation of the aerodynamics of flapping wings, where the wing motion and deformation patterns are extracted from high-speed video camera recordings of a locust in a wind tunnel. The computations demonstrate how the temporal NURBS basis functions are used in representation of the motion of the locust wings and the motion of the volume mesh as well as in remeshing, demonstrating that all this provides an accurate and efficient way of dealing with the wind tunnel data and the mesh.

7.1 2D Flow Past an Elastic Beam Attached to a Fixed, Rigid Block

Our first example, from Bazilevs *et al.* (2008), is 2D flow past a thin elastic beam attached to a fixed, rigid square block. This test problem was proposed in Wall (1999) to study the accuracy

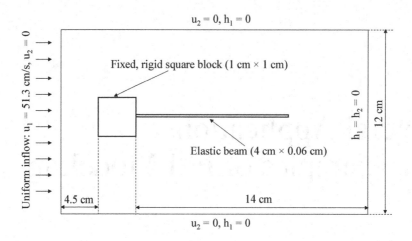

Figure 7.1 2D flow past an elastic beam attached to a fixed, rigid block. Problem setup

and robustness of FSI methods. The problem setup is shown in Figure 7.1. The flow is driven by a uniform inflow velocity of 51.3 cm/s. The lateral boundaries are assigned zero normal velocity and zero tangential stress. Zero-traction boundary condition is applied at the outflow.

The fluid density and viscosity are 1.18×10^{-3} g/cm^3 and 1.82×10^{-4} g/cm s, respectively, resulting in a Reynolds number of 100 based on the edge length of the block. The beam is modeled as a solid made of the neo-Hookean material described in Section 1.2. The density of the beam is 0.1 g/cm^3, and the Young's modulus and Poisson's ratio are 2.5×10^6 g/cm s^2 and 0.35, respectively. The problem dimensions, material properties, and boundary conditions are taken from the original reference.

The ALE FSI method is employed in this computation. The combined fluid and structure mesh is comprised of 6936 quadratic NURBS elements. The through-thickness discretization of the beam consists of two C^1-continuous quadratic elements and four basis functions. The mesh is allowed to move everywhere in the fluid mechanics domain except at the inflow and on the rigid block, where it is held fixed, and also at the lateral boundaries and outflow boundary, where the mesh is constrained not to move in the normal direction. The time-step size is 0.00165 s.

Figure 7.2 shows the velocity vectors and pressure at different instants. The flow features are characteristic of Re = 100. Vortices that are being shed from the square block are impinging on the bar, eventually forcing it into an oscillating motion. The bar experiences large deformations, requiring a robust mesh-moving algorithm. The Jacobian-based stiffening method is employed in this example (see Section 4.7). The computation was performed without remeshing and the good quality of the NURBS mesh is preserved as a result of the Jacobian-based stiffening method (see the smaller frames in Figure 7.2). Comparison with the data from Wall (1999) for a periodic flow regime is shown in Figure 7.3. The amplitude of the tip displacement is between 1.0 and 1.5 cm, and the period is approximately 0.33 s. The results are in good agreement with those from Wall (1999).

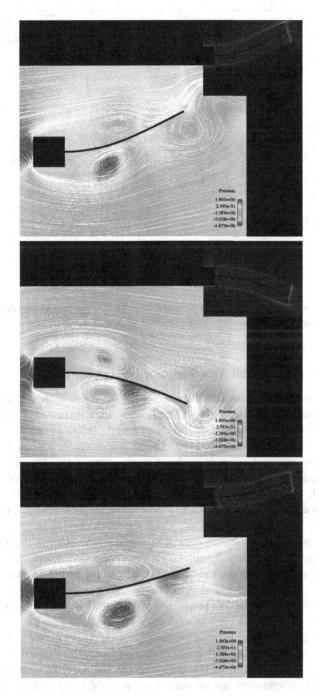

Figure 7.2 2D flow past an elastic beam attached to a fixed, rigid block. Larger frames: fluid velocity vectors superposed on the pressure. Smaller frames: deformed fluid mesh

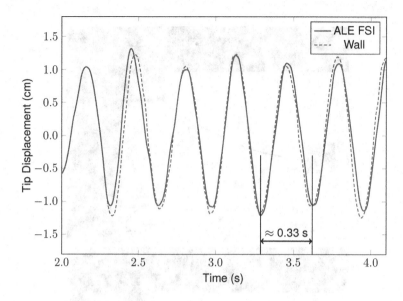

Figure 7.3 2D flow past an elastic beam attached to a fixed, rigid block. Tip displacement at the midplane as a function of time. Data from Wall (1999) is shown for comparison

7.2 2D Flow Past an Airfoil Attached to a Torsion Spring

This computation is from Takizawa and Tezduyar (2012b). The airfoil is NACA 64-618. The computational domain is $(-5, 10) \times (-5, 5)$. The leading edge is located at $(0, 0)$. The length, velocity, and density scales are based on the chord length, inflow velocity, and fluid density, respectively. The Reynolds number is 1000. The airfoil is attached to the torsion spring at $(0.9, 0)$. The spring has zero torsion when the angle of attack is $15°$. The inertia of the airfoil with respect to the attachment point of the spring is 0.5, and the spring coefficient is 10 rad^{-1}. The mesh is made of linear finite elements (see Figure 7.4); it has 1450 nodes and 2780 elements. As shown in the figure, the mesh in the core rectangular region is rotating with the airfoil, and the rest of the mesh is deforming with the mesh moving technique mentioned in Section 4.7. The boundary conditions consist of a uniform velocity at the inflow boundary, zero stress at the outflow boundary, no-slip conditions on the airfoil, and slip conditions at the top and bottom boundaries. We use the quasi-direct coupling technique (see Section 6.1.2). The predictors for the fluid and airfoil velocities are the velocities from the previous time step. This results in an interface movement when the airfoil velocity is nonzero. Therefore we solve the mesh moving block once as a "domain predictor."

Prior to the actual computation, to wipe out the initial shakeup, we compute 200 steps with a time-step size of 1.0, and to have a more time-accurate solution, we compute 200 steps more with a time-step size of 0.1, followed by another 200 steps with a time-step size of 0.01. These computations are all performed with the DSD/SST-SUPS technique. At this point we set $t = 0.0$, and compute the problem with both the DSD/SST-SUPS and the conservative form of DSD/SST-VMST. The stabilization parameters are those given by Equations (4.115)–(4.120) and (4.125). The time-step size is 0.01. In all stages of the computations

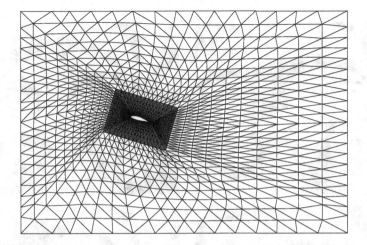

Figure 7.4 2D flow past an airfoil attached to a torsion spring. Mesh made of linear finite elements. The mesh has 1450 nodes and 2780 elements. The shaded region rotates with the airfoil

the number of nonlinear iterations per time step is 4. The number of GMRES iterations for the fluid+structure block is 30, 60, 270, and 270 for the first, second, third, and fourth nonlinear iterations, respectively. For the mesh moving block, including when used as the domain predictor, we have 30 GMRES iterations per nonlinear iteration. We use Selective Scaling (see Section 6.1.2), with the scale for the structure part set to 100.

Figure 7.5 shows the pressure coefficients at $t = 3.07$ and $t = 3.75$, when the angle of attack is approximately minimum and maximum. The angle of attack is measured with respect to the free-stream velocity. In both cases, the pressure coefficient is normalized using the stagnation-point pressure at $t = 3.75$. Figure 7.6 shows the time history of the angle of attack. Figure 7.7 shows the time history of the torque acting on the airfoil, which consists of the aerodynamic and spring torques. We see that the airfoil motion is less damped with DSD/SST-VMST than it is with DSD/SST-SUPS.

7.3 Inflation of a Balloon

This computation is from Tezduyar and Sathe (2007). A balloon, initially spherical, is inflated by pumping air through a circular hole as shown in Figure 7.8. The inflow is pulsating in the form of a cosine wave with a period of 2 s. The minimum and maximum values of the magnitude of the inflow velocity are 0.0 m/s and 2.0 m/s. Initially, the diameter of the balloon is 2 m and the diameter of the circular hole is 0.625 m. The thickness, density, and stiffness of the balloon are 2.0 mm, 100 kg/m^3, and 1.0×10^3 N/m^2, respectively. The mesh for the balloon consists of 1479 nodes and 2936 three-node triangular membrane elements. The fluid mechanics mesh for the air inside balloon contains 6204 nodes and 32 455 four-node tetrahedral elements. The computation is carried out with the SSTFSI-TIP1 technique (see Remarks 4.10 and 5.9) and the SUPG test function option WTSA (see Remark 4.4). The stabilization parameters used are those given by Equations (4.115)–(4.120) and (4.124). The quasi-direct coupling technique (see Section 6.1.2) is used. The GMRES search technique is

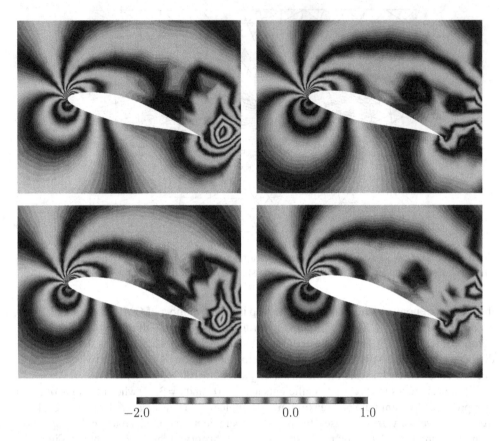

Figure 7.5 2D flow past an airfoil attached to a torsion spring. Pressure coefficient at $t = 3.07$ (left) and $t = 3.75$ (right). Computed with DSD/SST-SUPS (top) and DSD/SST-VMST (bottom)

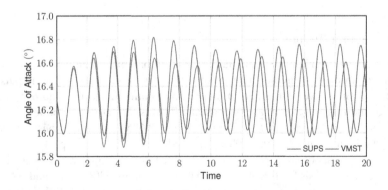

Figure 7.6 2D flow past an airfoil attached to a torsion spring. Time history of the angle of attack. Computed with DSD/SST-SUPS ("SUPS") and DSD/SST-VMST ("VMST")

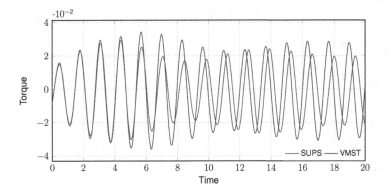

Figure 7.7 2D flow past an airfoil attached to a torsion spring. Total (aerodynamic and spring) torque acting on the airfoil. Computed with DSD/SST-SUPS ("SUPS") and DSD/SST-VMST ("VMST")

Figure 7.8 Inflation of a balloon. Problem setup

used with a diagonal preconditioner. The time-step size is 0.1 s, and the computation duration is 4 s. The number of nonlinear iterations per time step is 5, and the number of GMRES iterations per nonlinear iteration is 30. The entire computation was completed without any remeshing. Figure 7.9 shows that volumetric flow rate for the inflow matches the rate of change for the balloon volume. Figure 7.10 shows that the instantaneous balloon volume matches the initial balloon volume plus the volume of air added. Figures 7.11 and 7.12 show the flow field during the two periods of inflation.

7.4 Flow Through and Around a Windsock

This computation is from Tezduyar and Sathe (2007). The windsock has a length of 1.5 m and a diameter ranging from 0.25 m upstream to 0.15 m downstream (see Figure 7.13). Initially

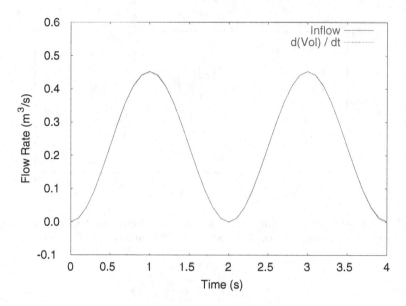

Figure 7.9 Inflation of a balloon. Volumetric flow rate for the inflow and the rate of change for the balloon volume

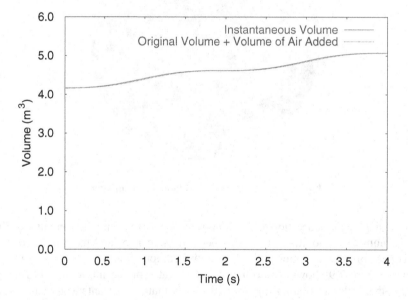

Figure 7.10 Inflation of a balloon. Instantaneous balloon volume compared with the initial balloon volume plus the volume of air added

the windsock is in a horizontal position, and the starting condition for the flow field is the developed flow field corresponding to a rigid windsock held in that horizontal position. Then the gravity is turned on for the windsock, the FSI starts, and the windsock starts falling down. The wind velocity is constant at 10 m/s. The thickness, density, and stiffness of the

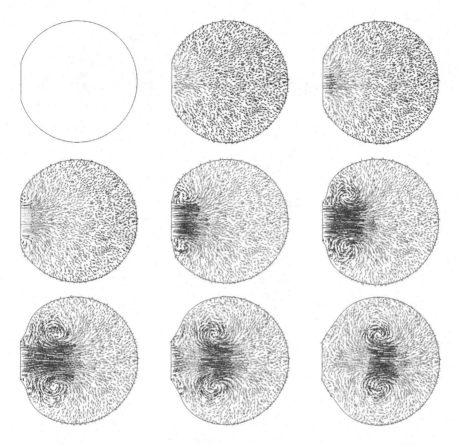

Figure 7.11 Inflation of a balloon. Velocity vectors colored by air pressure, from 0 to 2 s

windsock are 2.0 mm, 100 kg/m^3, and 1.0×10^6 N/m^2, respectively. The upstream edge of the structure is held fixed while the remaining structure is free and flaps in cycles. The mesh for the windsock is semi-structured and consists of 984 nodes and 1920 three-node triangular membrane elements (see Figure 7.14). The fluid mechanics mesh contains 19 579 nodes and 113 245 four-node tetrahedral elements. Initially, the fluid mesh at the interface is identical to the windsock mesh. The computation is carried out with the SSTFSI-SV technique (see Remarks 4.11 and 5.9) and the SUPG test function option WTSE (see Remark 4.4). The stabilization parameters used are those given by Equations (4.117)–(4.123) and (4.125). The quasi-direct coupling technique (see Section 6.1.2) is used. The GMRES search technique is used with a diagonal preconditioner. The time-step size is 0.0125 s, and the computation duration is two flapping cycles. The number of nonlinear iterations per time step is 5, and the number of GMRES iterations per nonlinear iteration is 30.

We expected the windsock to develop kinks as it flaps in the wind. Therefore we used the FSI-GST (see Section 5.5) for smoothing the fluid mesh at the interface. The nodes of the windsock mesh were generated on straight longitudinal gridlines, and with that we were able to use the directional version of the FSI-GST, that is, FSI-DGST (see Section 5.5).

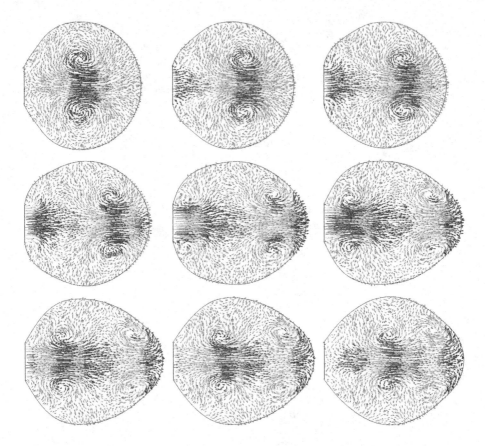

Figure 7.12 Inflation of a balloon. Velocity vectors colored by air pressure, from 2 to 4 s

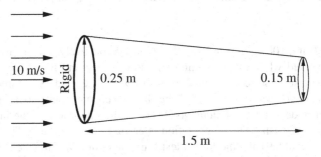

Figure 7.13 Flow through and around a windsock. Problem setup

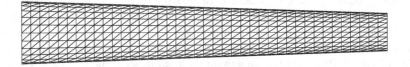

Figure 7.14 Flow through and around a windsock. The initial windsock mesh

General Applications and Examples of FSI Modeling

For a node A on such a gridline, we use a weighted averaging involving four nearby nodes on each side: $A\pm 1, A\pm 2, A\pm 3$ and $A\pm 4$. The weighted averaging formula is given as follows:

$$\mathbf{x}_A^{\text{SMOOTH}} = 0.2\mathbf{x}_A + 0.16(\mathbf{x}_{A-1} + \mathbf{x}_{A+1}) + 0.12(\mathbf{x}_{A-2} + \mathbf{x}_{A+2}) \\ + 0.08(\mathbf{x}_{A-3} + \mathbf{x}_{A+3}) + 0.04(\mathbf{x}_{A-4} + \mathbf{x}_{A+4}). \tag{7.1}$$

We note that this directional smoothing does not introduce any smoothing in the circumferential direction. During the FSI computations the structure develops kinks, which would make mesh updating more difficult and increase the frequency of remeshing. With the FSI-DGST, two flapping cycles were computed without any remeshing. Figure 7.15 shows the structural and fluid mechanics meshes at the interface, one with a kink and the other smooth. Figure 7.16 shows the zoomed (around the kink) versions of the pictures in Figure 7.15. Figure 7.17 shows the windsock and the flow field at various instants.

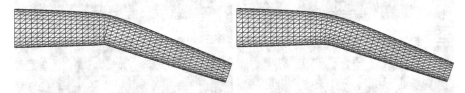

Figure 7.15 Flow through and around a windsock. Meshes at the interface: structure (left) and fluid (right)

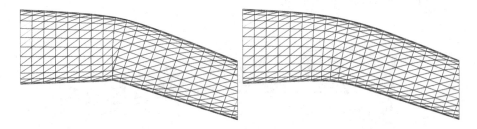

Figure 7.16 Flow through and around a windsock. Meshes at the interface: structure (left) and fluid (right)

7.5 Aerodynamics of Flapping Wings

This computation is from Takizawa *et al.* (2012d). The motion and deformation patterns of the wings are prescribed. They are based on data extracted from the high-speed, multi-camera video recordings of a locust in a wind tunnel.

7.5.1 Surface and Volume Meshes

Based on a digital, scanned copy of locust wings, we construct a surface mesh of the forewing (FW) and hindwing (HW) using NURBS. The FW is modeled with a single, degenerated

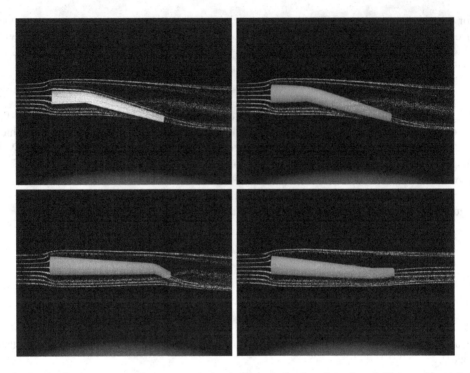

Figure 7.17 Flow through and around a windsock. The windsock and the flow field (traced particles) at various instants. Sectional view (left top)

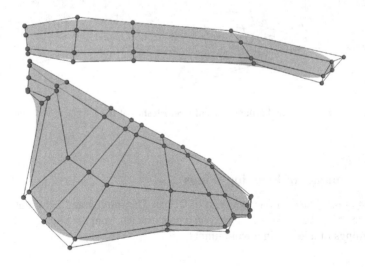

Figure 7.18 Forewing (FW) and hindwing (HW) surfaces represented by NURBS and the control points

Figure 7.19 Wing and body surface meshes with triangular elements

patch. The HW is modeled with two patches – one with and one without degeneration. There are 21 and 51 control points for the FW and HW, respectively (see Figure 7.18).

We also generate a surface mesh of the locust body using 16 NURBS patches. We base the mesh on empirical height and width measurements provided at five cross-sectional positions and estimate the axial curvature of the body using video of the locust flying. After the wing spatial-control meshes are deformed to the various positions in the flapping motion as we will describe in Section 7.5.2, we discretize them at each temporal-control point. The triangular surface mesh used in the computation is shown in Figure 7.19.

For automatic volume mesh generation, the wings have a finite thickness of 1% of the FW root chord, which tapers to zero thickness at the wing edges. We generate a one-layer refinement region near the wing surface. In this region, the element height is 10% of the FW root chord. In addition, we have a cylindrical region of increased refinement around the locust. We also specify increased volume mesh refinement in the region between the FW and HW. The volume mesh within this cylindrical region is then generated with tetrahedral elements using an automatic mesh generator. Next we define a box that contains the cylindrical region and generate a tetrahedral mesh in that, again using an automatic mesh generator. We rotate this mesh to an angle representing the approximate body angle of the locust within the full computational domain. A volume mesh within the full domain is then also automatically generated with tetrahedral elements. The number of nodes and elements in the volume mesh varies between each temporal patch. The average number of nodes and elements in these meshes are approximately 430 000 and 2.6 million, respectively. The volume mesh and refinement regions are shown in Figure 7.20.

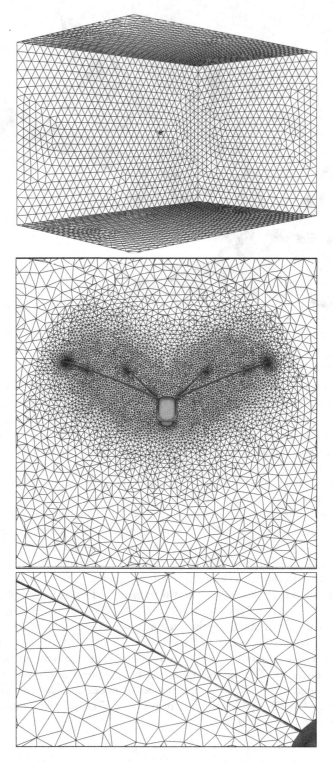

Figure 7.20 Volume mesh shown for the full computational domain (top), cylindral-refinement region (middle), and refinement region near the wing surface (bottom)

7.5.2 Flapping-Motion Representation

We must also provide a prescribed motion that is representative of straight-flight flapping. Here we face the challenge of reconciling data acquired through photogrammetry with data that is suitable as an input for computational analysis. We use 4, 1, and 10 wind-tunnel-acquired tracking points for the body, HW (wingtip) and FW (1 common with the body). This motion is reflected across the sagittal plane of the locust to create symmetric flapping motion. We generate additional HW "tracking points" to deform the wing in a similar manner to that observed in wind tunnel videos. In total, we use 76 tracking points to represent the desired wing motion.

Next, we temporally interpolate our representative data set. For each tracking point, we apply a temporal NURBS representation, as discussed in Sections 6.4.2 and 6.4.3. We use quadratic B-splines with temporal-control points as defined by Equation (6.66), and we reduce 171 sampling points (about 7 flapping cycles) to 60 temporal-control points by using Equation (6.67).

Then, we spatially represent the tracking points at each temporal-control point. Spatial interpolation is accomplished using the SSDM described in Section 6.5. An illustration of this process for the left HW is shown in Figure 7.21. The FW and HW SS consist of 6 and 9 control points, respectively.

Figure 7.21 Deformed SS and associated control points along with the projected HW NURBS surface

In the least-squares fit from tracking points to SS control points, the control points nearest to the body are fixed. To minimize the effect of an unrealistic least-squares fit due to the single tracking point at the tip, additional points are generated using linear extrapolation between the outermost tracking points. These additional points are included in the 76 total points mentioned earlier. At each temporal-control point, a final least-squares projection is

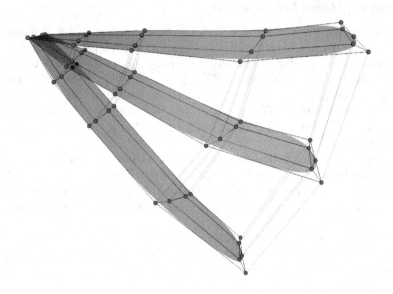

Figure 7.22 FW control mesh and corresponding surface at three temporal-control points

then performed between each SS and corresponding wing surface defined by the NURBS mesh. Now, we have a NURBS-represented data set in both space and time for each wing as illustrated for the FW in Figure 7.22.

The motion of the wings requires that we remesh at some time during the flapping cycle. To facilitate remeshing, we use temporal knot insertion (see Section 6.6.2) to create equal-duration temporal patches prior to volume meshing. Each temporal patch contains 5 control points. We note that the spatial position corresponding to the last control point of each temporal patch is identical to that of the first control point in the next patch. Within each temporal patch, we select the middle control point and generate a volume mesh.

7.5.3 Mesh Motion

To capture the wing motion and deformation within each temporal patch, the volume mesh inside the box must be deformed to the corresponding temporal-control surface mesh. Due to the relatively large change in deformation between each temporal-control point, we use subiterations for the mesh computation to divide the steps between temporal-control points into 20 smaller steps. We move the mesh, which corresponds to the middle control point, backward and forward through each smaller step using 1500 GMRES iterations. Using this approach, as shown in Figure 7.23, the worst mesh quality occurs at the beginning and end of each temporal patch.

To summarize, each patch has 5 control points (3 knot spans). For each patch, we generate the mesh only once (corresponding to the middle control point) and move the mesh 4 times (corresponding to the other 4 control points) using the mesh-moving equations. We can then, in the space–time computation, have as many space–time slabs associated with each knot

General Applications and Examples of FSI Modeling 187

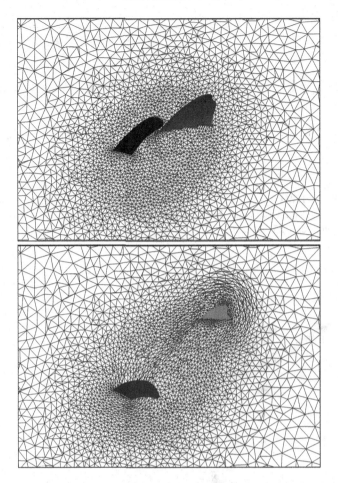

Figure 7.23 The volume mesh obtained by the automatic mesh generator (top) and after being moved to the first temporal-control point of that patch (bottom)

span as determined by the time-step size requirements. In the computation over each of those space–time slabs, we interpolate the mesh information needed from the temporal NURBS representation of the mesh.

We use a fluid dynamics starting condition as described in Section 6.7.2.1. We obtain the temporal-control meshes M_c^{-1} and M_c^{-2} by using Equations (6.88) and (6.89). This starting condition can be seen in the gray region of Figure 7.24.

7.5.4 Fluid Mechanics Computation

Prior to beginning the prescribed flapping motion, we compute for 200 time steps to develop the flow field. Over the first 100 time steps of this computation, we use a cosine form to smoothly increase the inflow velocity from 0 to 2.4 m/s, which represents the average wind tunnel velocity. In this flow-development computation, the time-step size is 2.2×10^{-4} s, with

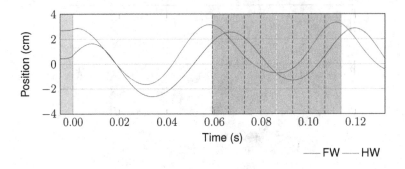

Figure 7.24 FW and HW tip position in time with the shaded regions showing the extrapolation region (gray) and section used for visualizing the results (green). Dashed, vertical lines indicate the points in the cycle used in Figures 7.25, 7.26, and 7.27

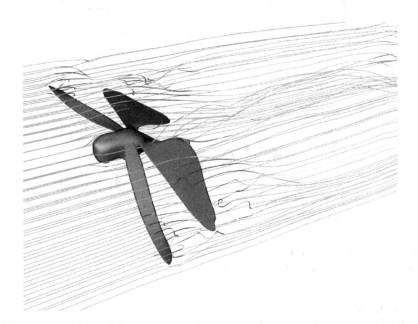

Figure 7.25 Streamlines colored by velocity at the instant indicated by the vertical, fifth dashed line in Figure 7.24

3 nonlinear iterations per time step. We use the DSD/SST-SUPS technique. The stabilization parameters are those given by Equations (4.115)–(4.120) and (4.127). We compute the stabilization parameters after the predictor step, and use the same values at all three nonlinear iterations. The number of GMRES iterations for the nonlinear iterations are 30, 60, and 90.

In the computation with flapping, we use 25 space–time slabs (with linear basis functions) for each of the 3 knot spans in the temporal representation of the mesh, which results in a remeshing frequency of every 75 time steps. The time-step size is 2.2×10^{-4} s. We use 4 nonlinear iterations per time step. The DSD/SST-SUPS and the conservative form of

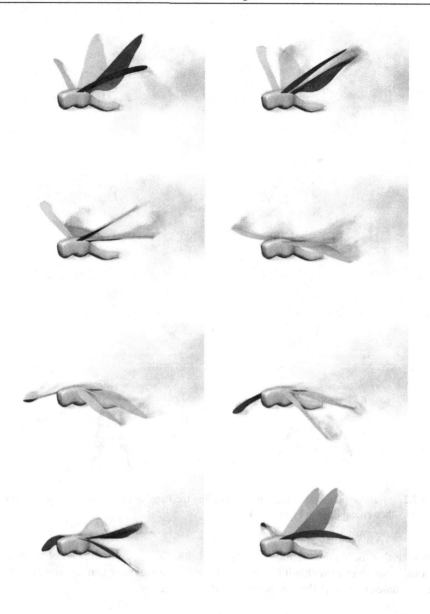

Figure 7.26 Vorticty at eight instants during the flapping cycle (left to right, top to bottom) indicated by the vertical lines in Figure 7.24

DSD/SST-VMST are used for the first two and last two nonlinear iterations, resepectively. The stabilization parameters are those given by Equations (4.115)–(4.120) and (4.127).

We compute the stabilization parameters after the predictor step and after the first two nonlinear iterations. The number of GMRES iterations for the nonlinear iterations are 30, 60, 60, and 120.

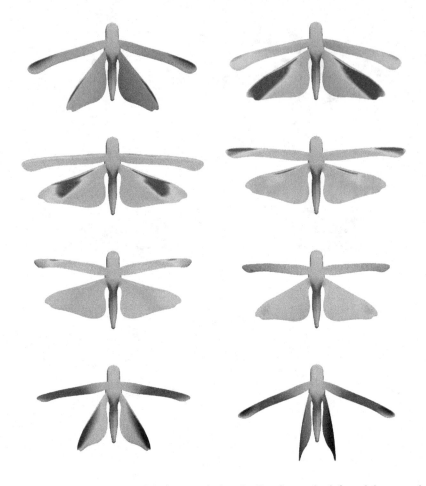

Figure 7.27 Surface pressures at eight instants during the flapping cycle (left to right, top to bottom) indicated by the vertical lines in Figure 7.24

We note that the fluid dynamics starting condition, described at the end of Section 7.5.3, is accomplished over a patch with single knot span, with the other computational parameters being the same as those in the computation with flapping.

REMARK 7.1 *We have seldom observed close-to-zero or negative diagonal terms when the time-step size is large. This occurs when the fine-scale velocity is much larger than the coarse-scale (discrete) velocity. We believe this is due to the predictor, which assumes all velocities are the same from the previous time step except those with Dirichlet conditions.*

Because the wing velocity varies during the stroke, the Reynolds number range is 1000–2500 as calculated at 75% of the wing span (measured from root chord to tip). Figures 7.25–7.27 show the results form the preliminary computations.

8

Cardiovascular FSI

Application of computational fluid dynamics (CFD) (without FSI) to the simulation of blood flow in patient-specific geometries, first reported by Taylor *et al.* (1998a), has led to significant advances in our understanding of how the mechanics of blood flow and the hemodynamic quantities of interest are affected by the vessel wall geometry and boundary conditions. Over the last 10–15 years CFD was successfully used for a variety of clinical applications, including blood flow studies in major vessels, comparison of rest and exercise conditions, examination of surgical treatment options, simulation of medical devices, such as blood pumps and drug-eluting stents, and many more (see, e.g., a review article Bazilevs *et al.*, 2010b and references therein). Just CFD modeling without FSI continues to dominate blood flow computations to this day primarily because of the availability of general-purpose commercial CFD software. The Reynolds number at the peak systole, which characterizes the flow regime and complexity, is of the order of several hundreds in cerebral arteries and a few thousands in the aortic arch. This range of Reynolds numbers corresponds to complex 3D flows, which are, nevertheless, laminar. Mild turbulence is seen only under rare circumstances, if at all. Consequently, state-of-the-art commercial CFD software can deliver stable and reasonably accurate solutions for patient-specific hemodynamics on meshes with a reasonable number of degrees-of-freedom.

Nevertheless, just CFD modeling of blood flow invokes the simplifying assumption that the blood vessel walls are stationary and rigid, which is not physiologically realistic. Vascular walls are (visco)elastic and undergo large deformations due to hemodynamic forces. Wall deformations alter the blood flow patterns, which, in turn, alter the hemodynamics. Therefore, for the computational modeling to be physiologically realistic, the blood flow and wall deformation need to be treated in a coupled fashion, and that makes it an FSI problem, where the blood is the fluid and the vascular wall is the structure.

The coupled problem is formulated as follows. The blood flow is governed by the Navier–Stokes equations of incompressible flows. The vascular wall is typically assumed to behave as an elastic material that is allowed to undergo large deformations (Humphrey, 2002). For simplicity, a hyperelastic framework is typically employed for this purpose (see Holzapfel, 2000; Humphrey, 2002 and references therein). Hyperelasticity makes use of a stored energy

---

Computational Fluid–Structure Interaction: Methods and Applications, First Edition.
Yuri Bazilevs, Kenji Takizawa and Tayfun E. Tezduyar.
© 2013 John Wiley & Sons, Ltd. Published 2013 by John Wiley & Sons, Ltd.

functional, which depends on the strain experienced point-wise by the vascular wall relative to a convenient reference configuration, often assumed to be unloaded. Different forms of the stored energy functionals give rise to different constitutive models for the wall, which may vary in complexity depending on the phenomena one is trying to represent. The subject of constitutive modeling of vessel wall tissue is rich and may be found in a recent review article (Holzapfel and Ogden, 2010 and references therein). At the luminal surface, the interface between the blood-filled lumen and vascular wall, kinematic and traction compatibility conditions are assumed to hold point-wise. That is, the blood flow velocity must equal that of the wall and the tractions associated with the fluid and structural domains must balance at every point on the luminal surface. These conditions are both physically meaningful and lead to a mathematically well-posed problem. To complete the formulation of the FSI problem, the motion of the blood vessel, which includes the vascular wall and the blood-filled lumen, must be computed point-wise. This added complexity of the FSI modeling relative to just CFD modeling creates a new set of computational challenges, which are discussed in this chapter. The poor performance of loosely-coupled procedures and the lack of commercial software allowing monolithic fluid–structure coupling make FSI modeling less accessible for hemodynamic research than just CFD modeling.

Although introducing FSI for blood flow increases modeling and simulation complexity, the computations produce physiologically more realistic results than those generated by just CFD. Effects of including wall elasticity in vascular simulations have been examined, for example, for carotid artery (Torii *et al.*, 2004, 2006a), for cerebral aneurysms (Torii *et al.*, 2007a, 2008; Bazilevs *et al.*, 2010c), and for the total cavopulmonary connection (Bazilevs *et al.*, 2009b). The rigid-wall assumption consistently shows an overestimation of the wall shear stress (WSS) compared to the flexible wall, in some cases by as much as 50%. Some qualitative and quantitative differences between the rigid- and flexible-wall simulations were also observed for the blood flow patterns. We note that the blood vessels of young children are significantly more flexible that those of adults, making FSI modeling especially important for pediatric cardiology (Bazilevs *et al.*, 2010b).

Unlike the rigid-wall assumption, FSI enables simulation of the complete mechanical environment of the vascular wall, both the loads acting on the wall due to blood flow and the loads acting within the wall. The latter loads are particularly important because they act on the cells that control wall structure and function, which in turn may change the bulk elastic properties of the wall and hence the hemodynamics.

The preferred method of handling the moving interfaces involved in FSI modeling of patient-specific cardiovascular hemodynamics has mostly been the ALE finite element formulation. ALE methods in the context of Galerkin formulations with BB-stable finite element pairs for the velocity and pressure variables were employed in Gerbeau *et al.* (2005) and Fernandez and Moubachir (2005). An ALE-VMS formulation (see Section 4.6.1), which is a more robust CFD technology suitable for equal-order velocity-pressure interpolation, was employed in cardiovascular FSI simulations of a number of patient-specific models of major blood vessels (see Bazilevs *et al.*, 2006a, 2008, 2009a,b, 2010c,d; Zhang *et al.*, 2007, 2009; Isaksen *et al.*, 2008; Hsu and Bazilevs, 2011).

Patient-specific arterial FSI modeling with the DSD/SST formulation was first reported in Torii *et al.* (2004). Over the years following, Torii *et al.* conducted one of the most extensive series of patient-specific arterial FSI modeling of cerebral aneurysms (Torii *et al.*, 2006a,b, 2007a,b, 2008, 2009, 2010a,b, 2011). The cases studied in these articles by Torii *et al.* were almost all for middle cerebral arteries, and the geometries were constructed from computed tomography (CT) images. In these arterial FSI computations the DSD/SST formulation was used together with advanced mesh update methods (Tezduyar *et al.*, 1992b, 1993; Johnson and Tezduyar, 1994; Tezduyar, 2001b) and was implemented with block-iterative coupling (Tezduyar, 2004a). The inflow boundary condition used in the computations is a pulsatile velocity profile, which closely represents the measured flow rate during a heartbeat cycle. A brief, chronological review of the computations reported in Torii *et al.* (2004, 2006a,b, 2007a) was provided in Tezduyar *et al.* (2007b). The SSTFSI technique was extended in Tezduyar *et al.* (2007b, 2008b, 2009a,b, 2010a) and Takizawa *et al.* (2010a,b) to cardiovadcular FSI modeling, with emphasis on arteries with aneurysm.

A number of special techniques for arterial FSI were developed in conjunction with the SSTFSI technique. These include techniques for calculating an estimated zero-pressure (EZP) arterial geometry (Tezduyar *et al.*, 2007a, 2008b, 2009b; Takizawa *et al.*, 2010a,b), a special mapping technique for specifying the velocity profile at an inflow boundary with non-circular shape (Takizawa *et al.*, 2010a; Tezduyar *et al.*, 2009b), techniques for using variable arterial wall thickness (Takizawa *et al.*, 2010a,b; Tezduyar *et al.*, 2009b), mesh generation techniques for building layers of refined fluid mechanics mesh near the arterial walls (Tezduyar *et al.*, 2009a,b; Takizawa *et al.*, 2010a,b), a recipe for pre-FSI computations that improves the convergence of the FSI computations (Tezduyar *et al.*, 2007b, 2008b), the Sequentially-Coupled Arterial FSI (SCAFSI) technique (Tezduyar *et al.*, 2007c, 2008b, 2009a,b, 2010a) and its multiscale versions (Tezduyar *et al.*, 2009a,b, 2010a), and techniques (Takizawa *et al.*, 2010b) for the projection of fluid–structure interface stresses, calculation of the WSS, and calculation of the oscillatory shear index (OSI). In FSI modeling of three cerebral artery segments with aneurysm reported in Takizawa *et al.* (2011c), the arterial geometries came from 3D rotational angiography (3DRA). In Takizawa *et al.* (2011c), the computational challenges related to extraction of the arterial-lumen geometry from 3DRA, generation of a mesh for that geometry, and building a good starting point for the FSI computations were also addressed. In Tezduyar *et al.* (2011) and Takizawa *et al.* (2012b), new techniques were presented for determining the shrinking amount in the EZP process, the arterial wall thickness, and the thickness of the layers of refined fluid mechanics volume mesh near the arterial walls. These techniques were originally proposed in Remark 3 of Takizawa *et al.* (2011c), but the description was very brief. In Tezduyar *et al.* (2011) and Takizawa *et al.* (2012b), a new scaling technique was also introduced for specifying a more realistic volumetric flow rate.

In Bazilevs *et al.* (2009b), a technique was proposed to use the Laplace's equation for specifying a variable vessel wall thickness, and in Bazilevs *et al.* (2010c) and Hsu and Bazilevs (2011) a prestressing technique was developed for blood vessels. The former addressed the challenge of how to specify spatially varying vessel wall thickness. It inspired the idea of using the Laplace's equation over the surface mesh covering the lumen to specify a variable vessel wall thickness (Takizawa *et al.*, 2011c, 2012b; Tezduyar *et al.*, 2011). The latter addressed the challenge presented by the fact that patient-specific blood vessel geometry data comes from a configuration that is not stress-free, the same fact that earlier motivated the development of methods for calculating an EZP arterial geometry (Tezduyar *et al.*, 2007a, 2008b). Both techniques are quite general and, most importantly, are independent of the details of the patient-specific blood vessel geometry.

In this chapter we provide a description of these special space–time and ALE techniques developed by the authors' research teams for patient-specific cardiovascular FSI modeling, and present results from earlier computations. These special techniques were developed to enable the application of the core FSI methods presented in the earlier chapters of this book to patient-specific cardiovascular FSI.

8.1 Special Techniques

8.1.1 Mapping Technique for Inflow Boundaries

The special mapping technique for inflow boundaries was introduced in Takizawa et al. (2010a). We repeat here from Takizawa et al. (2010a) the need for such a mapping technique and how the technique works.

Some inflow profiles require the inlet to be circular, however the inlets in many of the geometries we encounter are not circular. Furthermore, as the artery deforms, the inlet shape changes. Thus, even if the inlet is initially circular, it will not remain so. The technique introduced in Takizawa et al. (2010a) to meet this requirement maps the inflow boundaries from non-circular shapes to circular shapes. The actual inflow profile $U(\mathbf{z}, t)$, where $\mathbf{z}$ is the coordinate vector in the inflow plane, is obtained by mapping from a preferred inflow profile $U^P(r, t)$. Here r is the circular coordinate and $0 \leq r \leq r_B$, where r_B is the average radius of the inflow cross-sectional area, which comes from the image-based data. It is calculated by dividing that area by π and taking the square-root of that.

The technique involves two steps:

1. Map $\mathbf{z}$ to r and calculate a "trial" velocity:

$$r(\mathbf{z}) = \frac{\|\mathbf{z} - \mathbf{z}_C\|}{\|\mathbf{z} - \mathbf{z}_B\| + \|\mathbf{z} - \mathbf{z}_C\|} r_B, \tag{8.1}$$

$$U^T(\mathbf{z}, t) = U^P(r, t), \tag{8.2}$$

where subscripts "C" and "B" denote the centroid and the closest boundary, respectively, as shown in Figure 8.1, and the superscript "T" stands for "trial."

2. Adjust the velocity:

$$U(\mathbf{z}, t) = \frac{Q(t)}{\int_{\Gamma_{IN}} U^T(\mathbf{z}, t) \, d\Gamma} U^T(\mathbf{z}, t), \tag{8.3}$$

where Q is the flow rate and Γ_{IN} is the discretized inflow area; that is, the integration area in the finite element space.

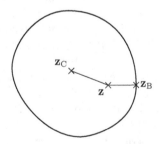

Figure 8.1 Special mapping technique

8.1.2 Preconditioning Technique

In computations with hyperelastic materials, we do not compute the diagonal of the tangent stiffness matrix. Therefore, as proposed in Tezduyar *et al.* (2009a), we use a diagonal preconditioner based on the assembly of only the element-level lumped mass matrices $\mathbf{m}^e_{\text{LUMP}}$, but after being multiplied by a factor that, to some extent, takes into account the material stiffness. In computations with the Fung material, for the multiplication factor, as proposed in Tezduyar *et al.* (2009a), we use $\left(C^e_{\text{HYFU}}\right)^2$, where

$$C^e_{\text{HYFU}} = \max\left(\frac{\sqrt{\lambda^{\text{FP}}/\rho}\, \Delta t}{h^e}, \frac{\sqrt{\mu^{\text{FP}}/\rho}\, \Delta t}{h^e}\right). \tag{8.4}$$

Here h^e is the cube-root of the element volume, and λ^{FP} and μ^{FP} are given as

$$\lambda^{\text{FP}} = \frac{6 D_1 D_2 \nu_{\text{PEN}}}{(1 + \nu_{\text{PEN}})(1 - 2\nu_{\text{PEN}})}, \tag{8.5}$$

$$\mu^{\text{FP}} = \frac{3 D_1 D_2}{(1 + \nu_{\text{PEN}})}. \tag{8.6}$$

In Tezduyar *et al.* (2009a), $\left[1 + (1 - \alpha)\beta \left(2 C^e_{\text{HYFU}}\right)^2\right]$ was proposed as an alternative multiplication factor, where α and β are the parameters of the Hilber–Hughes–Taylor time integration scheme (Hilber *et al.*, 1977).

REMARK 8.1 *We use the Selective Scaling (see Section 6.1.2) to shift the emphasis between the fluid and structure parts. As pointed out in Takizawa et al. (2010a), the preconditioner described in this section improves, within the fluid+structure block, the relative scaling between the fluid and structure parts and provides a better beginning point for selective scaling, if we still find a need to use such a scaling.*

8.1.3 Calculation of Wall Shear Stress

A new technique for calculating the WSS was proposed in Takizawa *et al.* (2010b). We provide the description of the technique from Takizawa *et al.* (2010b).

We first decompose the spatial version of $(\mathbf{w}^h_{1\text{I}})^-_{n+1}$ into its two components:

$$\mathbf{w}^h_{1\text{I}} = \left(\mathbf{w}^h_{1\text{I}}\right)^{\text{W}} + \left(\mathbf{w}^h_{1\text{I}}\right)^{\text{R}}, \tag{8.7}$$

where $\left(\mathbf{w}^h_{1\text{I}}\right)^{\text{R}}$ is the part associated with the rim nodes at the lumen ends, and $\left(\mathbf{w}^h_{1\text{I}}\right)^{\text{W}}$ is the part associated with the rest of the fluid mechanics nodes at the arterial wall. We then calculate $\left(\mathbf{h}^h_\nu\right)_{1\text{I}}$ as follows:

$$\int_{\Gamma_h} \left(\mathbf{w}^h_{1\text{I}}\right)^{\text{W}} \cdot \left(\mathbf{h}^h_\nu\right)_{1\text{I}} \, d\Gamma = \int_\Omega 2\mu\boldsymbol{\varepsilon}\left(\left(\mathbf{w}^h_{1\text{I}}\right)^{\text{W}}\right) : \boldsymbol{\varepsilon}(\mathbf{u}^h) \, d\Omega$$

$$+ \sum_{e=1}^{(n_{\text{el}})_n} \int_{\Omega^e} \left(\mathbf{w}^h_{1\text{I}}\right)^{\text{W}} \cdot \boldsymbol{\nabla} \cdot \left(2\mu\boldsymbol{\varepsilon}(\mathbf{u}^h)\right) d\Omega, \tag{8.8}$$

$$\int_{\Gamma_h} \left(\mathbf{w}_{II}^h\right)^R \cdot \left((\mathbf{n} \times \mathbf{e}^R) \cdot \nabla\right) \left(\mathbf{h}_v^h\right)_{II} \, d\Gamma = 0, \qquad (8.9)$$

where $\mathbf{e}^R$ is the unit vector along the rim.

8.1.4 Calculation of Oscillatory Shear Index

The OSI is a measure of the degree to which WSS oscillates during a heart beat cycle. It is defined (see Taylor et al., 1998b) as follows:

$$\text{OSI} = \frac{1}{2}\left(1 - \frac{\left(\mathbf{h}_v^h\right)_{II}^{\text{NM}}}{\left(\mathbf{h}_v^h\right)_{II}^{\text{MN}}}\right), \qquad (8.10)$$

where, following the notation from Takizawa et al. (2010b), "NM" and "MN" stand for "norm of the mean" and "mean of the norm," and

$$\left(\mathbf{h}_v^h\right)_{II}^{\text{NM}} = \frac{1}{T}\left\|\int_0^T \left(\mathbf{h}_v^h\right)_{II} dt\right\|, \qquad (8.11)$$

$$\left(\mathbf{h}_v^h\right)_{II}^{\text{MN}} = \frac{1}{T}\int_0^T \left\|\left(\mathbf{h}_v^h\right)_{II}\right\| dt. \qquad (8.12)$$

Here T is the period of the cardiac cycle. Higher OSI indicates larger flow direction variation in a cardiac cycle. As pointed out in Takizawa et al. (2010b), calculating the OSI based on a fixed reference frame is not the best way, because, for example, if an artery segment undergoes rigid-body rotation, that should not influence the OSI. Two methods that exclude rigid-body rotation from the OSI calculation were proposed in Takizawa et al. (2010b).

Method 1

$$\left(\mathbf{h}_v^h\right)_{II}^{\Delta} = J\mathbf{F}^{-1}\left(\mathbf{h}_v^h\right)_{II}, \qquad (8.13)$$

where $\mathbf{F}$ is the deformation gradient tensor associated with the deformation of the fluid–structure interface (not the volumetric deformation gradient of the fluid-domain motion), and $J = \det \mathbf{F}$.

Method 2

$$\left(\mathbf{h}_v^h\right)_{II}^{\Delta} = \mathbf{R}^T\left(\mathbf{h}_v^h\right)_{II}, \qquad (8.14)$$

where $\mathbf{R}$ is the rotation tensor coming from the decomposition of $\mathbf{F}$ as

$$\mathbf{F} = \mathbf{R}\mathbf{U}, \qquad (8.15)$$

and $\mathbf{U}$ is the right stretch tensor.

For both methods, $\left(\mathbf{h}_v^h\right)_{1\mathrm{I}}^\Delta$ is calculated as follows:

$$\int_{(\Gamma_{1\mathrm{I}})_{\mathrm{ROSI}}} \mathbf{w}_{1\mathrm{I}}^h \cdot \left(\mathbf{h}_v^h\right)_{1\mathrm{I}}^\Delta \, \mathrm{d}\Gamma = \int_{(\Gamma_{1\mathrm{I}})_{\mathrm{ROSI}}} \mathbf{w}_{1\mathrm{I}}^h \cdot \mathcal{R}\left(\mathbf{h}_v^h\right)_{1\mathrm{I}} \, \mathrm{d}\Gamma, \tag{8.16}$$

where $\mathcal{R} = J\mathbf{F}^{-1}$ or $\mathcal{R} = \mathbf{R}^T$, and $(\Gamma_{1\mathrm{I}})_{\mathrm{ROSI}}$ is a reference configuration of the fluid–structure interface used in the OSI calculations. In Equations (8.11) and (8.12), we replace $\left(\mathbf{h}_v^h\right)_{1\mathrm{I}}$ with $\left(\mathbf{h}_v^h\right)_{1\mathrm{I}}^\Delta$.

REMARK 8.2 *A similar concept can be found in Green and Naghdi (1976) as the corotated Cauchy stress, $\mathbf{R}^T \boldsymbol{\sigma} \mathbf{R}$.*

REMARK 8.3 *The OSI calculations reported in this book are based on Equation (8.13).*

REMARK 8.4 *As pointed out in Takizawa et al. (2010b), the reference configuration used in Equation (8.16) is not necessarily the unstressed configuration of the fluid–structure interface. For the calculations reported in this book, it is the configuration corresponding to the instant when the pressure is at its time-averaged value (on the way up, i.e., at the ascending part of the pressure curve).*

8.1.5 Boundary Condition Techniques for Inclined Inflow and Outflow Planes

In earlier arterial FSI computations (Tezduyar *et al.*, 2008b, 2009a; Takizawa *et al.*, 2010a; Tezduyar *et al.*, 2010a; Takizawa *et al.*, 2010b) with the SSTFSI technique, the inflow and outflow planes were parallel to the Cartesian coordinate planes, and slip boundary conditions were imposed on those planes for the structural mechanics and mesh-moving equations. With the techniques introduced in Takizawa *et al.* (2011c), such slip boundary conditions were extended to inclined inflow and outflow planes. Here we describe those techniques from Takizawa *et al.* (2011c).

8.1.5.1 Structural Mechanics Equations

The unknown space for the structural mechanics nodes at the inflow and outflow planes is rotated in such a way that one of the directions is perpendicular to the plane. The normal vector of the plane, $\mathbf{n}_{\mathrm{S}2}$, is calculated for each arterial end by the area-weighted average of the normal vectors of the element surfaces at that end, where the subscripts "S" and "2" refer to the slip plane and the structural mechanics equations. With that, the normal component of the structural displacement is set to zero.

8.1.5.2 Mesh-Moving Equations

The fluid mechanics nodal positions calculated with Equation (5.112) include the positions of the rim nodes at the lumen ends. However, as pointed out in Takizawa *et al.* (2011c), there is no guarantee that these nodes will all be on the same plane. As proposed in Takizawa et al. (2011c), we bring them all to the same plane by adjusting their positions as follows:

$$(\mathbf{x}_{II}^h)_{n+1} \leftarrow (\mathbf{x}_{II}^h)_{n+1} \left(\left((\mathbf{x}_{S1}^h)_{n+1} - (\mathbf{x}_{II}^h)_{n+1} \right) \cdot \frac{\mathbf{n}_{S1}}{\|\mathbf{n}_{S1}\|} \right) \frac{\mathbf{n}_{S1}}{\|\mathbf{n}_{S1}\|}, \quad (8.17)$$

where $\mathbf{x}_{S1}^h$ is the centroid of the set of fluid element edges coinciding with the rim, and the normal vector of the plane, $\mathbf{n}_{S1}$, is calculated by using the following expression:

$$\mathbf{n}_{S1} = \sum_{k=1}^{n_{S1}} \left((\mathbf{x}_L^k)_{n+1} - (\mathbf{x}_{S1}^h)_{n+1} \right) \times \left((\mathbf{x}_R^k)_{n+1} - (\mathbf{x}_{S1}^h)_{n+1} \right). \quad (8.18)$$

Here n_{S1} is the number of element edges coinciding with the rim, and $(\mathbf{x}_L^k)_{n+1}$ and $(\mathbf{x}_R^k)_{n+1}$ are the positions of the left and right nodes of the kth edge. We apply the adjustment of Equation (8.17) also to the other nodes of the inflow and outflow boundaries. With $\mathbf{n}_{S1}$ given for each inflow and outflow boundary, the fluid mechanics mesh-moving equations are solved with slip condition at the inflow and outflow planes, in the same way the structural mechanics equations are solved.

REMARK 8.5 *It was pointed out in Takizawa et al. (2011c) that it is also possible to use at each arterial end the cutting plane used in the arterial-surface extraction (see Section 8.2.1) but that that plane is represented by the structural mechanics mesh after the extraction and is not kept as separate data in its originally-defined way.*

REMARK 8.6 *The boundary conditions for inclined inflow and outflow planes are handled in our most current implementation as described in this subsection. However, the results reported in Takizawa et al. (2011c) were obtained with an earlier implementation where the only thing that was left out of what is described here was applying the adjustment given by Equation (8.17) to the rim nodes.*

8.2 Blood Vessel Geometry, Variable Wall Thickness, Mesh Generation, and Estimated Zero-Pressure (EZP) Geometry

8.2.1 Arterial-Surface Extraction from Medical Images

In the most recent arterial FSI research of the Team for Advanced Flow Simulation and Modeling (T★AFSM) (tafsm.org) the arterial geometries came as voxel data from 3D rotational angiography (3DRA) performed at one of the neuroangiography suites at the Memorial Hermann Hospital at the Texas Medical Center. This was done on a biplane neuroangiographic unit (Allura FD20/10; Philips Medical System, Best, the Netherlands). Adjusting the contrast ratio for this voxel data allows us to visualize and create a triangular surface mesh using a marching cubes algorithm. The vertices of the surface mesh are then passed through a Gaussian smoothing filter to eliminate any high frequency noise and obtain a smooth surface. At the artery inlets and outlets, we select cutting planes that are approximately perpendicular to the flow direction. As pointed out in Takizawa et al. (2011c), this provides better inflow and outflow planes for specifying the fluid mechanics boundary conditions and is also important for imposing proper slip boundary conditions at the inlets and outlets for the structural mechanics and fluid mesh motion (see Section 8.1.5). This entire process is carried out using software originally designed by Warren and McPhail for the purpose of interactively imaging the pulmonary structure of the human lung McPhail and Warren (2008). Similar procedures for the luminal surface extraction are employed for ALE-VMS computations.

8.2.2 Mesh Generation and EZP Arterial Geometry

We use the arterial lumen geometry as input to ANSYS Meshing Tools to generate a quadrilateral surface mesh. As mentioned in Takizawa *et al.* (2011c), at locations where the arteries have large curvature we use more mesh refinement. Based on the surface mesh, we go through a process of determining the arterial wall thickness, generating a hexahedral structural mechanics mesh for the arterial wall (typically with two layers of elements across the arterial wall), and calculating the EZP arterial geometry (Tezduyar *et al.*, 2008b, 2009b; Takizawa *et al.*, 2010a,b).

The concept of EZP geometry was introduced in Tezduyar *et al.* (2008b). Quite often, the image-based geometries are used as arterial geometries corresponding to zero blood pressure. As pointed out in Tezduyar *et al.* (2008b), it is more realistic to use that image-based geometry as the arterial geometry corresponding to the time-averaged value of the blood pressure. Given that arterial geometry at the time-averaged pressure value, an estimated arterial geometry corresponding to zero blood pressure needs to be constructed. This is where the need for an EZP arterial geometry comes from. In estimating that geometry, the time-averaged value of the blood pressure, obtained by averaging over a cardiac cycle, is 92 mm Hg.

In Takizawa *et al.* (2011c), different wall-thickness ratios are tried with the zero-pressure shape until, approximately, a 10% wall-thickness ratio (relative to the diameter of the arterial lumen) is obtained at the inflow. At each iteration, the trial wall-thickness ratio is globally uniform (which comes out to be in the range 12–13% when the iterations end), but the base length scales for the "patches" are defined individually, with a smooth transition between the patches. The patches are identified as the regions associated with the inflow trunk, each of the outflow branches, and the aneurysm/bifurcation area. The length scales for the inflow and outflow patches are the lumen diameters at those ends. The length scale for the aneurysm/bifurcation patch is a factor times the lumen diameter at the inflow, where the factor was less than one and varied between the three different patient-specific artery models used in Takizawa *et al.* (2011c). The zero-pressure shape at each EZP iteration is obtained by shrinking the surface mesh generated in the surface-extraction process (see Section 8.2.1) by an amount equal to the trial wall-thickness described above. It was pointed out in Takizawa *et al.* (2011c) that this was a simplified implementation and it was proposed to calculate the shrinking amount not with such direct dependence on the trial wall-thickness, but based on a more sophisticated rule of dependence or based on an independent trial objective.

In Takizawa *et al.* (2010a,b) and Tezduyar *et al.* (2009b, 2010a) the EZP geometry was calculated in a simpler way. The zero-pressure shape used at each EZP iteration was simply the surface mesh generated in the surface-extraction process, without any shrinking. The calculation was even simpler in Tezduyar *et al.* (2008b, 2009a), where the entire artery segment was treated as a single patch.

Following the calculation of the EZP geometry, the structure is inflated to a pressure corresponding to the pressure at the start of our computation cycle (cardiac cycle). After that, we generate, with ANSYS Meshing Tools, a fluid mechanics surface mesh associated with the inflated arterial-wall structure. Then, using that surface mesh, we generate a desired number of layers of refined fluid mechanics volume mesh near the arterial walls. The rest of the fluid mechanics volume mesh is generated with the T★AFSM automatic mesh generator. Layers of refined fluid mechanics volume mesh near the arterial walls were used in T★AFSM computations as early as the computations reported in Tezduyar et al. (2009a), followed by the computations reported in Takizawa et al. (2010a,b, 2011c) and Tezduyar et al. (2009b, 2010a).

In Takizawa et al. (2011c), the layers of refined mesh have locally variable thickness (with smooth transition between areas of different thickness), because some artery branches have very small diameters. The thickness of the layers of refined mesh is determined basically in the same way as the arterial wall thickness is determined in Takizawa et al. (2011c). The layers of refined mesh were generated in a simpler way in Tezduyar et al. (2009a,b, 2010a) and Takizawa et al. (2010a,b), where the entire artery segment was treated as a single patch. The number of layers was 6 in Tezduyar et al. (2009a), with a progression factor of approximately 1.25, and 4 in Takizawa et al. (2010a,b, 2011c) and Tezduyar et al. (2009b, 2010a), with a progression factor of 1.75.

New techniques were proposed in Remark 3 of Takizawa et al. (2011c) for determining the shrinking amount in the EZP process, the arterial wall thickness, and the thickness of the layers of refined fluid mechanics volume mesh near the arterial walls. More detailed descriptions of these techniques were presented in Tezduyar et al. (2011) and Takizawa et al. (2012b), and we repeat them here. Instead of using (nearly) patch-wise constant values (with a smooth transition between the patches) for the EZP shrinking amounts, the wall thicknesses, and the thickness of the layers of refined mesh, we determine the local values of all three based on the solution of the Laplace's equation over the surface mesh covering the lumen. In each of the three cases, the Laplace's equation is solved with values specified at the inflow and outflow boundaries and for the shrinking amount and wall thickness, as needed,[1] at a set of inter-patch points (i.e., points that are considered to be at the boundaries between the patches). The trial ratios for the shrinking amount and wall thickness are no longer globally uniform but are defined individually for the inflow and outflow boundaries (which still come out to be in the range 12–13% for the wall thickness when the iterations end), and the values specified at the inter-patch points are not directly related to these ratios. Furthermore, instead of targeting just a 10% wall-thickness ratio at the inflow, we take some additional considerations into account, such as targeting a 10% wall-thickness ratio also at the outflow boundaries, targeting a wall-thickness for the aneurysm or a set inter-patch points, reasonableness of the aneurysm size and overall shape, and the mesh quality. The trial shrinking is applied, as needed, in multiple steps, with surface remeshing between the steps. Because the parameter space is wider and the targets are multiple, the process involves more user experience, intuition, and judgment. Still, of course, the objective in iterating on the values for the shrinking amount and wall thickness is to have an EZP geometry that after inflation to average pressure gives us a shape that closely resembles the lumen geometry from the 3DRA. We note that the trial ratios specified at the boundaries for the shrinking amount and wall thickness are not independent quantities, but related by the incompressibility constraint. In generating the refined fluid mechanics volume mesh near the arterial walls, the number of layers is 4 and the progression factor is 1.75.

---

[1] In some cases where the outflow diameters significantly differ, the solution obtained from the Laplace's equation for shrinking amount and wall thickness for the aneurysm/bifurcation area could have an undesirable distribution. The need for specifying values at a set of inter-patch points comes from seeking a better distribution in that area.

REMARK 8.7 *The original version of the technique for calculating an EZP geometry was introduced in a 2007 conference paper (Tezduyar et al., 2007a) and the 2008 journal paper (Tezduyar et al., 2008b) as "a rudimentary technique" for addressing the issue. Newer techniques have been introduced since then, such as the version introduced in Tezduyar et al. (2011) and the approach proposed in Bazilevs et al. (2010c) and further refined in Hsu and Bazilevs (2011). In the approach given in Bazilevs et al. (2010c) and Hsu and Bazilevs (2011), which is presented in Section 8.3, the geometry of the vessel is left unchanged and a state of pre-stress is found, which puts the artery in equilibrium with the cardiac-cycle-averaged pressure (and viscous forces). The pre-stress is then directly employed for the blood vessel wall tissue modeling in the FSI computations.*

REMARK 8.8 *A technique for wall-thickness prescription, based on the solution of the Laplace's equation over the fluid volume mesh, was developed in Bazilevs et al. (2009b). It is presented in detail in the next section. The idea of using the Laplace's equation over the surface mesh covering the lumen to determine the local values of the EZP shrinking amount, the arterial wall thickness, and the thickness of the layers of refined mesh was motivated by this earlier wall-thickness determination work.*

8.2.3 Blood Vessel Wall Thickness Reconstruction

Modeling and discretization of the blood vessel wall requires information not only about the material properties, but also local wall thickness. In the literature, wall thickness is often reported as a percentage of the blood vessel radius. This definition is meaningful for blood vessels that are straight and circular, however, the meaning is lost for the case of real patient-specific vasculature due to the presence of local curvature, vessel branching, and geometric anomalies, such as aneurysms. CT imaging is able to produce accurate blood volume data, yet blood vessel wall thickness information is not easily accessible. However, local wall thickness is a critical parameter for generating structural mechanics meshes in the case of 3D continuum modeling or for using directly in the case of membrane and shell formulations of structural mechanics.

The need to specify variable wall thickness and ways of doing this were discussed in Section 8.2.2. In this section, we describe a general method for blood vessel wall thickness specification, which is independent of the blood vessel type and geometric complexity. The method was developed in Bazilevs *et al.* (2009b).

Let $\Omega \in \mathbb{R}^3$ be the blood vessel domain occupied by the blood, and Γ be its boundary. Let Γ_a, $a = 1, 2, \ldots, n_{\mathrm{srf}}$, denote a^{th} inlet or outlet surface, and n_{srf} be the total number of inlets and outlets in a given patient-specific model. We introduce a *volumetric* thickness function $h_{\mathrm{th}} : \Omega \to \mathbb{R}$, whose restriction to the boundary defines the blood vessel wall thickness at every point on the surface. The thickness function h_{th} is assumed to satisfy the following boundary value problem:

$$-\Delta h_{\mathrm{th}} = 0 \quad \text{in} \quad \Omega, \tag{8.19}$$

$$h_{\mathrm{th}} = \left(\frac{\int_{\Gamma_a} d\Gamma}{\pi}\right)^{1/2} \times x\% \quad \text{on} \quad \Gamma_a, \tag{8.20}$$

$$\mathbf{n} \cdot \nabla h_{\mathrm{th}} = 0 \quad \text{on} \quad \Gamma \setminus \cup_{a=1}^{n_{\mathrm{srf}}} \Gamma_a, \tag{8.21}$$

which is the Laplace's (or heat) equation with prescribed essential boundary conditions at the model inlets and outlets, and homogeneous natural boundary conditions at the luminal surface. Here, x is the wall thickness expressed as a percentage of the effective radius of a given inlet or outlet.

The method effectively collects the wall thickness information at the inlets and outlets of the patient-specific model and propagates it into the domain interior. Smooth distribution of wall thickness is expected everywhere in the domain, including geometrically complex branching regions, due to the favorable properties of the Laplace operator. The method can be applied to any patient-specific model independent of its complexity, and guarantees that the wall thickness at all inlets and outlets is exactly x% of the area-averaged radius. The formulation given by Equations (8.19)–(8.21) is amenable to a heat transfer interpretation, where inlets and outlets correspond to regions of prescribed temperature with no heat exchange on the rest of the boundary.

We tested the proposed thickness reconstruction method on an idealized bifurcation model as well as patient-specific Fontan surgery configurations. In both cases, inlet and outlet vessel wall thickness was assumed to be 10% of the effective radii. Figures 8.2 and 8.3 show the resultant wall thickness distribution for the bifurcation and Fontan models, respectively. In both cases a very reasonable smooth distribution of wall thickness is attained, especially considering how little information was taken as input data. In particular, the results of the Fontan configuration show a physiologically-realistic, gradual thinning of the vessel wall from larger to smaller branches.

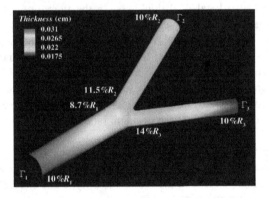

Figure 8.2 Reconstructed thickness distribution from inlet and outlet data for an idealized bifurcation model. The radii of the arterial branches are $R_1 = 0.31$ cm, $R_2 = 0.22$ cm, and $R_3 = 0.175$ cm. Near the bifurcation, the largest branch thins to 8.7% of its radius, while the two smaller ones thicken to 11.5% and 14% of their respective radii

REMARK 8.9 *As mentioned in Section 8.2.2, the thickness boundary condition specification is not restricted to the inlets and outlets. This information, if available from measurements or other sources, may be incorporated in other parts of the patient-specific model domain. In situations where the geometry is locally complex (such as extreme stenosis or aneurysm), additional constraints on the thickness may be imposed at specific locations within the domain.*

REMARK 8.10 *Here we do not imply that vessel wall thickness is distributed according to the Laplace's equation. The proposed method is an approximate technique that allows for*

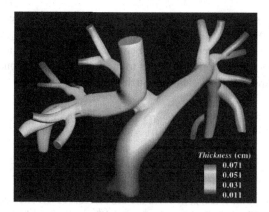

Figure 8.3 Reconstructed thickness distribution from inlet and outlet data for a patient-specific Fontan surgery configuration

incorporation of a reasonably realistic variable wall thickness in the simulations that make use of limited input data. This approach gives physiologically more realistic results than the constant-wall-thickness assumption, which is employed in many patient-specific vascular FSI computations reported in the literature by other researchers.

8.3 Blood Vessel Tissue Prestress

The structural mechanics formulation for the blood vessel wall assumes the reference configuration to be stress free. However, as pointed out in Section 8.2.2, the blood vessel configuration coming from imaging data is typically not stress free. It is subjected to blood pressure and viscous traction and develops an internal stress state to resist these external loads. In Section 8.2.2, we described the approach for calculating the EZP blood vessel geometry. Here, we propose an alternative approach, which is based on the idea of first computing the state of stress (i.e., prestress) of the blood vessel wall, which puts it in the equilibrium with the cardiac-cycle-averaged tractions coming from blood flow, and then using this directly in the FSI simulations of patient-specific hemodynamics. The method was first proposed in Bazilevs et al. (2010c), and further refined and studied in Hsu and Bazilevs (2011). Here we provide the description of the method and postpone the presentation of the numerical results until later sections.

8.3.1 Tissue Prestress Formulation

We propose to modify the structural mechanics formulation as follows: find $\mathbf{y}^h \in \mathcal{S}_y$ such that $\forall\, \mathbf{w}_2^h \in \mathcal{V}_y$:

$$\int_{(\Omega_2)_t} \mathbf{w}_2^h \cdot \rho_2 \frac{d^2 \mathbf{y}^h}{dt^2}\, d\Omega + \int_{(\Omega_2)_0} \delta \mathbf{E}^h : \left(\mathbf{S}^h + \mathbf{S}_0^h\right) d\Omega - \int_{(\Omega_2)_t} \mathbf{w}_2^h \cdot \rho_2 \mathbf{f}_2^h\, d\Omega$$

$$- \int_{\Gamma_{2E}} \mathbf{w}_{2E}^h \cdot \mathbf{h}_{2E}^h\, d\Gamma - \int_{\Gamma_{2I}} \mathbf{w}_{2I}^h \cdot \mathbf{h}_{2I}^h\, d\Gamma = 0. \qquad (8.22)$$

The modification consists of adding an a priori specified symmetric prestress tensor $\mathbf{S}_0^h$ in the stress term (written in the reference configuration) as seen in the above equation. The prestress tensor is designed such that for zero displacement the blood vessel is in equilibrium with the blood flow forces. This design condition leads to the following variational problem, obtained by setting $\mathbf{y}^h = \mathbf{0}$ in Equation (8.22): find $\mathbf{S}_0^h$ such that $\forall \, \mathbf{w}_2^h \in \mathcal{V}_y$:

$$\int_{(\Omega_2)_0} \nabla_X \mathbf{w}_2^h : \mathbf{S}_0^h \, d\Omega - \int_{(\Omega_2)_0} \mathbf{w}_2^h \cdot (\rho_2)_0 \mathbf{f}_2^h \, d\Omega$$
$$- \int_{(\Gamma_{2E})_0} \mathbf{w}_{2E}^h \cdot \hat{\mathbf{h}}_{2E}^h \, d\Gamma - \int_{(\Gamma_{2I})_0} \mathbf{w}_{2I}^h \cdot \hat{\mathbf{h}}_{2I}^h \, d\Gamma = 0. \tag{8.23}$$

In Equations (8.22) and (8.23), $(\Omega_2)_0$ denotes the blood vessel reference configuration, coming from the imaging data. The interface traction vector $\hat{\mathbf{h}}_{2I}^h$ is obtained from a separate rigid-wall blood flow simulation on the reference domain with steady inflow and resistance outflow boundary conditions. The latter guarantees a physiological intramural pressure level in the blood vessel. The inflow flow rate for the prestress problem is selected so that the intramural pressure corresponds to the cardiac-cycle-averaged pressure (about 85 mm Hg in this case).

Because Equation (8.23) is a vector equation with a tensor unknown $\mathbf{S}_0^h$, it may, in principle, have an infinite number of solutions. We obtain a particular solution for the state of prestress by means of the procedure outlined below.

Starting with step $n = 0$, and setting $\mathbf{S}_0^h = \mathbf{0}$, we repeat the following steps:

1. We zero out the displacement $\mathbf{y}^h = \mathbf{0}$, which gives $\mathbf{S}^h = \mathbf{0}$.

2. We integrate Equation (8.22) from t_n to t_{n+1}, and use the computed displacement field $\mathbf{y}^h$ to calculate the deformation gradient, the strain, and the stress $\mathbf{S}^h$, which is consistent with a chosen constitutive model (see Section 1.2).

3. We update the prestress as $\mathbf{S}_0^h = \mathbf{S}^h + \mathbf{S}_0^h$, and increment n.

We continue the above three-step iteration until $\mathbf{y}^h \to \mathbf{0}$, $\mathbf{S}^h \to \mathbf{0}$, and we arrive at the solution of Equation (8.23). Once the prestress is computed, we use the modified structural mechanics formulation given by Equation (8.22) in the FSI simulations. In Step 2 of the above prestress procedure, we move from t_n to t_{n+1} using the generalized-α time integration method (see Section 5.2).

8.3.2 Linearized Elasticity Operator

Linearization of the stress terms with respect to the displacement increment $\Delta \mathbf{y}$ in the conventional structural mechanics formulation without the prestress leads to the following bilinear form (see Equation (1.151)):

$$\int_{(\Omega_2)_0} \nabla_X \mathbf{w}_2 : \mathbf{D} \nabla_X \Delta \mathbf{y} \, d\Omega, \tag{8.24}$$

where

$$\mathbf{D} = [\bar{D}_{iJkL}], \tag{8.25}$$
$$\bar{D}_{iJkL} = \bar{F}_{iI}\bar{\mathbb{C}}_{IJKL}\bar{F}_{kK} + \delta_{ik}\bar{S}_{JL}. \tag{8.26}$$

When the current and reference configurations coincide,

$$\bar{D}_{IJKL} = \mathbb{C}_{IJKL}, \tag{8.27}$$

and only the material contribution remains in the tangent stiffness. When the prestress tensor $\mathbf{S}_0$ is present in the structural mechanics equations, the tangent stiffness is modified as

$$\bar{D}_{iJkL} = \bar{F}_{iI}\bar{\mathbb{C}}_{IJKL}\bar{F}_{kK} + \delta_{ik}(\bar{S}_{JL} + (S_0)_{JL}), \tag{8.28}$$

where $(S_0)_{JL}$ are the components of $\mathbf{S}_0$. For the case when the reference and current configurations coincide, we obtain

$$\bar{D}_{IJKL} = \mathbb{C}_{IJKL} + \delta_{IK}(S_0)_{JL}, \tag{8.29}$$

which states that the prestress is correctly accounted for in the geometric stiffness terms.

8.4 Fluid and Structure Properties and Boundary Conditions

8.4.1 Fluid and Structure Properties

As it was done for the computations reported in Torii et al. (2004, 2006a,b, 2007a,b), the blood is assumed to behave like a Newtonian fluid (see Section 2.1 in Tezduyar et al., 2008b). The density and kinematic viscosity are set to 1000 kg/m³ and 4.0×10⁻⁶ m²/s. The material density of the arterial wall is known to be close to that of the blood and therefore set to 1000 kg/m³. The arterial wall is modeled with the continuum element made of hyperelastic (Fung) material. The Fung material constants D_1 and D_2 (from Huang et al., 2001) are 2.6447×10³ N/m² and 8.365, and the penalty Poisson's ratio is 0.45. Cerebral arteries are surrounded by cerebrospinal fluid, and we expect that to have a damping effect on the structural dynamics of the arteries. Therefore we add a mass-proportional damping, which also helps in removing the high-frequency modes of the structural deformation. The damping coefficient is chosen in such a way that the structural mechanics computations remain stable at the time-step size used. It is 1.5×10^4 s⁻¹.

8.4.2 Boundary Conditions

On the arterial walls, we specify no-slip boundary conditions for the flow. In the structural mechanics part, as boundary conditions at the ends of the arteries, we set the normal component of the displacement to zero (see Section 8.1.5), and for one of those nodes we also set to zero the tangential displacement component that needs to be specified to preclude rigid-body motion.

In what follows, we give a detailed description of inflow and outflow boundary conditions for the fluid mechanics part, which we employ in our simulations.

8.4.2.1 Inflow Boundary Conditions

At the inflow boundary we specify the velocity profile as a function of time, by using the technique introduced in Takizawa et al. (2010a). Here we describe the technique from Takizawa et al. (2010a). We use a velocity waveform which represents the cross-sectional maximum velocity as a function of time. Assuming that the maximum velocity occurs at $r = 0$, the artery is rigid and the cross-sectional shape is a perfect circle, we can apply the Womersley Womersley, 1955 solution as follows:

$$U^P(r,t) = A_0 \left(1 - \left(\frac{r}{r_B}\right)^2\right) + \sum_{k=1}^{N} A_k \frac{J_0(\alpha \sqrt{k}\imath^{\frac{3}{2}}) - J_0(\alpha \sqrt{k}\left(\frac{r}{r_B}\right)\imath^{\frac{3}{2}})}{J_0(\alpha \sqrt{k}\imath^{\frac{3}{2}}) - 1} \exp\left(\imath 2\pi k \frac{t}{T}\right), \quad (8.30)$$

where N is the number of Fourier coefficients (we use $N = 20$), $A_k \in \mathbb{C}$ are the Fourier coefficients of the waveform, T is the period of the cardiac cycle, J_0 is the Bessel functions of the first kind of order 0, $\imath$ is the imaginary number, and α is the Womersley parameter:

$$\alpha = r_B \sqrt{\frac{2\pi}{\nu T}}. \quad (8.31)$$

We use the special mapping technique described in Section 8.1.1 for non-circular shapes. Figure 8.4 shows a sample volumetric flow rate as a function of time.

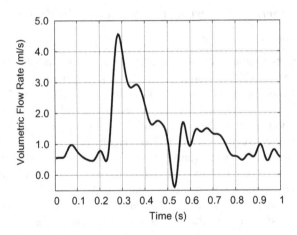

Figure 8.4 Sample volumetric flow rate

REMARK 8.11 *In the current T⋆AFSM computations, the volumetric flow rate (which was calculated based on a velocity waveform that represents the cross-sectional maximum velocity) is scaled by a factor. The scaling factor is determined in such a way that the scaled flow rate, when averaged over the cardiac cycle, yields a target WSS for Poiseuille flow over an equivalent cross-sectional area. The target WSS is 10 dyn/cm^2 in the current T⋆AFSM computations. This technique was introduced in Tezduyar et al. (2011) and Takizawa et al. (2012b).*

8.4.2.2 Explicit Outflow Boundary Conditions

At all outflow boundaries of an artery segment, we specify the same traction boundary condition. The traction boundary condition is based on a pressure profile computed as described in Takizawa *et al.* (2010a). In that computation, the pressure profile, as a function of time, is determined based on the flow rate and by using the Windkessel model Frank, 1899. From Equation (8.30), we obtain the flow rate as follows:

$$Q(t) = \int_0^{r_B} 2\pi r U^P(r,t) dr \tag{8.32}$$

$$= \pi r_B^2 \frac{A_0}{2} + \pi r_B^2 \sum_{n=1}^{N} A_k \frac{J_0(\alpha \sqrt{k\imath^{\frac{3}{2}}}) - 2\left(\alpha \sqrt{k\imath^{\frac{3}{2}}}\right)^{-1} J_1(\alpha \sqrt{k\imath^{\frac{3}{2}}})}{J_0(\alpha \sqrt{k\imath^{\frac{3}{2}}}) - 1} \exp\left(\imath 2\pi k \frac{t}{T}\right) \tag{8.33}$$

$$= \sum_{k=0}^{N} B_k \exp\left(\imath 2\pi k \frac{t}{T}\right), \tag{8.34}$$

where J_1 is the Bessel functions of the first kind of order 1, and for notational convenience we introduce another set of coefficients, $B_k \in \mathbb{C}$. The pressure, based on the Windkessel model, can be written as follows:

$$p(t) = p_0 + \exp\left(-\frac{t}{RC}\right) \int_0^t \frac{1}{C} Q(\tau) \exp\left(\frac{\tau}{RC}\right) d\tau, \tag{8.35}$$

where C and R are the compliance and resistance of the distal arterial networks, and p_0 is a constant of integration. Substituting Equation (8.34) into Equation (8.35), we obtain the following:

$$p(t) = p_0 + \sum_{k=0}^{N} \frac{B_k}{\imath 2\pi k \frac{C}{T} + \frac{1}{R}} \left[\exp\left(\imath 2\pi k \frac{t}{T}\right) - \exp\left(-\frac{t}{RC}\right)\right]. \tag{8.36}$$

After a sufficient number of periods, the $\exp\left(-\frac{t}{RC}\right)$ term in Equation (8.36) goes to 0:

$$p(t) = p_0 + \frac{T}{C} \sum_{k=0}^{N} \frac{B_k}{\imath 2\pi k + \frac{T}{RC}} \exp\left(\imath 2\pi k \frac{t}{T}\right). \tag{8.37}$$

Here $\frac{T}{RC}$ is only a profile factor, because it is a parameter that only acts on each Fourier coefficient. We set $\frac{T}{RC}$ to 18.2, and the other parameters, $\frac{T}{C}$ and p_0, are set in such a way that the range for the pressure profile is from 80 to 120 mm Hg for normal blood pressure. Figure 8.5 shows the pressure profile corresponding to the sample flow rate shown in Figure 8.4.

8.4.2.3 Implicit Outflow Boundary Conditions

As opposed to prescribing traction boundary conditions explicitly, we consider a class of outflow boundary conditions in which the outlet traction is a function of the flow rate passing through the boundary.

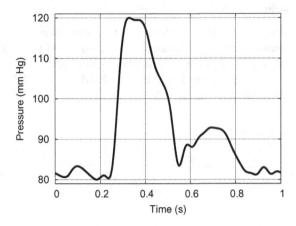

Figure 8.5 Outflow pressure profile corresponding to the sample flow rate shown in Figure 8.4

We assume that the outlet faces are denoted by Γ_a, where a is the outflow boundary index. At every outlet face Γ_a we set

$$\mathbf{n} \cdot \tilde{\sigma}(\mathbf{u}, p)\mathbf{n} + f(Q_a) = 0, \qquad (8.38)$$

$$\mathbf{t}_1 \cdot \tilde{\sigma}(\mathbf{u}, p)\mathbf{n} = 0, \qquad (8.39)$$

$$\mathbf{t}_2 \cdot \tilde{\sigma}(\mathbf{u}, p)\mathbf{n} = 0, \qquad (8.40)$$

where

$$\tilde{\sigma}(\mathbf{u}, p)\mathbf{n} = -p\mathbf{n} + 2\mu\varepsilon(\mathbf{u})\mathbf{n} - \beta\rho\{(\mathbf{u} - \hat{\mathbf{u}}) \cdot \mathbf{n}\}_- \mathbf{u} \qquad (8.41)$$

is a modified traction vector, $\{A\}_-$ denotes the negative part of A, that is,

$$\begin{cases} \{(\mathbf{u} - \hat{\mathbf{u}}) \cdot \mathbf{n}\}_- = (\mathbf{u} - \hat{\mathbf{u}}) \cdot \mathbf{n} & \text{if } (\mathbf{u} - \hat{\mathbf{u}}) \cdot \mathbf{n} < 0 \\ \{(\mathbf{u} - \hat{\mathbf{u}}) \cdot \mathbf{n}\}_- = 0 & \text{otherwise} \end{cases}, \qquad (8.42)$$

where $\mathbf{n}$ is the unit outward normal to the outlet face, $\mathbf{t}_1$ and $\mathbf{t}_2$ are mutually orthogonal unit tangent vectors on the outlet face, and β is a positive constant. The last term in Equation (8.41) is active only in the case of flow reversal through the outflow boundary. Otherwise, Equation (8.41) coincides with the usual definition of the fluid traction vector.

In Equation (8.38), Q_a denotes the volumetric flow rate through the outlet face Γ_a,

$$Q_a = \int_{\Gamma_a} \mathbf{u} \cdot \mathbf{n} \, d\Gamma, \qquad (8.43)$$

and $f(Q_a)$ represents any functional dependence. The following choice for $f(Q_a)$ results in a well-known affine resistance boundary condition (see, e.g., Bazilevs et al., 2009a):

$$f(Q_a) = C_a Q_a + p_0, \qquad (8.44)$$

where C_a is the positive resistance constant and p_0 sets the physiological value of intramural blood pressure. More complicated choices for $f(Q_a)$ include impedance and RCR-type boundary conditions; $f(Q_a)$ may also be defined implicitly as a solution of a system of ordinary differential equations modeling the behavior of downstream vasculature (see, e.g., Formaggia *et al.*, 2001; Vignon-Clementel *et al.*, 2006).

We impose boundary conditions given by Equation (8.38) weakly by adding the following terms to the fluid mechanics part of the FSI problem at all the model outlet faces:

$$\sum_a \left(-\beta \int_{\Gamma_a} \mathbf{w}_1^h \cdot \rho \left\{ (\mathbf{u}^h - \hat{\mathbf{u}}^h) \cdot \mathbf{n} \right\}_- \mathbf{u}^h \, d\Gamma + \left(\int_{\Gamma_a} \mathbf{w}_1^h \cdot \mathbf{n} \, d\Gamma \right) f(Q_a) \right). \quad (8.45)$$

Imposition of outflow boundary conditions often calls for implicit treatment of these terms to maintain robustness of the computational procedure. Linearizing Equation (8.45) gives

$$\sum_a \left(-\beta \int_{\Gamma_a} \mathbf{w}_1^h \cdot \rho \left\{ (\mathbf{u}^h - \hat{\mathbf{u}}^h) \cdot \mathbf{n} \right\}_- \Delta \mathbf{u}^h \, d\Gamma + \left(\int_{\Gamma_a} \mathbf{w}_1^h \cdot \mathbf{n} \, d\Gamma \right) f'(Q_a) \left(\int_{\Gamma_a} \Delta \mathbf{u}^h \cdot \mathbf{n} \, d\Gamma \right) \right), \quad (8.46)$$

where $\Delta \mathbf{u}^h$ is the fluid velocity solution increment (we kept the convective velocity "frozen" during the linearization). Note that the second term in Equation (8.46) contains a product of surface integrals and leads to nonstandard sparsity structure of the left-hand-side matrix. The numerical treatment of such terms is discussed in detail in Bazilevs *et al.* (2009a).

REMARK 8.12 *The term* $-\beta \int_{\Gamma_a} \mathbf{w}_1^h \cdot \rho \left\{ (\mathbf{u}^h - \hat{\mathbf{u}}^h) \cdot \mathbf{n} \right\}_- \mathbf{u}^h \, d\Gamma$ *adds stability to the formulation in the presence of locally-reversed flow through the outflow boundaries. Such reverse flows occur due to the fact that velocity fluctuations are convected out of the computational domain by the mean flow or due to the presence of flow recirculation zones near the outlets. We found that the addition of this term is important for the overall stability of the computations, especially in the case when large blood vessels (e.g., thoracic aorta) with complicated, near turbulent blood flow features are simulated.*

REMARK 8.13 *It can be shown analytically that $\beta > 1/2$ for stability. However, in practice, lower values of β (0.2 or so) may be used to obtain stable solutions (see Moghadam et al., 2011).*

REMARK 8.14 *It was shown in Moghadam et al. (2011) that among several options available to prevent outflow divergence, the proposed approach is the most stable and cost effective, and the least invasive in that the effect on the blood flow solution away from the outlets is minimized.*

8.5 Simulation Sequence

Recipes for pre-FSI computations that provide a good starting point for the FSI computations and improve their convergence were introduced in Tezduyar *et al.* (2007b). Now, in all arterial simulations carried out by the T⋆AFSM, the FSI computations are preceded by a set of pre-FSI computations. These pre-FSI computations include fluid-only and structure-only computations. The recipes introduced in Tezduyar *et al.* (2007b) were used also in Tezduyar *et al.* (2008b, 2009a). A slightly modified recipe was introduced in Takizawa *et al.* (2010a), resulting in a simulation sequence called "S→F→S→FSI," and this is the one that the T⋆AFSM has been using in its arterial simulations since then.

Structure→Fluid→Structure→FSI sequence (S→F→S→FSI)

Step 1: Generate the structure mesh based on the shape of the unstressed structure.

Step 2: Compute the structural deformation with a uniform fluid pressure held steady at a value close to 80 mm Hg (100 mm Hg for high blood pressure).

- Structural deformation can be determined with a steady-state computation or a time-dependent computation that eventually yields a steady-state solution.

Step 3: Generate the fluid mesh based on the shape of the deformed structure.

Step 4: Compute a developed flow field while holding the structure from Step 2 rigid.

- The outflow traction is set to a value close to 80 mm Hg.
- The inflow velocity is set to a value corresponding to the outflow traction.

Step 5: Recompute the structural deformation, with the fluid stresses at the interface held steady at their values from Step 4, and simultaneously update the fluid mesh.

- Structural deformation can be determined with one of the two choices we had in Step 2.

Step 6: Compute the FSI with the same inflow and outflow conditions used in Step 4, with the initial condition for the flow velocity coming from Step 4.

Step 7: Compute the FSI with the inflow and outflow conditions pulsating.

8.6 Sequentially-Coupled Arterial FSI (SCAFSI) Technique

The SCAFSI technique was first proposed in Tezduyar *et al.* (2007c, 2009a) as an approximate FSI approach in arterial fluid mechanics. In the SCAFSI technique, first we compute a "reference" (i.e., "base") arterial deformation as a function of time, driven only by the blood pressure, which is given as a function of time by specifying the pressure profile in a cardiac cycle. Then we compute a sequence of updates involving mesh motion, fluid dynamics calculations, and recomputing the arterial deformation. The SCAFSI steps are preceded by a set of pre-FSI computation steps, which can be found in their originally-proposed form in Tezduyar *et al.* (2007b, 2008b) and most current form in Section 8.5. The details of the SCAFSI steps are described below.

Step 1

Compute the "reference" arterial displacement:
$(\mathbf{Y}_R)_n \quad n = 1, 2, \ldots, n_{ts}$
Driven only by the blood pressure: $p_R(t)$
Predictor options in moving from time level n to $n + 1$:

$$((\mathbf{Y}_R)_{n+1})^0 = (\mathbf{Y}_R)_n \qquad (8.47)$$

$$((\mathbf{Y}_R)_{n+1})^0 = 2(\mathbf{Y}_R)_n - (\mathbf{Y}_R)_{n-1} \qquad (8.48)$$

$$((\mathbf{Y}_R)_{n+1})^0 = 3(\mathbf{Y}_R)_n - 3(\mathbf{Y}_R)_{n-1} + (\mathbf{Y}_R)_{n-2} \qquad (8.49)$$

$$((\mathbf{Y}_R)_{n+1})^0 = (\mathbf{Y}_R)_n + \frac{(\mathbf{Y}_R)_n - (\mathbf{Y}_R)_{n-1}}{p_R(t_n) - p_R(t_{n-1})} (p_R(t_{n+1}) - p_R(t_n)) \qquad (8.50)$$

Nodal values of $p_R(t_n)$: $(\mathbf{P}_R)_n$
Nodal values of the interface stress: $(\mathbf{H}_R)_n$

Step 2

Compute the "reference" mesh motion:
$(\hat{\mathbf{Y}}_R)_n \quad n = 1, 2, \ldots, n_{ts}$
Predictor options:

$$\left((\hat{\mathbf{Y}}_R)_{n+1}\right)^0 = \mathbf{0} \tag{8.51}$$

$$\left((\hat{\mathbf{Y}}_R)_{n+1}\right)^0 = (\hat{\mathbf{Y}}_R)_n \tag{8.52}$$

$$\left((\hat{\mathbf{Y}}_R)_{n+1}\right)^0 = 2(\hat{\mathbf{Y}}_R)_n - (\hat{\mathbf{Y}}_R)_{n-1} \tag{8.53}$$

$$\left((\hat{\mathbf{Y}}_R)_{n+1}\right)^0 = 3(\hat{\mathbf{Y}}_R)_n - 3(\hat{\mathbf{Y}}_R)_{n-1} + (\hat{\mathbf{Y}}_R)_{n-2} \tag{8.54}$$

$$\left((\hat{\mathbf{Y}}_R)_{n+1}\right)^0 = (\hat{\mathbf{Y}}_R)_n + \frac{(\hat{\mathbf{Y}}_R)_n - (\hat{\mathbf{Y}}_R)_{n-1}}{p_R(t_n) - p_R(t_{n-1})} (p_R(t_{n+1}) - p_R(t_n)) \tag{8.55}$$

We note that here, and at Step 5, $\hat{\mathbf{Y}}$ is the mesh motion relative to the position at the time level n.

Step 3

For zero-stress conditions at the outflow boundaries, compute the time-dependent flow field and the corresponding interface stress: $(\mathbf{H}_1)_n \quad n = 1, 2, \ldots, n_{ts}$
Predictor options:

$$((\mathbf{P}_1)_{n+1})^0 = (\mathbf{P}_1)_n \tag{8.56}$$

$$((\mathbf{P}_1)_{n+1})^0 = 2(\mathbf{P}_1)_n - (\mathbf{P}_1)_{n-1} \tag{8.57}$$

$$((\mathbf{P}_1)_{n+1})^0 = 3(\mathbf{P}_1)_n - 3(\mathbf{P}_1)_{n-1} + (\mathbf{P}_1)_{n-2} \tag{8.58}$$

$$((\mathbf{P}_1)_{n+1})^0 = (\mathbf{P}_1)_n + \frac{(\mathbf{P}_1)_n - (\mathbf{P}_1)_{n-1}}{U(t_n) - U(t_{n-1})} (U(t_{n+1}) - U(t_n)) \tag{8.59}$$

Here $U(t)$ is the cross-sectional average of the inflow velocity. To enhance the stability of the computation in Step 4, smooth $(\mathbf{H}_1)_n$ by time averaging:

$$\begin{aligned}(\mathbf{H}_1)_n \leftarrow{}& \omega_0(\mathbf{H}_1)_n + \omega_{\pm 1}\left((\mathbf{H}_1)_{n+1} + (\mathbf{H}_1)_{n-1}\right) \\&+ \omega_{\pm 2}\left((\mathbf{H}_1)_{n+2} + (\mathbf{H}_1)_{n-2}\right) \\&+ \omega_{\pm 3}\left((\mathbf{H}_1)_{n+3} + (\mathbf{H}_1)_{n-3}\right) \\&+ \omega_{\pm 4}\left((\mathbf{H}_1)_{n+4} + (\mathbf{H}_1)_{n-4}\right)\end{aligned} \tag{8.60}$$

Options for time-averaging weights:

$$(\omega_0, \omega_{\pm 1}, \omega_{\pm 2}, \omega_{\pm 3}, \omega_{\pm 4}) = \frac{1}{9}(3, 2, 1, 0, 0) \tag{8.61}$$

$$(\omega_0, \omega_{\pm 1}, \omega_{\pm 2}, \omega_{\pm 3}, \omega_{\pm 4}) = \frac{1}{16}(4, 3, 2, 1, 0) \tag{8.62}$$

$$(\omega_0, \omega_{\pm 1}, \omega_{\pm 2}, \omega_{\pm 3}, \omega_{\pm 4}) = \frac{1}{25}(5, 4, 3, 2, 1) \tag{8.63}$$

Now the total interface stress: $(\mathbf{H}_R)_n + (\mathbf{H}_1)_n$

Step 4

Compute the updated arterial displacement:
$\mathbf{Y}_n \quad n = 1, 2, \ldots, n_{ts}$
Predictor options:

$$(\mathbf{Y}_{n+1})^0 = 2\mathbf{Y}_n - \mathbf{Y}_{n-1} \tag{8.64}$$

$$(\mathbf{Y}_{n+1})^0 = (\mathbf{Y}_R)_{n+1} + ((\mathbf{Y}_1)_{n+1})^0 \tag{8.65}$$

Displacement increment: $(\mathbf{Y}_1)_n = \mathbf{Y}_n - (\mathbf{Y}_R)_n$
Predictor options for the displacement increment:

$$((\mathbf{Y}_1)_{n+1})^0 = (\mathbf{Y}_1)_n \tag{8.66}$$

$$((\mathbf{Y}_1)_{n+1})^0 = 2(\mathbf{Y}_1)_n - (\mathbf{Y}_1)_{n-1} \tag{8.67}$$

$$((\mathbf{Y}_1)_{n+1})^0 = 3(\mathbf{Y}_1)_n - 3(\mathbf{Y}_1)_{n-1} + (\mathbf{Y}_1)_{n-2} \tag{8.68}$$

Step 5

Compute the updated mesh motion:
$\hat{\mathbf{Y}}_n \quad n = 1, 2, \ldots, n_{ts}$
Predictor options:

$$\left(\hat{\mathbf{Y}}_{n+1}\right)^0 = 0 \tag{8.69}$$

$$\left(\hat{\mathbf{Y}}_{n+1}\right)^0 = (\hat{\mathbf{Y}}_R)_{n+1} + \left((\hat{\mathbf{Y}}_1)_{n+1}\right)^0 \tag{8.70}$$

Mesh-motion increment: $(\hat{\mathbf{Y}}_1)_n = \hat{\mathbf{Y}}_n - (\hat{\mathbf{Y}}_R)_n$
Predictor options for the mesh-motion increment:

$$\left((\hat{\mathbf{Y}}_1)_{n+1}\right)^0 = 0 \tag{8.71}$$

$$\left((\hat{\mathbf{Y}}_1)_{n+1}\right)^0 = (\hat{\mathbf{Y}}_1)_n \tag{8.72}$$

$$\left((\hat{\mathbf{Y}}_1)_{n+1}\right)^0 = 2(\hat{\mathbf{Y}}_1)_n - (\hat{\mathbf{Y}}_1)_{n-1} \tag{8.73}$$

$$\left((\hat{\mathbf{Y}}_1)_{n+1}\right)^0 = 3(\hat{\mathbf{Y}}_1)_n - 3(\hat{\mathbf{Y}}_1)_{n-1} + (\hat{\mathbf{Y}}_1)_{n-2} \tag{8.74}$$

Step 6

For zero-stress conditions at the outflow boundaries, compute the time-dependent flow field and the corresponding interface stress: $(\mathbf{H}_2)_n \quad n = 1, 2, \ldots, n_{ts}$
Predictor options:

$$((\mathbf{P}_2)_{n+1})^0 = (\mathbf{P}_2)_n \tag{8.75}$$

$$((\mathbf{P}_2)_{n+1})^0 = 2(\mathbf{P}_2)_n - (\mathbf{P}_2)_{n-1} \tag{8.76}$$

$$((\mathbf{P}_2)_{n+1})^0 = 3(\mathbf{P}_2)_n - 3(\mathbf{P}_2)_{n-1} + (\mathbf{P}_2)_{n-2} \tag{8.77}$$

$$((\mathbf{P}_2)_{n+1})^0 = (\mathbf{P}_2)_n + \frac{(\mathbf{P}_2)_n - (\mathbf{P}_2)_{n-1}}{U(t_n) - U(t_{n-1})} (U(t_{n+1}) - U(t_n)) \tag{8.78}$$

$$((\mathbf{P}_2)_{n+1})^0 = (\mathbf{P}_1)_{n+1} \tag{8.79}$$

Now the total interface stress: $(\mathbf{H}_R)_n + (\mathbf{H}_2)_n$

The SCAFSI algorithm described above is based on the assumption that in computations with more than one outflow boundary, we specify the same traction condition for all. Versions of the SCAFSI technique that do not rely on that assumption were proposed in Tezduyar et al. (2009a). These versions are applicable even if the outflow traction conditions are not specified explicitly but are modeled as a function of the flow rate at each outflow boundary. We refer the interested reader to Tezduyar et al. (2009a).

REMARK 8.15 *Clearly, the SCAFSI technique results in savings in computer time compared to the (fully) coupled arterial FSI (CAFSI) technique. These savings come from various aspects of SCAFSI, which can be found in Tezduyar et al. (2009a).*

REMARK 8.16 *The predictors given by Equations (8.76) and (8.77) were written in Tezduyar et al. (2009a) with a subscript typo in each equation. Although the typos were obvious, they were pointed out in Tezduyar et al. (2010a).*

REMARK 8.17 *Due to a combination of publisher's typesetting errors and misinterpretation of what has been used in the computations, the predictor options identified in Tezduyar et al. (2009a) as those used in the test computations were not the ones that were actually used. The predictor options used, in reference to the equation numbers of this paper, were those given by Equations (8.48), (8.51), (8.56), (8.64), (8.69) and (8.79). This correction was pointed out in Tezduyar et al. (2010a).*

8.7 Multiscale Versions of the SCAFSI Technique

Temporally multiscale. A temporally multiscale version of the SCAFSI technique was proposed in Remark 1 of Tezduyar et al. (2009a), where different time-step sizes are used for the structural and fluid mechanics parts. This version was tested in Tezduyar et al. (2009a) on FSI modeling of a middle cerebral artery segment with aneurysm. The arterial geometry was a close approximation to the patient-specific image-based geometry used in Torii et al. (2007a). The geometry used in Torii et al. (2007a) was extracted from the CT model of an artery segment from a 57 year-old male. The arterial wall was modeled with the continuum element made of hyperelastic (Fung) material. The mesh for the artery had four-node tetrahedral elements, with two elements across the arterial wall. The time-step size for the structural mechanics part was twice that of the fluid mechanics part. The multiscale SCAFSI computation resulted in good mass balance, and the flow field obtained looked essentially the same as the flow field from the CAFSI computation. Time histories of the arterial volume and (spatially-averaged) interface stress obtained with the multiscale SCAFSI technique were also very close to those obtained with the CAFSI technique.

Spatially multiscale. Spatially multiscale versions of SCAFSI techniques were also proposed in Tezduyar et al. (2009a) (Remark 2), where the fluid mechanics meshes with different refinement levels are used at different stages of the SCAFSI computation. In the version called SCAFSI M1SC, a more refined fluid mechanics mesh is used at SCAFSI Steps 5 and 6 than the mesh used at Steps 2 and 3. With this approach we can increase the accuracy of the fluid mechanics solution at the final stage, just before we calculate the fluid mechanics quantities of interest, such as the WSS. As pointed out in Tezduyar et al. (2009a, 2010a), by using a relatively coarse fluid mechanics mesh at Steps 2 and 3, we avoid incurring high computational cost at stages where a highly-refined fluid mechanics mesh is not needed for accurately computing the arterial shape as a function of time. In the version called SCAFSI M1C, we first compute the arterial shape with the CAFSI technique and a relatively coarse fluid mechanics mesh, followed by mesh motion and fluid mechanics computations with a more refined mesh. Again, as pointed out in Tezduyar et al. (2009a, 2010a), by using a relatively coarse mesh at the stage where a highly-refined fluid mechanics mesh is not needed, we reserve our computational effort for the final stage, where we do need a highly-refined fluid mechanics mesh to calculate fluid mechanics quantities such as the WSS. Test computations carried out with the SCAFSI M1SC and SCAFSI M1C techniques were presented in Tezduyar et al. (2010a).

REMARK 8.18 *The original motivation in developing the SCAFSI technique was to have a computationally more economical alternative to the fully coupled FSI approach in arterial fluid mechanics. However, as pointed out in Tezduyar et al. (2009a), the flexibility of the technique, including its multiscale aspects, is even more appealing.*

REMARK 8.19 *Extending the multiscale sequentially-coupled FSI technique from arterial fluid mechanics to other classes of applications was proposed in Tezduyar et al. (2009b). The underlying concepts are still essentially the same as those described in the earlier part of this section. The name given in Tezduyar et al. (2009b), Multiscale Sequentially-Coupled FSI (SCFSI) Technique, however, no longer implies a functionality limited to arterial fluid mechanics. Specifically, the SCFSI M1C technique was proposed in Tezduyar et al. (2009b) as a way of reducing the FSI computational effort where we do not need it and increasing the accuracy of the fluid mechanics computations where we need accurate, detailed flow computation. We first compute the structural deformation with the (fully) coupled FSI (CFSI) technique and a relatively coarser fluid mechanics mesh, followed by mesh motion and fluid mechanics computations with a more refined mesh. A time integration version of this was also proposed in Tezduyar et al. (2009b), where the structural deformation would first be computed with the CFSI technique and a time-step size as small as it is needed in that computation, followed by mesh motion and fluid mechanics computations with a smaller time-step size that might be needed for more accurate, detailed flow computation. Test FSI computations for 2D flow past a flexible beam positioned perpendicular to the flow field were presented in Tezduyar et al. (2009b,c) to illustrate how the spatially multiscale SCFSI M1C technique works.*

REMARK 8.20 *Based on the spatially multiscale SCFSI techniques introduced in Tezduyar et al. (2009a,b), specifically the version called SCFSI M1C, a version that is spatially multiscale for the structural mechanics part was introduced in Tezduyar et al. (2009d) and designated with the acronym SCFSI M2C. In this technique, we first compute the time-dependent flow field with the CFSI technique and a relatively coarser structural mechanics mesh, followed by a structural mechanics computation with a more refined mesh, with the time-dependent interface stresses coming from the previously carried out CFSI computation. With this technique, we can reduce the FSI computational effort where we do not need it and increase the accuracy of the structural mechanics computation where we need accurate, detailed structural mechanics computations, such as computing the fabric stresses in a parachute FSI. Test computations for parachute modeling, with different purposes of using the SCFSI M2C technique, were presented in Tezduyar et al. (2009d, 2010b) and Takizawa et al. (2011f).*

8.8 Computations with the SSTFSI Technique

All computations were carried out in a parallel computing environment and were completed without any remeshing. The fully-discretized, coupled fluid and structural mechanics and mesh-moving equations are solved with the quasi-direct coupling technique (see Section 6.1.2). In solving the linear equation systems involved at every nonlinear iteration, the GMRES search technique (Saad and Schultz, 1986) is used with a diagonal preconditioner.

8.8.1 Performance Tests for Structural Mechanics Meshes

The arterial-modeling work presented in Takizawa *et al.* (2010a) included a performance evaluation for four different structural mechanics meshes: three hexahedral meshes, with one, two and three layers of elements across the arterial wall, and one tetrahedral mesh with two layers of elements across the arterial wall. We describe the performance evaluation tests from Takizawa *et al.* (2010a).

The geometry of the arterial lumen used in the tests is from Torii *et al.* (2007b, 2008, 2009), which was extracted from the CT model of a bifurcating segment of the middle cerebral artery of a 67 year-old female with aneurysm. The diameter of the arterial lumen is 2.39 mm at the inflow, and 1.53 and 1.73 mm at the two outflow ends. The cardiac cycle is 1.0 s. The arterial-wall thickness is based on using (nearly) patch-wise constant values (see Section 8.2.2). The length scale for the aneurysm/bifurcation patch is 0.67×(lumen diameter at the inflow). The zero-pressure shape used at each EZP iteration is simply the surface mesh generated in the surface-extraction process (see Section 8.2.1). Figure 8.6 shows, for the zero-pressure configuration, the wall thickness normalized by the wall thickness at the inflow. The "standard" structural mechanics mesh used by the T★AFSM for this model consists of 8067 nodes and 5316 eight-node hexahedral elements, with 2689 nodes and 2658 four-node

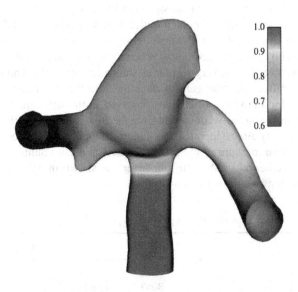

Figure 8.6 Zero-pressure surface configuration colored with normalized wall thickness

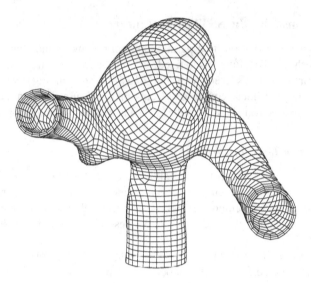

Figure 8.7 Standard structural mechanics mesh when the outflow pressure is maximum

quadrilateral elements on the fluid–structure interface and two layers of elements across the arterial wall. It is shown in Figure 8.7. The structure-mesh tests reported in Takizawa *et al.* (2010a) were performed under a prescribed traction over a cardiac cycle. The prescribed traction comes from the FSI computation described in Section 8.8.2 as the computation with the "medium" fluid mechanics mesh. The structural mechanics mesh properties are shown in Table 8.1. The only difference between the hexahedral meshes is the number of element layers across the arterial wall. The tetrahedral mesh is based on a triangular surface mesh, which is the same as the fluid interface mesh in the FSI computations. All meshes have the same number of nodes at the inlet and each outlet boundary. The results for all the hexahedral meshes are geometrically almost identical during the cardiac cycle, while the tetrahedral mesh results in a slightly different geometry. When the deformed tetrahedral mesh is rotated and translated, the geometries are very similar as shown in Figure 8.8. The least-squares projection of the traction from the triangular surface mesh to the quadrilateral surface mesh is the likely reason behind the small differences observed in the structure mesh deformations. The differences are insignificant when solving for the flow field within the artery. Since the results were geometrically almost identical, the lumen volume was provided in Takizawa *et al.* (2010a) as a quantitative measure of the differences. Each mesh volume, as a percentage of the volume of the hexahedral mesh with four layers, is shown in Table 8.2. These volume ratios remain almost constant throughout a cardiac cycle.

Table 8.1 Structural mechanics mesh properties

	Hex			Tet
Layers	1	2	4	2
Nodes	5378	8067	13445	9171
Elements	2658	5316	10632	36312

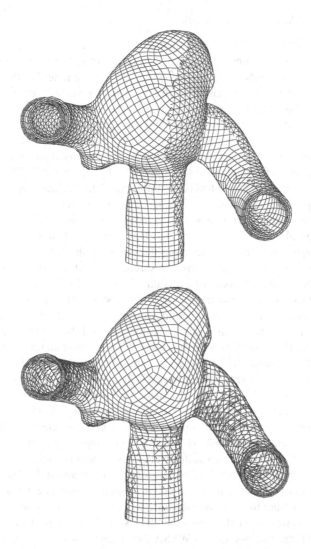

Figure 8.8 Structural mechanics meshes at the instant when the outflow pressure is at its maximum value. Blue represents the hexahedral mesh with four layers and red indicates the tetrahedral mesh with two layers. The top picture shows the original deformed geometries. The bottom picture shows the geometries after the tetrahedral mesh is rotated by 0.75° clockwise around an axis parallel to the inflow direction and translated such that the inlets are aligned

Table 8.2 Volume ratios for different structural mechanics meshes

	Hex			Tet
Layers	1	2	4	2
Lumen volume	97.6%	99.4%	100%	97.1%

8.8.2 Multiscale SCAFSI Computations

We describe the multiscale SCAFSI computations from Tezduyar *et al.* (2010a). In the spatially multiscale SCAFSI computations, Step 1–6 predictor options used are those given by Equations (8.48), (8.51), (8.56), (8.64), (8.69), and (8.75). We note that the pressure predictor option used at Step 6 is different from the one used in SCAFSI computations reported in Tezduyar *et al.* (2009a). This is because, as noted in Tezduyar *et al.* (2010a), in the spatially multiscale SCAFSI computations the fluid mechanics mesh at (Step 5 and) Step 6 is different from the fluid mechanics mesh used at the earlier stages, and therefore the pressure values obtained at the earlier stages cannot be directly used as pressure predictors at Step 6. The time-averaging weights are those given by Equation (8.63). The time-averaging, however, is not used in the SCAFSI M1C computation. The predictor options we are using are relatively simple ones among those proposed. As pointed out in Tezduyar *et al.* (2010a), using more sophisticated predictors is not expected to change the results that much, since the SCAFSI results are already quite close to the CAFSI results. Among the options proposed for the time-averaging weights, we are using the one with the largest spread. We have not experimented with the options with narrower spread. The arterial model used in Tezduyar *et al.* (2010a) is the same as the one used in Takizawa *et al.* (2010a), which was described in Section 8.8.1, and the structural mechanics mesh is the "standard" mesh described in the same section. The Womersley parameter, defined in Section 8.4.2, is 1.5. Figure 8.9 shows, for the test computations in this section, the volumetric flow rate and the outflow pressure profile.

The two different fluid mechanics meshes used in Tezduyar *et al.* (2010a) were a "coarse" mesh with 15 850 nodes and 88 573 four-node tetrahedral elements, and a "fine" mesh with 22 775 nodes and 128 813 four-node tetrahedral elements. The same meshes were called "coarse" and "medium" in Takizawa *et al.* (2010b) because a third, more refined mesh was introduced. We will therefore call them "coarse" and "medium" in this book. The medium mesh has four layers of elements with higher refinement near the arterial wall, with the thickness of the first layer being approximately 0.02 mm. The progression factor is 1.75. The coarse mesh has one layer of elements with a uniform thickness of approximately 0.2 mm. The coarse and medium meshes have the same number of nodes and elements at the fluid–structure interface: 3057 nodes and 6052 three-node triangular elements. Figures 8.10 and 8.11 show the mesh at the fluid–structure interface and the inflow plane for the coarse and medium meshes.

The computations were carried out with the SSTFSI-TIP1 technique (see Remarks 4.10 and 5.9) and the SUPG test function option WTSA (see Remark 4.4). The stabilization parameters used are those given by Equations (4.115)–(4.120) and (4.125). The time-step size is 3.333×10^{-3} s. In the CAFSI computations, the number of nonlinear iterations per time step is 6, and the number of GMRES iterations per nonlinear iteration is 300 for the fluid+structure block, and 30 for the mesh moving block. For all six nonlinear iterations the fluid scale is 1.0 and the structure scale is 50. In the SCAFSI M1SC computation, the mesh used at Step 2 and Step 3 is the coarse mesh, which reduces the cost of the computations, and the mesh used at Step 5 and Step 6 is the medium mesh, which increases the accuracy of the flow field computed. The number of nonlinear iterations per time step is 5 for the fluid mechanics part and 4 for the structural mechanics and mesh moving parts. The number of GMRES iterations per nonlinear iteration is 150, 50, and 30 for the fluid mechanics, structural mechanics, and mesh moving parts, respectively. In the SCAFSI M1C computation, Steps 1–4 are replaced with a CAFSI computation with the coarse mesh. The arterial shape obtained from

Cardiovascular FSI

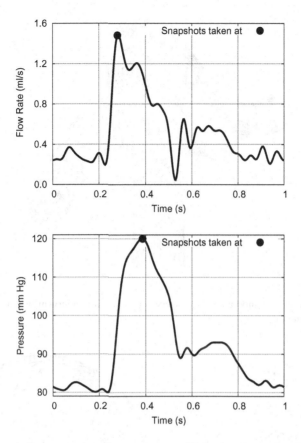

Figure 8.9 Volumetric flow rate and outflow pressure profile for the test computations in Section 8.8.2, with the maximum value marked

the CAFSI computation is used at Step 5 and Step 6 with the medium mesh. As pointed out in Tezduyar *et al.* (2010a), Step 4 arterial shape comes from the reduced-cost CAFSI computation with the coarse mesh, and the medium mesh used at Step 5 and Step 6 increases the accuracy of the flow field computed. The number of nonlinear iterations per time step is 5 for the fluid mechanics part and 4 for the structural mechanics and mesh moving parts. The number of GMRES iterations per nonlinear iteration is 150, 50, and 30 for the fluid mechanics, structural mechanics, and mesh moving parts, respectively.

REMARK 8.21 *The results reported in Tezduyar et al. (2010a), and described here, were computed based on one of the earlier T★AFSM implementations. In that implementation, the structural mechanics and mesh moving parts are not separable. Therefore the mesh motion can only be computed while computing the arterial deformation. Because of that, Step 4 arterial geometry does not actually come directly from the CAFSI computation with the coarse mesh, but it is recomputed with the interface stresses obtained from that CAFSI computation. The differences are very minor.*

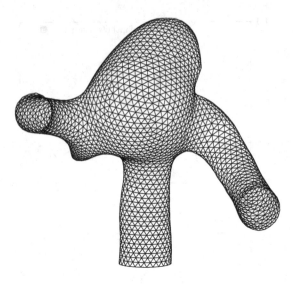

Figure 8.10 A bifurcating middle cerebral artery segment with aneurysm. Fluid mechanics mesh at the fluid–structure interface

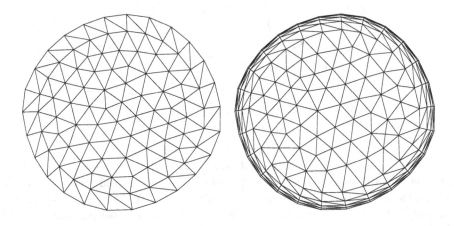

Figure 8.11 A bifurcating middle cerebral artery segment with aneurysm. Fluid mechanics mesh at the inflow plane for the coarse and medium meshes

REMARK 8.22 *The fluid mechanics computation at Step 6 requires an initial flow field. For SCAFSI versions that are not spatially multiscale, this initial flow field comes directly from Step 3. For the spatially multiscale versions, it was proposed in Tezduyar et al. (2010a) to use an initial flow field obtained from Step 3 by projection. However, the results reported in Tezduyar et al. (2010a), and described here, were computed based on one of the earlier T★AFSM implementations, which did not have that projection capability. Instead, a very brief fluid mechanics computation was carried out to produce a divergence-free flow field. The inflow velocity for this brief computation is the velocity at the beginning of Step 6. The initial condition consists of an essentially-zero velocity field.*

Good mass balance is achieved in all computations. This was verified by comparing the rate of change for the artery volume and the difference between the volumetric inflow and outflow rates. The pictures showing this can be found in Tezduyar *et al.* (2010a). Figure 8.12 illustrates the WSS for the CAFSI computations with the medium mesh. The WSS for the CAFSI computations with the coarse mesh can be found in Tezduyar *et al.* (2010a). Figures 8.13 and 8.14 illustrate the WSS for the SCAFSI M1SC and SCAFSI M1C computations. Figure 8.15 shows the time-averaged WSS for the CAFSI computations with the coarse and medium meshes. Figure 8.16 shows the time-averaged WSS for the SCAFSI M1SC and SCAFSI M1C computations. Table 8.3 shows the maximum, mean, and minimum values of the WSS for the CAFSI computations with the coarse and medium meshes. Table 8.4 shows the maximum, mean, and minimum values of the WSS for the CAFSI computation with the medium mesh and the SCAFSI M1SC and SCAFSI M1C computations.

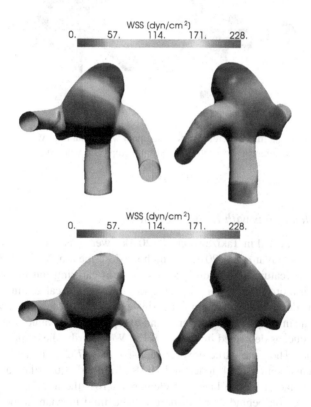

Figure 8.12 A bifurcating middle cerebral artery segment with aneurysm. WSS for the CAFSI computation with the medium mesh when the volumetric flow rate is maximum (top) and when the outflow pressure is maximum (bottom)

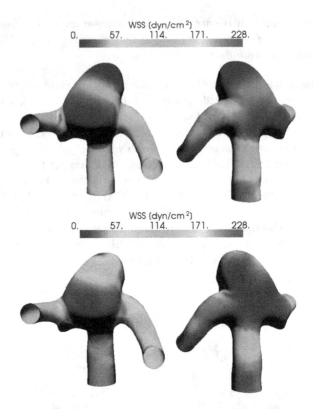

Figure 8.13 A bifurcating middle cerebral artery segment with aneurysm. WSS for the SCAFSI M1SC computation when the volumetric flow rate is maximum (top) and when the outflow pressure is maximum (bottom)

8.8.3 WSS Calculations with Refined Meshes

The computations reported in Takizawa *et al.* (2010b) were based on the same artery model as the one used in Tezduyar *et al.* (2010a), which we described in Section 8.8.2, and involved basically three new features. The new features were a) carrying out higher-resolution FSI computations with more refined fluid and structure meshes, b) calculating the WSS with a new technique, as described in Section 8.1.3 (instead of using the spatial version of Equation (5.119) as done in the preceding papers), and c) reporting OSI values that were calculated with a new technique, as described in Section 8.1.4. We describe the computations from Takizawa *et al.* (2010b). The "fine" structure mesh consists of 30 732 nodes and 20 366 eight-node hexahedral elements, with 10 244 nodes and 10 183 four-node quadrilateral elements on the fluid–structure interface and two layers of elements across the arterial wall. It is shown in Figure 8.17. The reason behind using a more refined fluid mechanics mesh in the higher-resolution FSI computations reported in Takizawa *et al.* (2010b) was to increase the accuracy of the WSS calculations by increasing the fluid mechanics mesh refinement also on the arterial wall, not just in the normal direction near the arterial wall. As pointed out in Takizawa *et al.* (2010b), this would normally make sense only if the structural mechanics mesh has

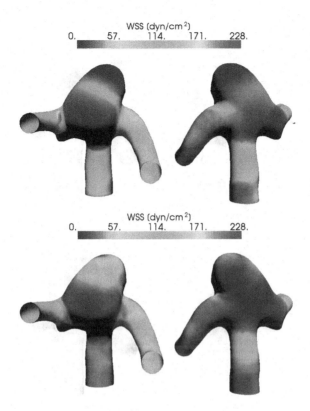

Figure 8.14 A bifurcating middle cerebral artery segment with aneurysm. WSS for the SCAFSI M1C computation when the volumetric flow rate is maximum (top) and when the outflow pressure is maximum (bottom)

comparable refinement. The refined structural mechanics mesh introduced in Takizawa *et al.* (2010b), which is shown in Figure 8.17, served that purpose. The "fine" fluid mechanics mesh has 138 713 nodes and 823 756 four-node tetrahedral elements. It has 11 713 nodes and 23 304 three-node triangular elements at the fluid–structure interface, which is shown in Figure 8.18. The fine mesh, just like the medium mesh, has four layers of elements with higher refinement near the arterial wall. The thickness of the first layer is approximately 0.02 mm and the progression factor is 1.75. Figure 8.19 shows the inflow plane for the fine mesh.

The computations were carried out with the SSTFSI-TIP1 technique (see Remarks 4.10 and 5.9) and the SUPG test function option WTSA (see Remark 4.4). The stabilization parameters used are those given by Equations (4.115)–(4.120) and (4.125). The time-step size is 3.333×10^{-3} s for the coarse and medium meshes and 1.667×10^{-3} s for the fine mesh. For all three meshes the number of nonlinear iterations per time step is 6. For the fluid+structure block the number of GMRES iterations per non-linear iteration is 300 for the coarse and medium meshes and 600 for the fine mesh. For all six nonlinear iterations the fluid scale is set to 1.0 and the structure scale to 50. For the mesh moving block the number of GMRES iterations is 30. Good mass balance is achieved in all computations. This was verified by comparing the rate of change for the artery volume and the difference between the volumetric

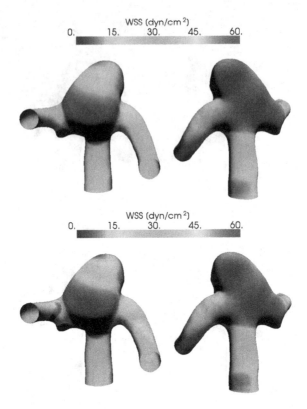

Figure 8.15 A bifurcating middle cerebral artery segment with aneurysm. Time-averaged WSS for the CAFSI computations with the coarse (top) and medium (bottom) meshes

inflow and outflow rates. The pictures showing this can be found in Takizawa et al. (2010b). Figure 8.20 shows the WSS for the three meshes when the volumetric flow rate is maximum. Figure 8.21 shows the time-averaged WSS for the three meshes. Table 8.5 shows the maximum, mean, and minimum values of the WSS for the three meshes.

REMARK 8.23 *Numbers shown in Table 8.5, which come from Takizawa et al. (2010b), are slightly different from the numbers in Section 8.8.2, which come from Tezduyar et al. (2010a). This is because, as mentioned at the beginning of this section, the WSS was calculated with a new technique in Takizawa et al. (2010b).*

Figure 8.22 shows the OSI for the three meshes. The higher OSI region indicates flow direction changes over the cardiac cycle. The medium and fine mesh results are in good agreement. Figure 8.23 shows typical streamlines around the higher OSI region at $t = 0.268$ s (accelerating flow) and $t = 0.448$ s (decelerating flow). When the flow accelerates, a vortex forms, which results in a downward WSS. Conversely, when the flow decelerates, the vortex dissipates and the flow creates an upward WSS. As pointed out in Takizawa et al. (2010b), one of the reasons behind this change in flow characterstics is the motion of the aneurysm. We observe an aneurysm movement towards the left in Figure 8.23 when the flow accelerates.

Cardiovascular FSI

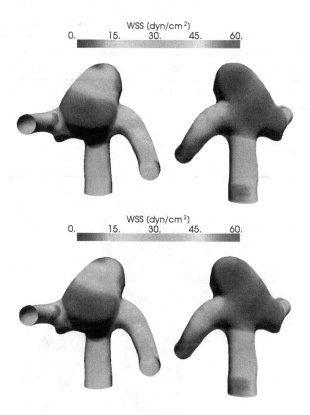

Figure 8.16 A bifurcating middle cerebral artery segment with aneurysm. Time-averaged WSS for the SCAFSI M1SC (top) and SCAFSI M1C (bottom) computations

Table 8.3 A bifurcating middle cerebral artery segment with aneurysm. WSS (dyn/cm^2) for the CAFSI computations with the coarse and medium meshes. Spatial maximum and mean at the peak systole, and the spatial maximum, mean, and minimum of the time-averaged values

Mesh	Peak Systole		Time Average		
	Max	Mean	Max	Mean	Min
Coarse	127	47	39	15	0.50
Medium	227	55	58	17	0.27

8.8.4 Computations with New Surface Extraction, Mesh Generation, and Boundary Condition Techniques

Three patient-specific cerebral artery segments with aneurysm were computed in Takizawa et al. (2011c) with the arterial-surface extraction, mesh generation, and boundary condition techniques introduced in Takizawa et al. (2011c). We describe the computations from Takizawa et al. (2011c). Figure 8.24 shows the geometries of the three models, obtained with the arterial-surface extraction technique introduced in Takizawa et al. (2011c), which was

Table 8.4 A bifurcating middle cerebral artery segment with aneurysm. WSS (dyn/cm^2) for the CAFSI computation with the medium mesh and the SCAFSI M1SC and SCAFSI M1C computations. Spatial maximum and mean at the peak systole, and the spatial maximum, mean, and minimum of the time-averaged values

Computation	Peak Systole		Time Average		
	Max	Mean	Max	Mean	Min
CAFSI	227	55	58	17	0.27
SCAFSI M1SC	227	55	60	17	0.29
SCAFSI M1C	225	55	60	17	0.30

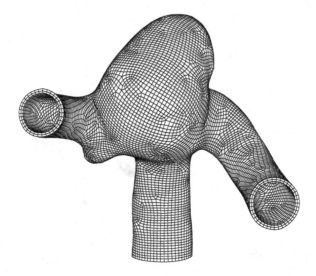

Figure 8.17 A bifurcating middle cerebral artery segment with aneurysm. Fine structural mechanics mesh when the outflow pressure is maximum

described in Section 8.2.1. These three models were called Model 1, Model 2, and Model 3 in Takizawa et al. (2011c). For Model 1, the diameter of the arterial lumen at the inflow and outflow ends is 4.61 and 2.67 mm. For Model 2, it is 3.30 and 2.51 mm. For Model 3, it is 2.97 mm at the inflow end, and 1.57, 1.54, 0.44, and 0.39 mm at the outflow ends. Figure 8.25 shows, for the three models, the mesh representing the arterial lumen, obtained with the mesh generation technique described in Takizawa et al. (2011c), which was highlighted in Section 8.2.2. Based on the surface mesh, the steps taken next in Takizawa et al. (2011c) were determining the arterial wall thickness, generating a hexahedral structural mechanics mesh with two layers of elements across the arterial wall, and calculating the EZP geometry. The techniques used for that were introduced in Takizawa et al. (2011c), which were described in Section 8.2.2. The length scales for the inflow and outflow patches are the lumen diameters at those ends. The length scale for the aneurysm/bifurcation patch is 0.80×(lumen diameter at the inflow) for Model 1. The factor is 0.89 for Model 2 and 0.43 for Model 3. Following that, the fluid mechanics meshes were generated with four layers of refined mesh near the arterial walls, where the layers were built with the techniques introduced in Takizawa et al.

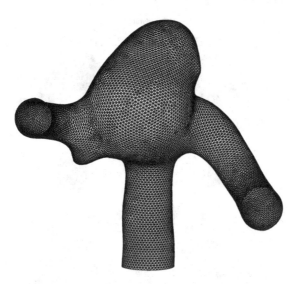

Figure 8.18 A bifurcating middle cerebral artery segment with aneurysm. Fine fluid mechanics mesh at the fluid–structure interface

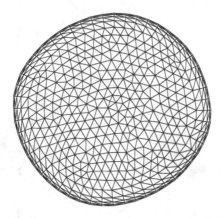

Figure 8.19 A bifurcating middle cerebral artery segment with aneurysm. Fine fluid mechanics mesh at the inflow plane

(2011c), which were described in Section 8.2.2. The thickness of the first layer for the inlet and outlet patches is 0.007×(lumen diameter at the ends associated with those patches), with a smooth interpolation between the patches. The progression factor is 1.75. Figures 8.26 and 8.27 show the fluid mechanics meshes for the three models. The node and element numbers for the models are given in Table 8.6. As pointed out in Takizawa *et al.* (2011c), Model 1 offers a geometry that is quite suitable for comparing the flow field obtained with what we get for the same artery segment with the aneurysm removed by virtual "surgery." Model 2, because of its long, thin geometry, poses computational challenges in iterative solution of the linear equation systems. The multiple outlets of Model 3, together with an adjacent sizable

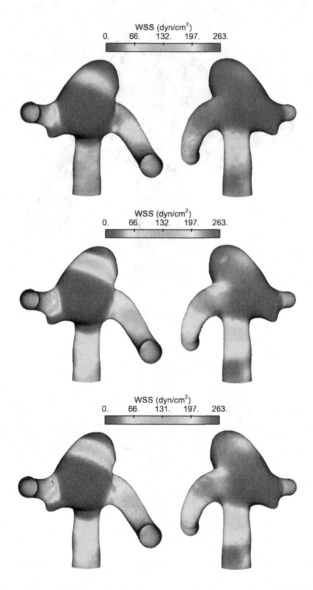

Figure 8.20 A bifurcating middle cerebral artery segment with aneurysm. WSS for the coarse (top), medium (middle), and fine (bottom) meshes when the volumetric flow rate is maximum

aneurysm, also pose computational challenges. The Womersley parameter (defined in Section 8.4.2) is 2.9, 2.1, and 1.9 for Model 1, Model 2, and Model 3, respectively. These are based on the duration of one cardiac cycle (1 s) and each representative diameter is calculated from the inflow area corresponding to the shape when inflated to the average pressure.

The computations were carried out with the SSTFSI-TIP1 technique (see Remarks 4.10 and 5.9) and the SUPG test function option WTSA (see Remark 4.4). The stabilization parameters used are those given by Equations (4.115)–(4.120) and (4.125). The time-step size is

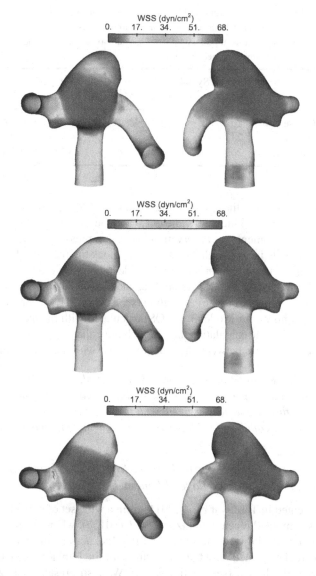

Figure 8.21 A bifurcating middle cerebral artery segment with aneurysm. Time-averaged WSS for the coarse (top), medium (middle), and fine (bottom) meshes

3.333×10^{-3} s. The number of nonlinear iterations per time step is 6. The number of GMRES iterations per nonlinear iteration for the fluid+structure block was chosen such that mass balance is satisfied to within at most 5% for each case. The number of GMRES iterations used in Model 1 and Model 3 computations was 300, which was sufficient for obtaining good mass balance. Model 2 required 500 GMRES iterations for good mass balance, at least partly because there are significantly more elements along the flow direction. As pointed out in Takizawa et al. (2011c), this is a significant increase in computational cost, so we are exploring

Table 8.5 A bifurcating middle cerebral artery segment with aneurysm. WSS (dyn/cm^2) for the coarse, medium, and fine meshes. Spatial maximum and mean at peak systole, and spatial maximum, mean, and minimum of time-averaged values

Mesh	Peak Systole		Time Average		
	Max	Mean	Max	Mean	Min
Coarse	102	37	32	12.53	0.16
Medium	237	54	60	16.76	0.32
Fine	263	53	68	16.53	0.24

better preconditioners, such as the one in Manguoglu et al. (2008). For all six nonlinear iterations, for Model 1 and Model 3 the fluid scale is 1.0 and the structure scale is 50. For Model 2, the fluid scales for the momentum conservation and incompressibility constraint are 1.0 and 10, and the structure scale is 50. For the mesh moving block the number of GMRES iterations is 30. Figure 8.28 shows the streamlines for each model when the volumetric flow rate is maximum. Figures 8.29–8.31 show the OSI for all three models, calculated with the technique that excludes rigid-body rotations from the calculation (see Section 8.1.4). For the purpose of comparison, Figure 8.32 shows, for Model 3, the OSI calculated with the technique that excludes rigid-body rotations from the calculation and the technique that does not. This is a case where we can clearly see differences between the OSI calculated with the two techniques.

REMARK 8.24 *As mentioned originally in Takizawa et al. (2011c) and also at the beginning of this section, Model 1 geometry is quite suitable for comparing the flow field obtained with what we get for the same artery segment with the aneurysm removed by virtual surgery. Figure 8.33 shows the streamlines for Model 1 with the aneurysm removed by virtual surgery.*

8.8.5 *Computations with the New Techniques for the EZP Geometry, Wall Thickness, and Boundary-Layer Element Thickness*

A sample was presented in Tezduyar et al. (2011) from a wide set of patient-specific cerebral-aneurysm models computed recently Takizawa et al. (2012b), where the shrinking amount in the EZP process, the arterial wall thickness, and the thickness of the layers of refined fluid mechanics mesh are determined based on the solution of the Laplace's equation over the surface mesh covering the lumen (see Section 8.2.2). We also present that sample here. The length scales used in conjunction with the trial ratios for the inflow and outflow boundaries are the lumen diameters at those ends. The value specified for the thickness of the first layer of elements at the inflow and outflow boundaries is 0.007×(lumen diameter at those ends). In these computations, the volumetric flow rate is specified by using the scaling technique described in Remark 8.11. Figure 8.34 shows the EZP shrinking amount, wall thickness, and structure mesh for the arterial model, which we call Model M6Acom. The diameter of the arterial lumen is 3.13 mm at the inflow end and 2.12 mm at both outflow ends. The structure mesh has two layers of elements across the arterial wall. For the layers of refined fluid mechanics mesh near the arterial wall, the progression factor is 1.75. Figure 8.35 shows the

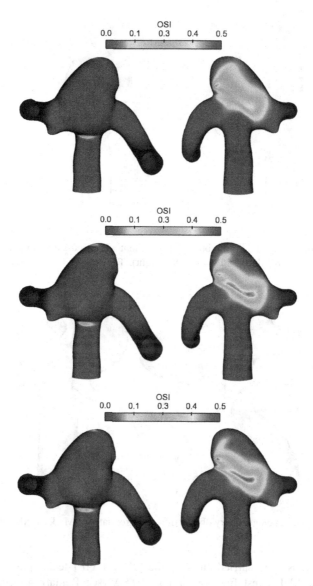

Figure 8.22 A bifurcating middle cerebral artery segment with aneurysm. OSI for the coarse (top), medium (middle), and fine (bottom) meshes

fluid mechanics mesh at the lumen, thickness of the first layer of elements near the arterial wall, and the mesh at the inflow plane. The node and element numbers for the model are given in Table 8.7. The Womersley parameter (defined in Section 8.4.2) is 1.96 and the peak volumetric flow rate is 1.2 ml/s. This is based on the duration of one cardiac cycle (1 s) and the representative diameter is calculated from the inflow area corresponding to the shape when inflated to the average pressure.

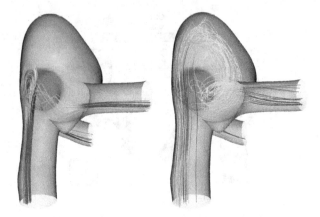

Figure 8.23 A bifurcating middle cerebral artery segment with aneurysm. Streamlines computed with the fine mesh at $t = 0.268$ s (left) and $t = 0.448$ s (right). The streamlines illustrate the WSS direction changes

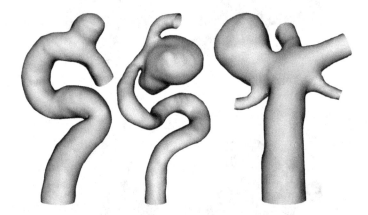

Figure 8.24 Arterial lumen geometry obtained from voxel data for Model 1, Model 2, and Model 3

The computations were carried out with the SSTFSI-TIP1 technique (see Remarks 4.10 and 5.9) and the SUPG test function option WTSA (see Remark 4.4). The stabilization parameters used are those given by Equations (4.115)–(4.120) and (4.125). The SSP option is used fully (see Remarks 5.15 and 5.16). The time-step size is 3.333×10^{-3} s. The number of nonlinear iterations per time step is 6. The number of GMRES iterations per nonlinear iteration for the fluid+structure block was chosen such that mass balance is satisfied to within at most 5% for each case. The number of GMRES iterations is 300, and this was sufficient for obtaining good mass balance. For all six nonlinear iterations the fluid scale is 1.0 and the structure scale is 100. For the mesh moving block the number of GMRES iterations is 30. Figure 8.36 shows the WSS when the volumetric flow rate is maximum. Figure 8.37 shows the OSI, calculated with the technique that excludes rigid-body rotations from the calculation (see Section 8.1.4).

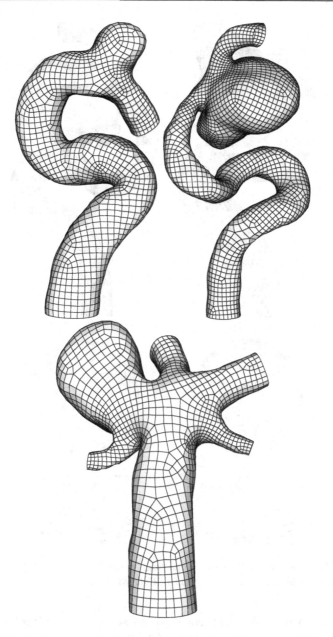

Figure 8.25 Mesh representing the arterial lumen for Model 1, Model 2, and Model 3

8.9 Computations with the ALE FSI Technique

In this section we present three examples of patient-specific cardiovascular FSI computations, which include cerebral aneurysms, the total cavopulmonary connection (TCPC), and the left ventricular assist device (LVAD). The cerebral aneurysm results are taken from Hsu

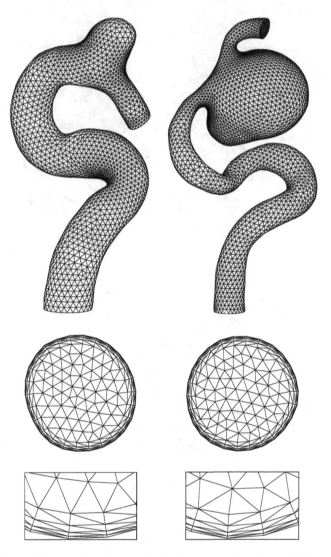

Figure 8.26 Top: fluid mechanics mesh at the fluid–structure interface. Middle and bottom: fluid mechanics mesh at the inflow plane. Left: Model 1, Right: Model 2

and Bazilevs (2011), while the TCPC computations were reported in Bazilevs *et al.* (2009b), and the LVAD computations in Bazilevs *et al.* (2009a).

All computations were carried out in a parallel computing environment and were completed without any remeshing. The FSI equations are advanced in time using the generalized-α method. The meshes for the blood volume and the vessel wall are matching at the interface, which simplifies the computational procedures.

In all cases a Neo-Hookean material with dilatational penalty is employed to model the response of the vascular wall. The cerebral aneurysm cases were computed using linear

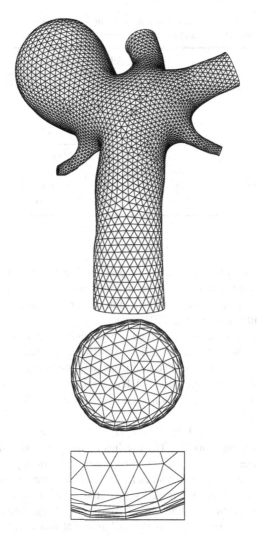

Figure 8.27 Top: fluid mechanics mesh at the fluid–structure interface. Middle and bottom: fluid mechanics mesh at the inflow plane. Model 3

tetrahedral elements for both fluid and structure discretizations. In the case of TCPC, the structural equations were discretized with a shell-like formulation using triangular elements (see Bazilevs *et al.*, 2009b for details). In the case of LVAD, the FSI computations were performed using NURBS-based IGA.

A quasi-direct solution strategy is used (see Section 6.1.2). As a result, the effect of the mesh motion on the fluid equations is omitted from the tangent matrix for efficiency, as advocated in Bazilevs *et al.* (2009a) for cardiovascular FSI applications. In solving the linear equation systems involved at every nonlinear iteration, the GMRES search technique Saad and Schultz (1986) is used with a block-diagonal preconditioner.

Table 8.6 Number of nodes and elements for the three models. Here *nn* and *ne* are number of nodes and elements, respectively

			Model 1	Model 2	Model 3
Structure	Volume	nn	8715	14454	8115
		ne	5760	9588	5318
	Interface	nn	2905	4818	2705
		ne	2880	4794	2659
Fluid	Volume	nn	21 628	58 988	47 719
		ne	122 241	343 552	278 650
	Interface	nn	2902	6377	4882
		ne	5754	12 706	9672

8.9.1 Cerebral Aneurysms: Tissue Prestress

In this section we focus on the importance of prestressing of arterial tissue and its effect on the key quantities of interest in cardiovascular FSI computations. For other results obtained in ALE FSI simulations of cerebral aneurysms the reader is referred to Isaksen *et al.* (2008), Zhang *et al.* (2009) and Bazilevs *et al.* (2010c,d).

We apply the prestress procedure described in Section 8.3 to two cerebral aneurysm models shown in Figure 8.38. The inflow and outflow branches are labeled as M1 and M2, respectively, in the same figure. Both models come from patient-specific imaging data and exhibit significant geometrical differences. Model 1 has a relatively small aneurysm dome and an inlet branch of large radius. The situation is reversed for Model 2. Table 8.8 shows the inlet cross-sectional area for both models.

Meshing techniques developed by Zhang *et al.* (2009) are used for generating linear tetrahedral elements for both models. The meshes contain both the blood flow volume and solid vessel wall. Meshes with boundary layer resolution are employed to ensure high fidelity of the computational results. The mesh sizes for both models are summarized in Table 8.9.

Figure 8.39 shows the final prestressed state for both models, which also demonstrates the applicability of the method to different vascular geometries. The models are colored by the isocontours of wall tension, which is defined as the absolute value of the first principal in-plane stress of $\mathbf{S}_0$.

To assess the influence of the prestress, we perform a coupled FSI simulation of both models and compare the results with and without prestress. Figure 8.40 shows the relative wall displacement between the deformed configuration and reference configuration coming from imaging data. The deformed configuration corresponds to the time instant when the fluid traction vector is closest to the averaged traction vector used for the prestress problem given by Equation (8.23). Almost no difference between the reference and deformed configurations is seen in the case of the prestressed-artery simulation, as expected. However, in the case of non-prestressed simulation, the differences between the two configurations are significant. This indicates that the FSI problem is not being solved on the correct geometry. Furthermore, the relative geometry error is larger for Model 2, which has a larger aneurysm dome and a thinner wall. Figure 8.41 shows the relative wall displacement between the deformed configuration at

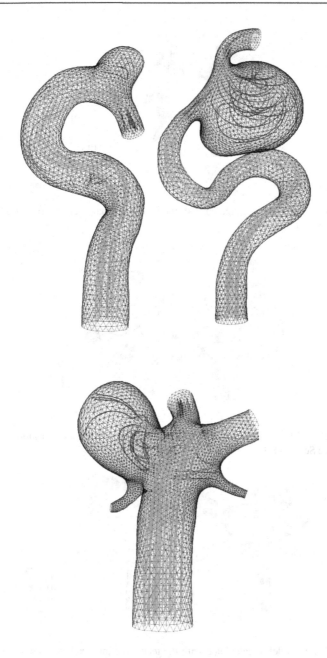

Figure 8.28 Streamlines when the volumetric flow rate is maximum. Model 1, Model 2, and Model 3. The streamlines are colored by the velocity magnitude

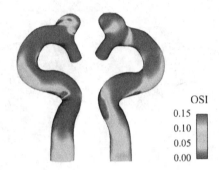

Figure 8.29 OSI for Model 1, calculated with the technique that excludes rigid-body rotations from the calculation (see Section 8.1.4)

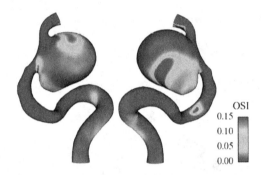

Figure 8.30 OSI for Model 2, calculated with the technique that excludes rigid-body rotations from the calculation (see Section 8.1.4)

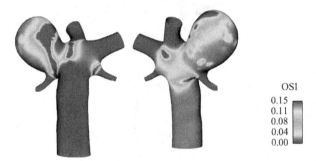

Figure 8.31 OSI for Model 3, calculated with the technique that excludes rigid-body rotations from the calculation (see Section 8.1.4)

peak systole and low diastole. In both the prestressed and non-prestressed cases the relative displacement is fairly small, yet non-negligible. The non-prestressed case, however, makes use of the geometry that is significantly more "inflated" compared to the prestressed case and the imaging data.

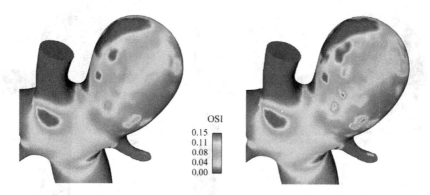

Figure 8.32 OSI for Model 3, calculated with the technique that excludes rigid-body rotations from the calculation (left) and the technique that does not (right)

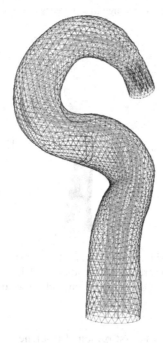

Figure 8.33 Model 1 with the aneurysm removed by virtual surgery. Streamlines when the volumetric flow rate is maximum. The streamlines are colored by the velocity magnitude

Figure 8.42 shows a comparison of the blood flow speed near peak systole for the simulations with and without prestress. The results between prestressed and non-prestressed cases are very similar, although some differences in the flow structures are visible, especially for Model 2. Figure 8.43 shows a comparison of the WSS near peak systole for both cases. The WSS, unlike blood flow velocity, exhibits significant differences in magnitude and spatial distribution. The wall tension results are shown in Figure 8.44. The wall tension is

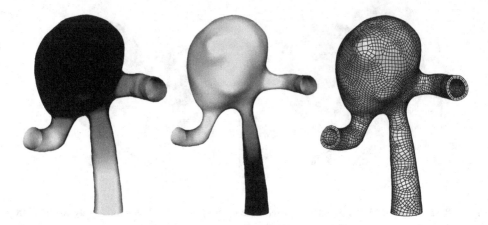

Figure 8.34 Model M6Acom. EZP shrinking amount over the surface (lumen) extracted from the medical image (left), wall thickness over the shrunk lumen (middle), and structure mesh at zero pressure (right). The color range represents a value range that increases from light to dark

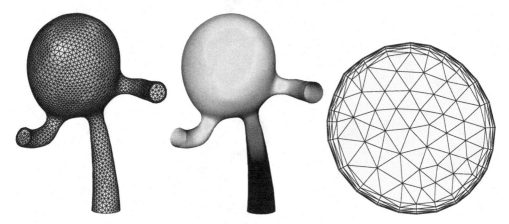

Figure 8.35 Model M6Acom. Fluid mechanics mesh at the lumen and outflow planes (left), thickness of the first layer of elements near the arterial wall (middle), and the mesh at the inflow plane (right). All pictures are from the starting point of our computation cycle. The color range represents a value range that increases from light to dark

defined as the absolute value of the first principal in-plane stress (see Bazilevs *et al.*, 2010c for details). Again, significant differences in magnitude and spatial distribution are observed. The comparisons clearly show the importance of considering prestress in the patient-specific vascular FSI simulations for accurate prediction of hemodynamic phenomena and vessel wall mechanics.

8.9.2 Total Cavopulmonary Connection

Congenital heart defects are among the most prevalent form of birth defects, occurring in roughly 1% of births. "Single ventricle"-type defects refer to cases where the heart has only

Cardiovascular FSI

Table 8.7 Model M6Acom. Number of nodes and elements. Here *nn* and *ne* are number of nodes and elements, respectively

			M6Acom
Structure	Volume	nn	17 574
		ne	11 650
	Interface	nn	5858
		ne	5825
Fluid	Volume	nn	33 040
		ne	192 112
	Interface	nn	3528
		ne	6996

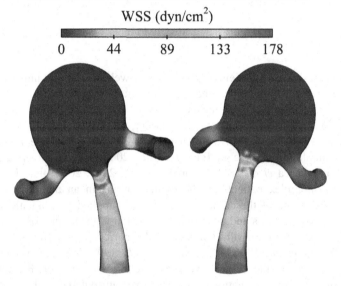

Figure 8.36 Model M6Acom. WSS when the volumetric flow rate is maximum

one effective or functional pumping chamber, and are usually fatal shortly after birth if left untreated. Single ventricle patients usually require a staged surgical approach which culminates with a Fontan procedure Fontan and Baudet (1971). There are two variants of the Fontan procedure, the extra-cardiac conduit (ECC) and the lateral tunnel (LT) Petrossian *et al.* (2006). In both cases, the superior vena cava (SVC) is connected to the right pulmonary artery. In the ECC variant, a baffle is also constructed to connect the inferior vena cava (IVC) to the pulmonary arteries, resulting in a modified T-shaped junction. In the LT variant, a tunnel-like patch is placed inside the atrium so that blood returning from the IVC is directed through this tunnel. A connection is then made between the end of the tunnel/top of the right atrium and the underside of the pulmonary artery. As a result of both the ECC and the LT, the circulation becomes a single pump system, and the heart contains only oxygenated blood. A surgical connection of the SVC and IVC directly to the left and right pulmonary arteries is referred to as the TCPC.

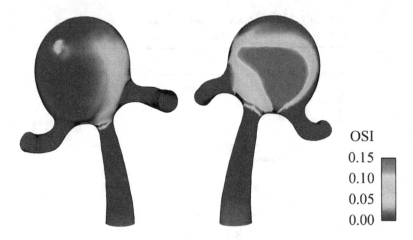

Figure 8.37 Model M6Acom. OSI

Congenital heart disease is a field that lends itself well to study by numerical techniques due to the wide range of anatomies and variations among patients. Numerical techniques allow us to examine the effects of the geometry of the Fontan connection that plays an important role in the overall success of the surgery, and assess blood flow characteristics and energy losses associated with a given surgical design. There are numerous articles on the TCPC simulation (see, e.g., Ensley *et al.*, 2000; Khunatorn *et al.*, 2002; Bove *et al.*, 2003; Migliavacca *et al.*, 2003; Marsden *et al.*, 2007, 2008) that do a very careful CFD analysis on complex patient-specific configurations in an attempt to answer some of these questions. Some of the earliest work in CFD applied to congenital heart disease compared energy loss in the standard T-junction Fontan with the proposed "offset" model, and led to the adoption of the offset model as the currently preferred method (de Leval *et al.*, 1996; Dubini *et al.*, 1996; Migliavacca *et al.*, 1997, 2003). However, despite the abundance of articles on the subject, very little clinical impact on Fontan surgery has been derived directly from simulations. This is in part attributable to the limitations of the simulation methods used for this application. The work of Bazilevs *et al.* (2009b) addressed one of these shortcoming by introducing flexible wall modeling in Fontan surgery simulations.

The Fontan surgery model that is used in the computations is shown in Figure 8.45. The model is comprised of two inlets, corresponding to the IVC and SVC, for which the time-periodic flow rate is prescribed. The model also has 20 outlet branches corresponding to the pulmonary circulation. At each outlet, resistance boundary conditions are prescribed. The resistance data (see Bazilevs *et al.*, 2009b for details) is chosen to match cardiac catheterization pressure data for this patient, as described below, and corresponds to case when the patient is resting, which we refer to as "rest conditions."

Note that, on the venous side, the intramural pressure is significantly lower than on the arterial side, and, as a result, the resistance boundary condition does not have an ambient pressure component (i.e., $p_0 = 0$ in Equation (8.44)). Because the ambient pressure in the venous circulation is very low as compared to the arterial circulation, the vessel configuration taken from image data may be used to approximate the reference, zero-stress configuration.

Cardiovascular FSI

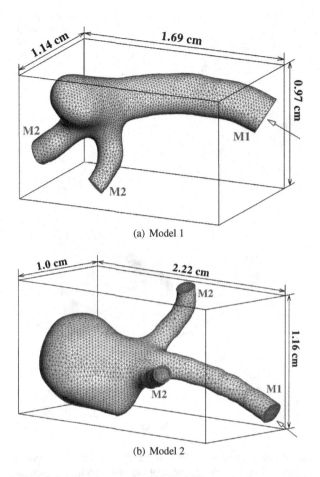

(a) Model 1

(b) Model 2

Figure 8.38 Tetrahedral finite element mesh of the middle cerebral artery (MCA) bifurcation with aneurysm. Inlet branches are labeled M1 and outlet branches are labeled M2 for both models. The arrows point in the direction of inflow velocity

Table 8.8 Inflow cross-sectional areas for the aneurysm models

Model	Inflow surface area (cm^2)
1	4.962×10^{-2}
2	2.102×10^{-2}

Table 8.9 Tetrahedral finite element mesh sizes for the aneurysm models

Model	nn	ne
1	30 497	164 140
2	30 559	167 563

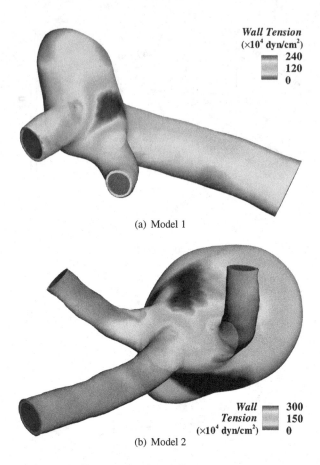

Figure 8.39 Final prestressed state for Model 1 and 2. The models are colored by the isocontours of wall tension, which is defined as the absolute value of the first principal in-plane stress of S_0

We use the following material properties in our computations. The fluid density and dynamic viscosity are $1.06 \, g/cm^3$ and $0.04 \, g/cm \, s$, respectively. The vessel wall has the density $1.00 \, g/cm^3$, and shear and bulk moduli of $1.72 \times 10^6 \, dyn/cm^2$ and $1.67 \times 10^7 \, dyn/cm^2$, respectively.

The tetrahedral mesh of the Fontan model is shown in Figure 8.46. The mesh is refined near boundary layers and regions of complex branchings based on the error indicators from a stand-alone fluid mechanics computation using vascular blood flow mesh adaption techniques in Sahni *et al.* (2006). The mesh has over 1M tetrahedral elements, which, in combination with boundary layer meshing, ensures high-fidelity simulation results.

We simulate rest and exercise conditions, and also compare rigid- and flexible-wall results in each case. Exercise flow conditions are generated by increasing the IVC flow rate three times, while keeping SVC flow fixed. These values are at or slightly above the typical range for a Fontan patient found in clinical exercise data, in which, on average, Fontan patients are able to approximately double their cardiac index at peak exercise (Shachar *et al.*, 1982; Giardini *et al.*, 2008).

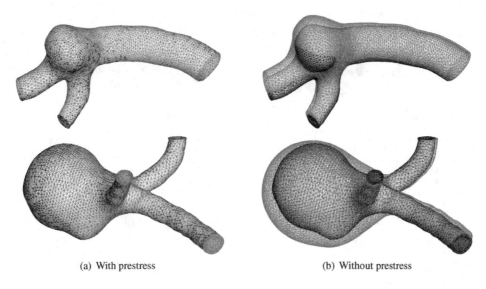

(a) With prestress (b) Without prestress

Figure 8.40 Relative wall displacement between the deformed configuration and reference configuration coming from imaging data. Top: Model 1; Bottom: Model 2. The deformed configuration corresponds to the time instant when the fluid traction vector is closest to the averaged traction vector used for the prestress problem given by Equation (8.23)

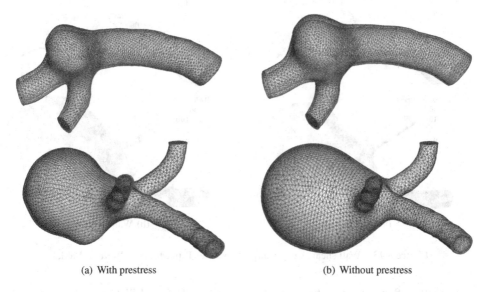

(a) With prestress (b) Without prestress

Figure 8.41 Relative wall displacement between the deformed configuration at peak systole and low diastole. Top: Model 1; Bottom: Model 2

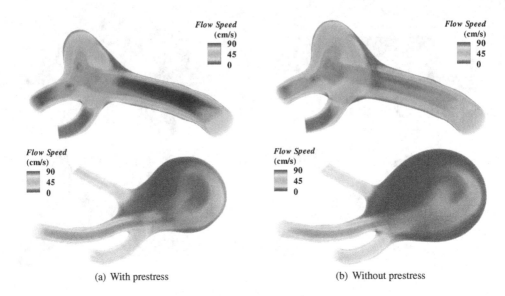

Figure 8.42 Volume-rendered blood flow velocity magnitude near peak systole. Top: Model 1; Bottom: Model 2

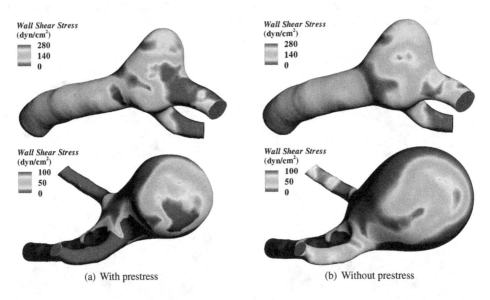

Figure 8.43 Wall shear stress near peak systole. Top: Model 1; Bottom: Model 2

The inflow flow rates as a function of time for both IVC and SVC branches are given in Figure 8.47. Note that the SVC flow rate is synchronized with the heart cycle, while the IVC flow rate is synchronized with the respiratory cycle. Cardiac catheterization pressure tracings, echocardiography, and magnetic resonance (MR) studies have all demonstrated that respiration significantly effects Fontan flow rates and pressures (Hjortdal *et al.*, 2003; Pedersen *et al.*,

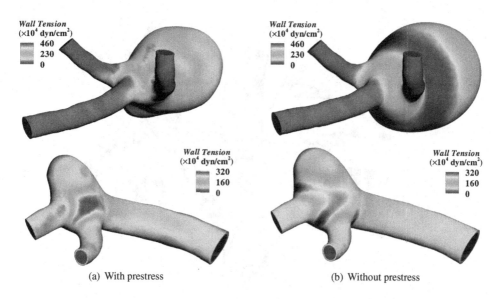

Figure 8.44 Wall tension near peak systole. Top: Model 1; Bottom: Model 2

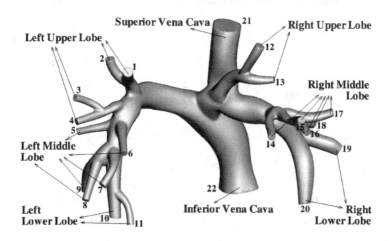

Figure 8.45 Patient-specific Fontan surgery model that includes the IVC, SVC, and pulmonary circulation represented by the left upper lobe, left middle lobe, left lower lobe, right upper lobe, right middle lobe, and right lower lobe. Image is rotated anterior to posterior for ease of viewing

2002). As seen with the echocardiographic tracings, quantitative real-time phase contrast MR measurements by Hjortdal et al. (2003) show that flow rates in the IVC vary significantly with respiration at rest (as much as 80%), with smaller cardiac pulsatility superimposed. Cardiac variations in the SVC were found to be small, with no significant respiratory variation. Based on this data, we impose a respiration model to model flow variations in the IVC following Marsden et al. (2007). This model assumes three cardiac cycles per respiratory cycle, and values of heart rate and respiratory rate are increased during exercise following the data

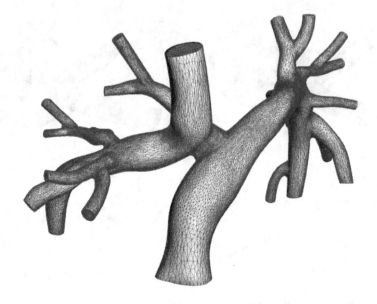

Figure 8.46 Tetrahedral mesh for the Fontan surgery model

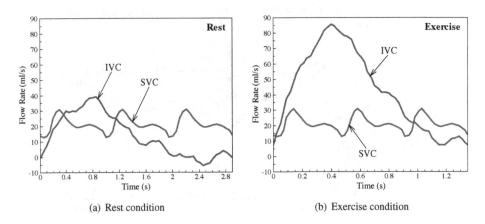

Figure 8.47 IVC and SVC inflow flow rates for rest and exercise conditions

of Hjortdal et al. (2003). Also note that the cycles are shorter and maximum flow rate is significantly higher for the exercise condition case.

Figure 8.48 shows isosurfaces of the vessel wall displacement magnitude for rest and exercise simulations. The inflow flow rate is higher for the case of exercise conditions leading to increased levels of intramural pressure and, as a result, larger magnitude of wall displacement.

Flow streamlines at low flow rate comparing rest and exercise conditions are shown in Figure 8.49. Under rest conditions, due to flow reversal during the later parts of the respiratory cycle, a swirl-like feature develops near the inflow of the IVC giving rise to a helical flow pattern in the branch. This feature is not present for the exercise condition case.

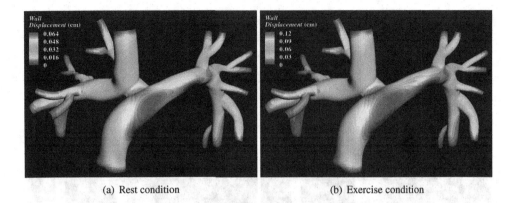

(a) Rest condition   (b) Exercise condition

Figure 8.48 Isosurfaces of vessel wall displacement magnitude

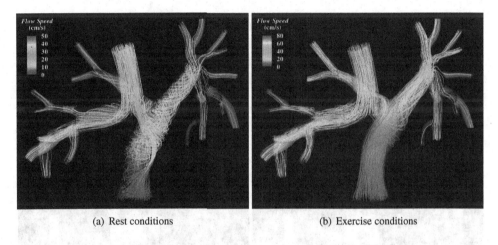

(a) Rest conditions   (b) Exercise conditions

Figure 8.49 Blood velocity streamlines at low flow rate

We next examine the WSS and compare the results of rigid-wall and flexible-wall simulations. In Figure 8.50, the rest conditions are compared at peak flow, while Figure 8.51 shows a comparison for the exercise case. The overall distribution of WSS is similar and WSS highs and lows tend to concentrate near the arterial branchings. A closer examination of the WSS data revealed that in the case of rest conditions, the WSS for the rigid-wall simulation overpredicts that of the flexible wall by as much 17% in the case of rest conditions and by as much as 45% in the case of exercise conditions. The data was taken at several discrete locations on the vascular wall and the locations are indicated by arrows in the figures. This data clearly shows that flexible-wall modeling is important for Fontan surgery simulations and has a greater effect on the outcomes of the simulations in the case of exercise conditions.

Computed pressure time histories for SVC, IVC, and selected outlet branches are shown in Figures 8.52 and 8.53. In all cases there is a distinct pressure time lag between rigid- and flexible-wall results. Furthermore, the flexible wall assumption produces a smoothing effect on the pressure output. In both rest and exercise cases, the pressure peak is always higher

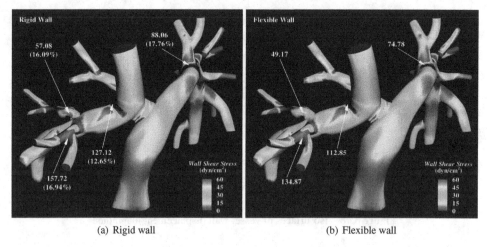

(a) Rigid wall

(b) Flexible wall

Figure 8.50 Comparison of WSS at rest conditions

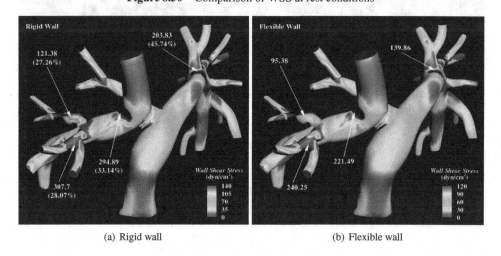

(a) Rigid wall

(b) Flexible wall

Figure 8.51 Comparison of WSS at exercise conditions

for the rigid-wall simulation and the overprediction is greater for the exercise conditions simulation, just as in the case of the WSS. In Figure 8.53, the outlet pressure data is presented in two ways: i. the pressure field is taken directly and averaged over the outlet cross-sections; ii. the flow rate is computed through the outlets and multiplied by the corresponding resistance constant according to Equation (8.44). The figure shows no visible differences between the two quantities, meaning the resistance BC, although satisfied only "weakly" in the variational formulation, actually holds in a strong sense.

8.9.3 Left Ventricular Assist Device

Cardiovascular diseases produce a number of physiological changes to the tissue of the cardiovascular system (e.g., loss of elasticity of the arteries as in arteriosclerosis, ischemic damage,

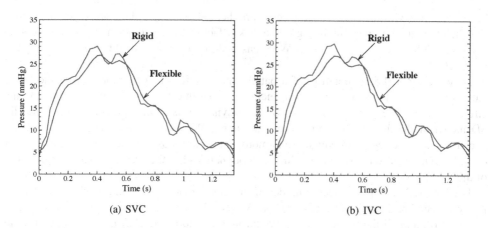

Figure 8.52 Comparison of blood pressure time histories at SVC and IVC at exercise conditions

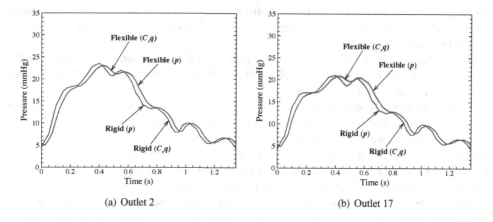

Figure 8.53 Comparison of blood pressure time histories at selected outlets at exercise conditions

and cardiomyopathies). These change the hemodynamics of the cardiovascular system with potentially disastrous consequences. When other treatments fail, implanted circulation support devices can be used to reestablish interrupted or inadequate flow. The emergence of axial flow assist devices has significantly advanced therapeutic options for patients with severe heart failure. These devices deliver continuous blood flow and provide distinct advantages with regard to reduction in size, weight, and energy demands, simplified implantation technique, and device control Hetzer et al. (2002).

New, small, efficient non-pulsatile axial flow LVADs are currently being studied as bridges to transplant, destination therapy, and recovery for congestive heart failure. These pumps are highly engineered, optimized devices, but the design of their most effective implant configurations and operating conditions has been more difficult. This is unfortunate because LVADs greatly alter the hemodynamics of the heart and aorta, which can be either helpful, as intended, or harmful, leading to significant complications. Tools to optimize LVAD device design and placement are notably lacking, though both have a significant effect on hemodynamics. Of particular concern regarding hemodynamics is the occurrence of regions in which the blood is stagnant, thought to be a key factor leading to thrombogenesis Wootton and Ku (1999). Flow stasis or mild WSS has been correlated with thrombotic events Liu et al. (2002).

Here we report FSI simulation of a patient-specific model of the aorta, from the aortic valve to the descending thoracic aorta, including flow into branch vessels, and include the effect of LVAD. The effect of an LVAD on hemodynamics is complex and demands a locally 3D model of the flow in the aortic valve and aorta. We focus on this section of the aorta because this is the region in which the hemodynamics are most affected by the introduction of an LVAD. It is also the region in which hemodynamics has the greatest effect on the health of the heart.

For this study a patient-specific model of the thoracic aorta with an added LVAD branch in the descending location was constructed. We consider three different flow conditions: 1) LVAD is off and all the blood flow occurs through the aortic root; 2) LVAD is operating in the regime where over one half of the blood supplied to the aorta comes from the pump; 3) LVAD is operating in the regime where nearly all the flow comes from the LVAD. Inflow data for the patient-specific model was obtained from a lumped-parameter closed-loop multiscale model of the cardiovascular system that was developed in Gohean (2007). The latter allows for the inclusion of assist devices.

We use NURBS-based IGA for this application. (See Bazilevs et al., 2006a, 2008, 2009a; Zhang et al., 2007; Calo et al., 2008; Isaksen et al., 2008 for the application of NURBS-based IGA to FSI modeling of vascular flows.)

Patient-specific geometry of the thoracic aorta of an over-30-year-old healthy volunteer was obtained from 64-slice CT angiography. The geometrical model is shown in Figure 8.54 (a). The computational mesh, consisting of 44 892 quadratic NURBS elements, is shown in Figure 8.54 (b), where an additional branch was added to the model to represent the inflow from the left ventricular assist device. Wall thickness for this model is taken to be 15% of the nominal radius of each cross-section of the fluid domain model. Two quadratic NURBS elements and four C^1-continuous basis functions are used for through-thickness resolution of the arterial wall.

We employ the following material properties in our computations. The fluid density and dynamic viscosity are $\rho = 1.06 \, \text{g/cm}^3$ and $\mu = 0.04 \, \text{g/cm s}$, respectively. The solid has the density $\rho = 1 \, \text{g/cm}^3$, Young's modulus, $E = 4.144 \times 10^6 \, \text{dyn/cm}^2$, and Poisson's ratio, $\nu = 0.45$. The solid model coefficients μ and κ are obtained using standard relationships for the Lamé parameters of elastostatics.

We fix the artery at the inlet and at all outlets. The top 50% of the right and left innominate and subclavian arteries are also constrained not to move. This is done so as to avoid the nonphysical swinging motion of the thoracic aorta during the simulation. Constraining these portions of these arteries mimics the effect of the surrounding tissue, albeit in a very crude way. Further research in accounting for the surrounding tissue is necessary, with respect to both mathematical modeling and computation. The LVAD branch is fixed in space.

Our model has two inflow boundaries, the inlet of the ascending aorta and the inlet of the LVAD branch, where we specify a periodic flow waveform (see Figure 8.55). Our model also has various outlets where resistance boundary conditions are applied. The resistance constants were set as $C_1 = (1\,500/A_1) \, \text{dyn s/cm}^5$, $C_2 = (2\,666/A_2) \, \text{dyn s/cm}^5$, $C_3 = (1\,400/A_3) \, \text{dyn s/cm}^5$, $C_4 = (1\,400/A_4) \, \text{dyn s/cm}^5$, where $A_a, a = 1, 2, 3, 4$, are surface

Cardiovascular FSI

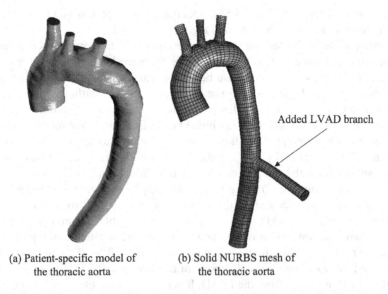

(a) Patient-specific model of the thoracic aorta

(b) Solid NURBS mesh of the thoracic aorta

Figure 8.54 Flow in a patient-specific thoracic aorta with LVAD. (a) Patient-specific model constructed from imaging data; (b) Smoothed solid NURBS model and mesh with the LVAD branch added. For more details of geometrical modeling for isogeometric analysis of blood flow, the reader is referred to Zhang et al. (2007)

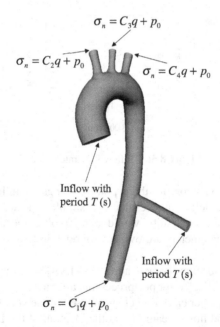

Figure 8.55 Flow in a patient-specific thoracic aorta with LVAD. Boundary conditions for the fluid mechanics domain. C_a, $a = 1, 2, 3, 4$, are the resistance constants, σ_n is the normal component of the traction vector, q is the volumetric flow rate, and p_0 is responsible for setting the physiological pressure level in the blood vessels

areas of the outlet faces (see Figure 8.55). This data was adapted from Olufsen (1998). The intramural pressure was set to $p_0 = 85$ mm Hg.

A lumped-parameter closed-loop multiscale model of the cardiovascular system has been developed in Gohean (2007) based on the framework of Olufsen (1998). Both the systemic and pulmonary circulation have been modeled to "close the loop" around the area of focus, the aorta and large arteries. The lumped-parameter cardiovascular model allows for the inclusion of the LVAD. In this particular case, we are studying the Jarvik 2000 model, which is a continuous-flow rotary blood pump. Because of the rotary nature of Jarvik 2000, commonly employed pressure-flow curves can be used to describe resultant flow rates over a range of pressures and pump speeds. The pump is introduced into the model as the additional outflow from the left ventricle and the additional inflow to the arterial tree. We consider three pump settings: (i) pump is off; (ii) pump is operating at the angular speed of 8000 rpm; (iii) pump is operating at the angular speed of 10 000 rpm. Periodic flow rates through the aorta and the LVAD inlets, with period $T = 0.6667$ s, are computed using the lumped-parameter cardiovascular model, and the results are illustrated in Figure 8.56. In the case when the pump is off, all the flow comes through the inflow of the aorta. In the case of 8000 rpm, the LVAD supplies more than 50% of the flow. Finally, in the case of 10 000 rpm, almost all of the flow comes from the LVAD. It is the latter case that is of most concern with respect to complications arising from flow stasis. We note that the case when the pump is off corresponds to the so-called "weak heart" condition, because the volume of blood supplied by the heart during the cycle is somewhat smaller, and the heart cycle is shorter than in the healthy case.

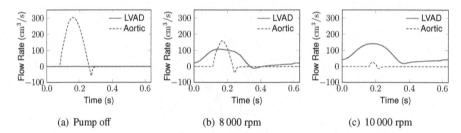

Figure 8.56 Inflow flow rate curves

We use the model outputs for the three pump settings as inflow boundary conditions for our patient-specific model, which results in a simple, one-way coupled simulation. A more faithful representation of reality would consist of embedding our patient-specific model inside the lumped-parameter cardiovascular model discussed here with a full two-way coupling.

The FSI of the patient-specific thoracic aorta with LVAD was computed for several heart cycles. Solution data was collected for postprocessing after a nearly periodic-in-time response was attained. 150 time steps per cardiac cycle were used in the simulations.

Figure 8.57 shows blood flow streamlines at the location of the LVAD branch attachment for the highest pump setting. Complex flow structures, thin boundary layers, as well as the presence of back flow up the descending part of the aorta are apparent in the figure. Aortic wall displacement at peak systole relative to low diastole is shown in Figure 8.58. The peak of the displacement occurs just below the descending part of the arch. This effect is due to the

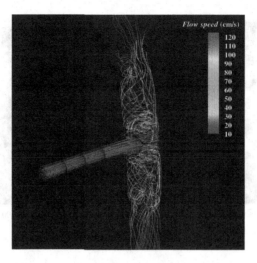

Figure 8.57 Flow in a patient-specific thoracic aorta with LVAD. Flow streamlines at peak systole in the LVAD attachment region. LVAD is operating at the highest setting

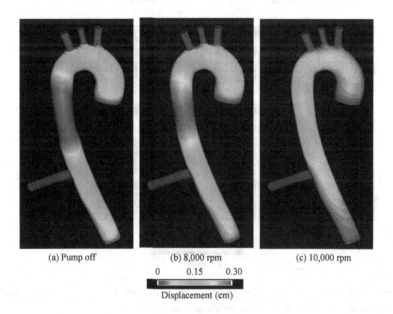

Figure 8.58 Flow in a patient-specific thoracic aorta with LVAD. Arterial wall displacement at peak systole

blood stream coming from the arch, impacting the arterial wall, and causing it to bend. This does not happen in the case of the highest LVAD setting, which is a direct consequence of the altered flow distribution in the aorta.

Figures 8.59 and 8.60 focus on the flow in the aortic arch. In the unassisted case, a helical flow pattern is obtained in early systole and late diastole, while in peak systole the flow vectors

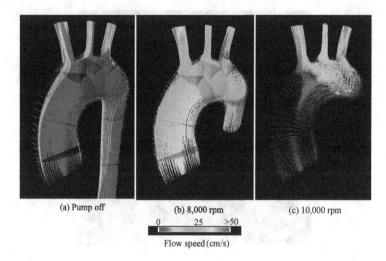

Figure 8.59 Flow in a patient-specific thoracic aorta with LVAD. Velocity vectors colored by the velocity magnitude in the ascending aorta and the arch at peak systole

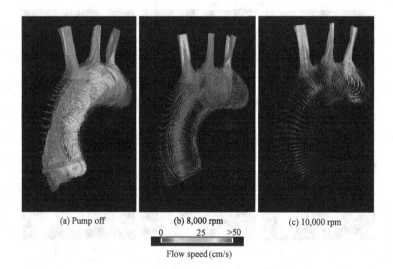

Figure 8.60 Flow in a patient-specific thoracic aorta with LVAD. Velocity vectors colored by the velocity magnitude in the ascending aorta and the arch in late diastole

are aligned with the arterial path. The helical flow pattern in the aortic arch is a well-known phenomenon. For example, in Kilner *et al.* (1993), helical flow structures were observed in a human thoracic aorta imaged by MR. The medium pump setting does not produce the helical flow patterns observed in the unassisted case and some flow stagnation is present in the ascending part of the aortic arch in early systole and late diastole. For the highest pump setting the flow in the ascending part of the arch is stagnant throughout the *entire* heart cycle. This is consistent with clinical observations and preliminary numerical simulations reported in Kar

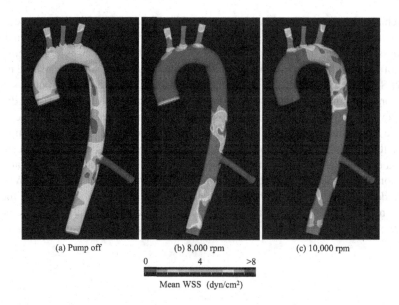

(a) Pump off   (b) 8,000 rpm   (c) 10,000 rpm

Mean WSS (dyn/cm²)

Figure 8.61 Flow in a patient-specific thoracic aorta with LVAD. Mean WSS

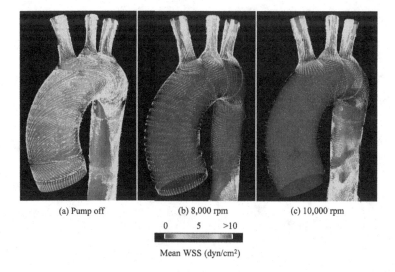

(a) Pump off   (b) 8,000 rpm   (c) 10,000 rpm

Mean WSS (dyn/cm²)

Figure 8.62 Flow in a patient-specific thoracic aorta with LVAD. Mean WSS vectors and the magnitude of the mean WSS

et al. (2005). Flow stagnation is known to increase uptake of atherogenic blood particles as a consequence of increased residence time, and is believed to be one of the mechanisms that is responsible for the development of atherosclerosis Glagov *et al.* (1988).

Figure 8.61 shows the cardiac-cycle-averaged ("mean") WSS magnitude on the luminal surface. In the arch, WSS is much lower for the cases when the pump is on. This observation is consistent with the predicted and observed flow stagnation in this area of the aorta. In contrast to the arch, in the descending branch near the LVAD, WSS significantly exceeds healthy levels. Figure 8.62 focuses on the arch and plots the WSS vectors on the luminal surface. In the unassisted case, the vectors follow the helical pattern, which is consistent with the behavior of the blood velocity in this case. For the pump-assisted simulations, the magnitude of the WSS is much lower in the arch, and, furthermore, for the highest pump setting, the WSS vectors point in the direction that is opposite to the conventionally assumed direction.

The relevance of WSS and its temporal oscillations (as measured by OSI) to atherosclerosis is an active area of medical research (see Shaaban and Duerinckx, 2000 for a comprehensive review of the subject). It was shown in Levesque and Nerem (1985) and Levesque et al. (1986) that endothelial cells subjected to elevated levels of WSS tend to elongate and align in the direction of flow, and that endothelial cells that experience low or oscillatory WSS remain more rounded and have no preferred alignment pattern. Moreover, exposure of the arterial wall to a relatively low WSS may increase intercellular permeability and consequently increase the vulnerability of these regions of the vessel to atherosclerosis Okano and Yoshida (1994).

9

Parachute FSI

Computer modeling of parachutes involves all the numerical challenges of FSI problems. The aerodynamics (fluid mechanics) of the parachute depends on the canopy shape and the deformation (structural mechanics) of the canopy depends on the aerodynamical forces, and the two systems need to be solved in a coupled fashion with proper interface conditions. Because the parachute FSI is in a category of problems where the structure is light (compared to the air masses involved in the parachute dynamics) and very sensitive to changes in the aerodynamical forces, the coupling technique, which determines how the coupling between the equation blocks representing the fluid mechanics, structural mechanics, and mesh moving equations is handled, requires extra care. Spacecraft parachutes are typically very large "ringsail" parachutes that are made of a large number of gores, where a gore is the slice of the canopy between two radial reinforcement cables running from the parachute vent to the skirt. Ringsail parachute gores are constructed from "rings" and "sails," resulting in a parachute canopy with hundreds of ring "gaps" and sail "slits." The complexity created by this geometric porosity makes FSI modeling inherently challenging. Spacecraft parachutes are also typically used in clusters of two or three parachutes, and the contact between the parachutes is one of the major challenges specific to FSI modeling of parachute clusters.

Parachute FSI modeling with the space–time FSI techniques started as early as 1997, with axisymmetric computation of the inflation of a parachute Stein *et al.* (1997). The 3D parachute computations with the space–time FSI techniques go as far back as 2000 (Kalro and Tezduyar, 2000; Stein *et al.*, 2000). Since then, many of the computational challenges involved in parachute FSI have been addressed (see Tezduyar and Osawa, 2001; Stein *et al.*, 2001a,b, 2003a,c; Tezduyar *et al.*, 2006a,b, 2008a,c, 2010b; Tezduyar and Sathe, 2007; Takizawa *et al.*, 2011d,e,f, 2012e; Takizawa and Tezduyar, 2012a) by developing both core space–time FSI techniques and special space–time FSI techniques specifically targeting parachutes, all implemented for large-scale parallel computing. Now, parachute modeling with the space–time FSI techniques is at a level of sophistication and power where it is helping with the design and testing of ringsail parachute clusters to be used with the next-generation NASA spacecraft. In this

---

Computational Fluid–Structure Interaction: Methods and Applications, First Edition.
Yuri Bazilevs, Kenji Takizawa and Tayfun E. Tezduyar.
© 2013 John Wiley & Sons, Ltd. Published 2013 by John Wiley & Sons, Ltd.

chapter we focus on spacecraft parachutes and special space–time FSI techniques developed in recent years.

The space–time FSI computation of parachutes started with the block-iterative FSI coupling technique (see Section 6.1.1) in the early years (see, for example, Kalro and Tezduyar, 2000; Stein et al., 2000, 2001a; Tezduyar and Osawa, 2001). Later the computations moved to a more robust version of the block-iterative coupling technique (see Section 6.1.1) that very much increased the coupling stability (see, Tezduyar, 2004a, 2007a; Tezduyar et al., 2006a,b). In 2004 and later, the space–time FSI computations were based on the quasi-direct coupling and direct coupling techniques (see Sections 6.1.2 and 6.1.3), which yield more robust algorithms for FSI computations where the structure is light and therefore more sensitive to the variations in the fluid dynamics forces. These techniques are for the general case of non-matching fluid and structure meshes at the interface, which is what we prefer in parachute computations, but reduce to monolithic techniques when the meshes are matching.

The SSTFSI technique is now the core technology used in parachute FSI modeling (see, for example, Tezduyar and Sathe, 2007; Tezduyar et al., 2008a,c, 2010b; Takizawa et al., 2011d,e,f, 2012e; Takizawa and Tezduyar, 2012a). A number of special FSI techniques were introduced in Tezduyar and Sathe (2007), Tezduyar et al. (2008a, 2010b) and Takizawa et al. (2010b, 2011d,e,f, 2012e) in conjunction with the SSTFSI technique. Many of these special techniques are in the category of interface projection techniques, such as the FSI-GST (see Section 5.5), SSP (see Section 5.3.2.3), Homogenized Modeling of Geometric Porosity (HMGP) (Tezduyar et al., 2008a), adaptive HMGP (Tezduyar et al., 2010b), "symmetric FSI" technique (Tezduyar et al., 2010b), accounting for fluid forces acting on structural components (such as parachute suspension lines) that are not expected to influence the flow (Tezduyar et al., 2010b), a new version of the HMGP that is called "HMGP-FG" (Takizawa et al., 2011d), and other interface projection techniques (Takizawa et al., 2010b). The special FSI techniques in other categories include the multiscale sequentially-coupled FSI techniques (Tezduyar et al., 2010b; Takizawa et al., 2011f) and rotational-periodicity techniques (Takizawa et al., 2011d,f).

Computer modeling of spacecraft parachutes with the space–time FSI techniques was first reported in Tezduyar et al. (2008a,c). With the HMGP, we bypass the intractable complexities of the geometric porosity by approximating it with an "equivalent," locally-varying "homogenized" porosity, which is obtained from an HMGP computation with an n-gore slice of the parachute canopy (for details, see Tezduyar et al., 2008a,c and Section 9.2). In the earlier HMGP computations with a four-gore slice, slip conditions were applied on the boundaries intersecting the canopy. With the rotational-periodicity techniques, less constraining conditions can be imposed on those boundaries (Takizawa et al., 2011d,f). The rotational-periodicity techniques can also be used for building starting conditions for parachute cluster computations (see Takizawa et al., 2011f).

The contact between the canopies of a spacecraft parachute cluster is a computational challenge that has been addressed recently (see Takizawa et al., 2011e,f) with the contact algorithm SENCT-FC (See Section 6.8). The SENCT-FC technique is now an essential technology in the parachute cluster computations (Takizawa et al., 2011e,f) carried out with the space–time FSI techniques.

Dynamical analysis of the data coming from parachute FSI computations requires an approach that helps us make sense out of a large volume of time-dependent information generated by the computations. We need to extract and present the information in a way that makes it easier for parachute design engineers to make use of. In addition to providing time histories of the aerodynamically significant quantities, such as the descent velocity and aerodynamical forces, we may sometimes find it useful to see the various contributors to these quantities separately. For this purpose, a special decomposition technique for parachute descent speed was introduced in Takizawa et al. (2011e). Another special technique presented in Takizawa et al. (2011e) is for extracting from a parachute FSI computation model parameters that can be used in fast, approximate engineering analysis models for parachute dynamics. The specific parameters targeted in Takizawa et al. (2011e) for extraction are the added mass and the coefficient for the velocity-proportional aerodynamical force.

In this chapter we describe many of the special space–time FSI techniques developed for parachute modeling. These include the FSI-GST and HMGP techniques, the technique that accounts for the fluid forces acting on the parachute suspension lines, the techniques for building a good starting point for the FSI computation, the symmetric-FSI technique, the multiscale sequentially-coupled FSI technique, and techniques for extracting model parameters. We also present in this chapter FSI computations for single and clusters of spacecraft parachutes.

9.1 Parachute Specific FSI-DGST

In a special version of the FSI-DGST introduced in Tezduyar et al. (2008a,c) for parachutes, the smoothing is done in the circumferential direction of the parachute canopy. This addresses the geometric complexities associated with the "peaks" and "valleys" of the parachute gores, which are formed by the inflation of a canopy with embedded reinforcement cables positioned longitudinally in the canopy structure. In this FSI-DGST for parachutes, certain nodes are picked from the structure interface mesh to generate the set of fluid interface nodes. Therefore the number of fluid interface nodes are smaller than the number of structure interface nodes. While generating the set of fluid interface nodes, the structure interface nodes from the valleys are picked. In picking these nodes circumferentially, a few valleys can be skipped, and in picking them longitudinally, inside a valley a few nodes can be skipped. The nodes are then connected with three-node triangular elements, resulting in a smooth fluid mesh along the circumferential direction. For parachutes with a large number of gores, where a gore is the slice of the canopy between two radial reinforcement cables running from the parachute vent to the skirt, the distance by which the gores bulge out is small compared to the parachute diameter. As pointed out in Tezduyar et al. (2008a,c), keeping the true shapes of the gores in the flow computations is not essential for calculating the fluid dynamics forces in this class of applications. A value (mesh coordinate or displacement rate) at a given fluid interface node is replaced by the value at the mapping structure interface node. When transferring the stress values from the smoothed interface to the structure, the values for the mapping nodes are transferred directly, and for the remaining nodes a weighted average is used. This transfer is based on the SSP (see Section 5.3.2.3).

Figures 9.1 and 9.2 show, as an example of how the FSI-DGST works, four-gore slices of the fluid and structure interface meshes used in the computations reported in Tezduyar et al. (2008a,c) for a ringsail parachute. The parachute has 80 gores and 4 rings and 9 sails, with 4 ring gaps and 8 sail slits (the terminology will become clearer in Section 9.2). We note that the fluid interface mesh does not have the gaps and slits the structure interface mesh has. This will be explained in Section 9.2. In generating and updating the fluid mechanics mesh at the interface, for the reasons explained in Tezduyar et al. (2008c), in the circumferential

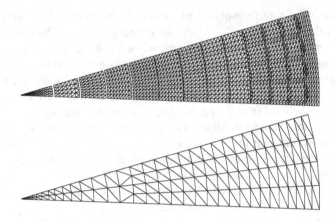

Figure 9.1 Four-gore structure (top) and fluid (bottom) meshes at the interface

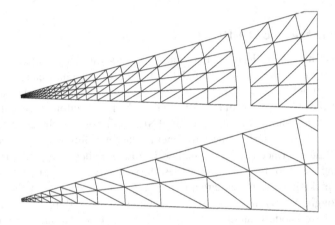

Figure 9.2 Structure (top) and fluid (bottom) meshes at the interface for the first ring

direction, for the rings every other valley node is picked, and for the sails every valley node. In the longitudinal direction, for the first ring every other valley node is picked. For the second ring three valley nodes are picked, and for each of the remaining rings and sails two valley nodes are picked. We note that the fluid mesh is sufficiently refined but has significantly less nodes and elements compared to the structure mesh. We also note that the surfaces curve into the paper as we move toward the skirt of the parachute, and therefore the aspect ratios for the meshes near the skirt are actually better than they appear to be in Figure 9.1.

9.2 Homogenized Modeling of Geometric Porosity (HMGP)

The HMGP was introduced in Tezduyar *et al.* (2008a,c) to address one of the computational challenges involved in the FSI modeling of ringsail parachutes. That particular challenge is the geometric porosity of the parachute canopy, a consequence of the many rings and sails used in the construction of the parachute canopy. Figure 9.3 shows, for an inflated ringsail

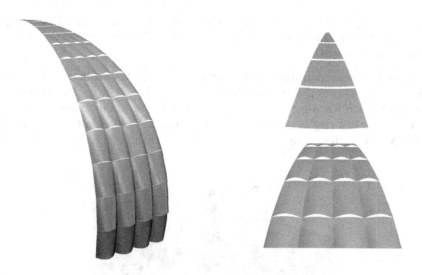

Figure 9.3 A four-gore slice of a ringsail parachute in its inflated form. The pictures on the right illustrate the ring and sail construction of the canopy and show the shapes of the ring gaps and sail slits

parachute from Takizawa *et al.* (2011d), the ring and sail construction and the ring gaps and sail slits.

In the STTFSI technique introduced in Tezduyar and Sathe (2007) the fabric porosity is accounted for (see Equation (5.115)), and one of test problems reported in Tezduyar and Sathe (2007) was the descent of a T–10 parachute with a realistic fabric porosity. With the HMGP, the intractable complexities of the geometric porosity are bypassed by approximating it with an equivalent, locally-varying, homogenized porosity. The fluid mechanics computations see a parachute canopy with no gaps or slits. The structural mechanics computations retain the ring and sail construction of the parachute canopy. The parachute canopy seen by the fluid mechanics computations is assigned a locally-varying homogenized porosity so that it closely approximates the parachute with gaps and slits. The equivalent homogenized-porosity coefficient is calculated from the cross-canopy pressure differentials and flow rates computed for the parachute with geometric porosity. This involves a one-time flow computation, holding the canopy rigid and using a thin slice of the canopy with a small number of gores, with all the rings, sails, gaps, and slits. Using only a four-gore slice, for example, keeps the problem size at a manageable level. From this, we calculate a locally-varying porosity coefficient. In our parachute computations the FSI-DGST and HMGP steps are combined into one single step.

REMARK 9.1 *In the SENCT technique (see Section 6.8), the objective, as it was originally envisaged, was to prevent the structural surfaces from coming closer than a predetermined minimum distance we would like to maintain. A new option for the SENCT technique was introduced in Tezduyar et al. (2008a). In this new option, which is denoted with the option key "-M1," the objective is preventing the fluid mechanics interface meshes from coming closer than a predetermined minimum distance. This differentiation becomes significant (and helpful) when smoothing and homogenization techniques are employed to shelter the fluid mechanics mesh from the consequences of the geometric complexities of the structural surfaces. In the*

SENCT-M1 technique, all the concepts and algorithms of the original SENCT technique are applied to the fluid mechanics interface meshes. The SENCT-FC technique (see Section 6.8) is also used with option M1.

In the computations reported in Tezduyar *et al.* (2008a,c), the parachute canopy is divided into 12 concentric patches and an equivalent homogenized-porosity coefficient is calculated for each. Each patch has a gap or slit, and part of a ring or sail on either side of the gap or slit. The entire first ring is in Patch 1, and the entire last sail is in Patch 12. Figure 9.4 shows Patch 4 of the four-gore slices of the fluid and structure interfaces. In the one-time

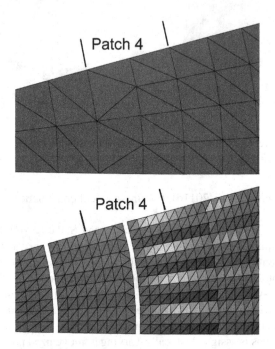

Figure 9.4 Patch 4 of the four-gore slices of the fluid (top) and structure (bottom) interfaces

flow computation, a four-gore canopy slice is used, with slip conditions on the boundaries of the slice intersecting the canopy. The details of the computational conditions can be found in Tezduyar *et al.* (2008c). Figure 9.5 shows the flow field, including the flow passing through the gaps and slits.

In earlier parachute FSI computations, calculation of the equivalent homogenized-porosity coefficient from the cross-canopy pressure differentials and flow rates was done by using the HMGP in its original form Tezduyar *et al.* (2008a,c). A new version, which was introduced in Takizawa *et al.* (2011d) and was named "HMGP-FG," is now used in T★AFSM parachute FSI computations. Computing with a periodic n-gore model (i.e., an n-gore canopy slice with rotational-periodicity conditions on the boundaries of the slice intersecting the canopy) was also introduced in Takizawa *et al.* (2011d). An n-gore periodic computation can be used for calculating the homogenized-porosity coefficient in HMGP (which would involve less constraining boundary conditions), for computing high-resolution flow fields (with less constraining boundary conditions) that the full-domain HMGP-based flow fields can be compared

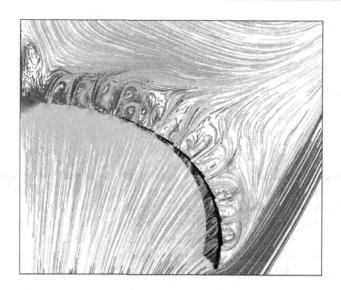

Figure 9.5 Flow field for the four-gore canopy slice with gaps and slits

to, or for other purposes such as investigating the dependence of the *n*-gore flow fields and homogenized-porosity coefficients on the number of gores used.

9.2.1 HMGP in its Original Form

In the original HMGP, the homogenized-porosity coefficient for a patch J is calculated by using the following expression:

$$\frac{\dot{V}_J}{(A_1)_J} = -(k_{\text{PORO}})_J \frac{\Delta F_J}{(A_2)_J}. \tag{9.1}$$

The area of the patch J calculated using the smoothed fluid interface is denoted by $(A_1)_J$, and the area calculated using the structure interface is denoted by $(A_2)_J$. With the additional notation of $(A_F)_J$ representing the fabric area and $(A_G)_J$ representing the gap (or slit) area, we can write $(A_2)_J = (A_F)_J$. The symbol $\dot{V}_J$ represents the volumetric flow rate crossing the patch J. It is the sum of the flow passing through the gap (or slit) and the flow through the fabric due to its porosity:

$$\dot{V}_J = (\dot{V}_F)_J + (\dot{V}_G)_J, \tag{9.2}$$

where $(\dot{V}_F)_J$ and $(\dot{V}_G)_J$ are calculated by integrating the flow over $(A_F)_J$ and $(A_G)_J$, respectively. The pressure differential seen when crossing the patch J is integrated over its area to yield a force differential denoted by ΔF_J:

$$\Delta F_J = \int_{(A_2)_J} \Delta p \, dA. \tag{9.3}$$

We define the average pressure differential across the structure for the patch J as

$$\Delta p_J = \frac{\Delta F_J}{(A_2)_J}, \qquad (9.4)$$

and from Equation (9.1) rewrite the expression used for calculating the homogenized porosity as

$$\frac{\dot{V}_J}{(A_1)_J} = -(k_{\text{PORO}})_J \Delta p_J. \qquad (9.5)$$

Table 9.1 shows the homogenized-porosity coefficients calculated as described above and used in the computations reported in Tezduyar et al. (2010b); Takizawa et al. (2011d). At the border between two patches, the average of the two porosity coefficients is used. If the parachute canopy has any missing sails (or rings) by design, the porosity at the edges facing a missing sail is calculated in the same way as the porosity at the edges of the vent and skirt. The porosity coefficient for the edge nodes is set to the fabric porosity, linearly progressing to the homogenized value for the adjacent patch.

Table 9.1 Homogenized-porosity coefficients for the 12 patches used in the computations reported in Tezduyar et al. (2010b) and Takizawa et al. (2011d)

Patch	1	2	3	4	5	6
CFM	816	627	449	364	116	135
Patch	7	8	9	10	11	12
CFM	130	146	182	288	303	300

9.2.2 HMGP-FG

In this new version Takizawa et al. (2011d) of the HMGP, instead of using a single expression for $\dot{V}_J$ as given by Equation (9.5), we use separate expressions for $(\dot{V}_F)_J$ and $(\dot{V}_G)_J$, with separate porosity coefficients $(k_F)_J$ and $(k_G)_J$ as follows:

$$\frac{(\dot{V}_F)_J}{(A_1)_J} = -(k_F)_J \frac{(A_F)_J}{(A_1)_J} \Delta p_J, \qquad (9.6)$$

$$\frac{(\dot{V}_G)_J}{(A_1)_J} = -(k_G)_J \frac{(A_G)_J}{(A_1)_J} \text{sgn}(\Delta p_J) \sqrt{\frac{|\Delta p_J|}{\rho}}. \qquad (9.7)$$

Then, the normal velocity crossing the fluid interface is modeled nodally by using the following expression:

$$u_n = -(k_F)_J \frac{A_F}{A_1} \Delta p - (k_G)_J \frac{A_G}{A_1} \text{sgn}(\Delta p) \sqrt{\frac{|\Delta p|}{\rho}}, \qquad (9.8)$$

where $(k_F)_J$, $(k_G)_J$, A_F, A_G and A_1 can be seen as "material properties," calculated for each node by area-weighted averaging of the material properties of the (triangular) fluid interface elements sharing that node. Each fluid interface element belongs to a material properties group. Each structural interface (fabric) element and each gap (or slit) also belongs to a material properties group. Each group is associated with a patch J. The values of $(k_F)_J$ and $(k_G)_J$ for a group come from the patch J that the group is associated with. The symbols A_F, A_G, and A_I represent for a group the total instantaneous area of the fabric, the sum of the instantaneous areas of the gap(s), and the sum of the areas of the fluid interface elements. In this new version of the HMGP we have 14 patches, with no gaps or slits in the first and last patches, and the groups are defined based on these 14 patches. Longitudinally, each group spans over one patch. Circumferentially, each group spans over 4 gores in Patch 1, 2 gores in Patches 2–5, and 1 gore in Patches 6–14.

9.2.3 Periodic n-Gore Model

In a periodic n-gore parachute computation Takizawa *et al.* (2011d), we use an n-gore slice of the parachute canopy and cylindrical domain, and apply rotational-periodicity conditions on the boundaries of the slice intersecting the canopy. Using rotational-periodicity conditions is less constraining than using slip conditions on those boundaries. Although we will not have a complete global representation of the flow field when we consider only a slice of the domain, by being able to afford a highly refined mesh for the slice, we can focus on local phenomena such as flow through the parachute gaps and slits.

In generating the mesh for an n-gore slice, first we extract one gore of the canopy. Figure 9.6 shows the one-gore surface mesh for the periodic four-gore HMGP computation reported in Takizawa *et al.* (2011d). The surface mesh has 2712 triangular elements and 1644 nodes, and there are 198 nodes along each valley. We are calling this "coarse mesh" because we will also present here results computed with a more refined mesh. Using this surface mesh, we generate a volume mesh with a pie-slice-shaped domain, which has an angle of

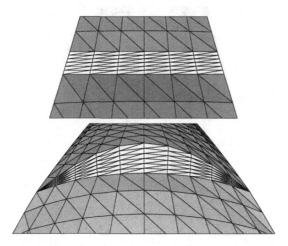

Figure 9.6 One-gore coarse surface mesh around the rings (top) and sails (bottom). The surface mesh is shaded and the mesh across the ring gaps and sail slits is unshaded. There are 6 elements across the ring gap and sail slit. The surface mesh has 2712 triangular elements and 1644 nodes, and there are 198 nodes along each valley

only 4.5°. Then, we repeat the one-gore model n times and merge them as an n-gore model. The four-gore mesh (coarse mesh) used in the HMGP computation reported in Takizawa *et al.* (2011d) consists of 953 884 tetrahedral elements and 165 400 nodes, with 16 896 triangular elements and 8730 nodes over the periodic boundary. Figure 9.7 shows (from Takizawa *et al.*, 2011d) the flow field computed with the four-gore coarse mesh.

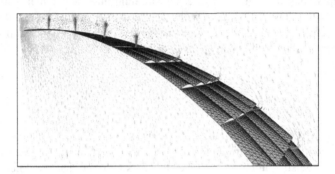

Figure 9.7 Flow field computed with the four-gore coarse mesh

The HMGP computations reported in Takizawa *et al.* (2011f) showed how we can make the rotational-periodicity conditions even less constraining by increasing the number of gores used and make the model more accurate by increasing the resolution of the fluid mechanics mesh used. We provide a summary of that here by starting with the increased mesh resolution. Figure 9.8 shows the one-gore fine surface mesh for the periodic four-gore HMGP computation reported in Takizawa *et al.* (2011f). The four-gore fine volume mesh generated

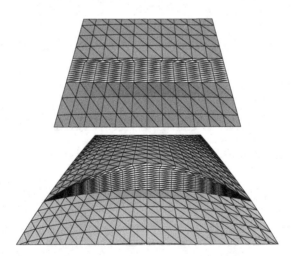

Figure 9.8 One-gore fine surface mesh around the rings (top) and sails (bottom). The surface mesh is shaded and the mesh across the ring gaps and sail slits is unshaded. There are 12 elements across the ring gap and sail slit. The surface mesh has 11 252 triangular elements and 6194 nodes, and there are 382 nodes along each valley

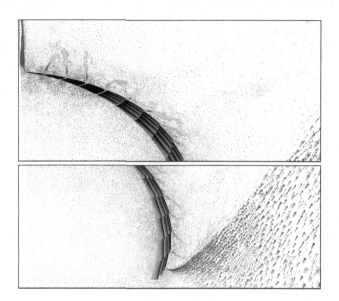

Figure 9.9 Flow field computed with the four-gore fine mesh. Flow around the rings (top) and sails (bottom)

from this surface mesh has 4 050 468 tetrahedral elements and 722 326 nodes, and there are 48 825 triangular elements and 24 954 nodes over the periodic boundary. Figure 9.9 shows (from Takizawa *et al.*, 2011f) the flow field computed with the four-gore fine mesh.

An 8-gore and a 16-gore models were used in the HMGP computations performed in Takizawa *et al.* (2011f) to show how we can make the rotational-periodicity conditions less constraining. The 8-gore model has 4 005 608 tetrahedral elements and 723 399 nodes, and the 16-gore model has 80 11 216 tetrahedral elements and 1 434 751 nodes. Both models have 23 124 triangular elements and 12 047 nodes over the periodic boundary. Figures 9.10 and 9.11 show (from Takizawa *et al.*, 2011f) the flow fields computed with the 8-gore and 16-gore models. As mentioned in Takizawa *et al.* (2011f), we did not observe much change in the homogenized-porosity values calculated as we increased the number of gores from 4 to 8 to 16.

9.3 Line Drag

In the parachute computations reported in Tezduyar *et al.* (2008a,c) and those reported earlier by the T★AFSM (see the references in Section 11 in Tezduyar and Sathe, 2007), the aerodynamic forces acting on the suspension lines of the parachutes were not accounted for. The suspension lines are very thin and are not expected to influence the flow field. However, not accounting for the fluid forces acting on them was based on guesstimating. A technique accounting for those forces was presented and tested in Tezduyar *et al.* (2010b). The technique is conceptually applicable to other classes of FSI problems where we are interested in accounting for or estimating the fluid forces acting on structural components that are not expected to influence the flow. We provide here an overview of the technique from Tezduyar *et al.* (2010b).

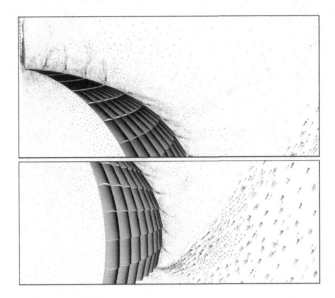

Figure 9.10 HMGP computation with the periodic 8-gore model. Flow around the rings (top) and sails (bottom)

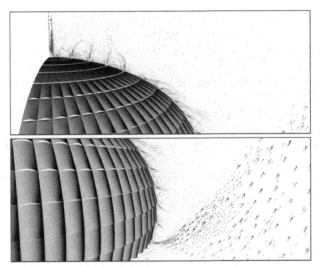

Figure 9.11 HMGP computation with the periodic 16-gore model. Flow around the rings (top) and sails (bottom)

Line (cable) drag is generated mainly in the cross-flow direction, and therefore flow parallel to the line creates negligible drag (see Hoerner, 1993). The relative velocity is $\mathbf{u}_R = \mathbf{u}_W - \dot{\mathbf{y}}$, where $\mathbf{u}_W$ is the wind velocity and $\dot{\mathbf{y}}$ represents the structural displacement rate (see Figure 9.12). The velocity $\mathbf{u}_W$ is evaluated at the centroid of each cable group, where a group consists of one or more cable elements. We can define, for example, each suspension line as a group or all the suspension lines as a single group or have a single cable element in each

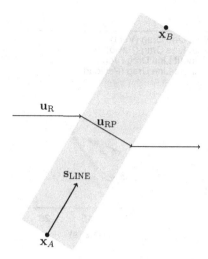

Figure 9.12 A cable element and relative flow directions

group. The relative velocity perpendicular to the line is $\mathbf{u}_{RP} = \mathbf{u}_R - (\mathbf{u}_R \cdot \mathbf{s}_{LINE})\mathbf{s}_{LINE}$, where $\mathbf{s}_{LINE} = (\mathbf{x}_B - \mathbf{x}_A)/\|\mathbf{x}_B - \mathbf{x}_A\|$, with "A" and "B" representing the cable-element nodes. For a cable with circular cross-section and radius R, the drag force (per unit length) is calculated as $f_D = \frac{1}{2}C_D\|\mathbf{u}_{RP}\|^2(2R)(\mathbf{u}_{RP}/\|\mathbf{u}_{RP}\|) = C_D\|\mathbf{u}_{RP}\|R\mathbf{u}_{RP}$. Here C_D is the drag coefficient, which is determined from tabulated experimental data for flow past a circular cylinder, based on the Reynolds number $\|\mathbf{u}_{RP}\|(2R)/\nu$ for each cable element.

Using the same conditions (except for the wind speed) used for the side-wind case described in Tezduyar *et al.* (2008c), FSI computations with and without line drag were carried out in Tezduyar *et al.* (2010b) to compare the results. The parachute diameter is about 120 ft, the descent speed is about 25 ft/s, and the side-wind speed is 6.25 ft/s. The 80 suspension lines each have a diameter of 0.3 inches. The information and the fluid and structural mechanics meshes and other computational conditions can be found in Tezduyar *et al.* (2010b). It was reported in Tezduyar *et al.* (2010b) that the difference in drag when we account for the line drag in addition to the canopy drag is less than 1% in the vertical direction and about 10% in the wind direction. There is almost no difference in the vertical positions we get with and without the line drag, and we see little difference in terms of the vent and payload positions in the wind direction (see Figure 9.13).

REMARK 9.2 *Because there is little added computational cost from the line drag algorithm, unless parachute performance or design comparison to an earlier computation without line drag is part of the objective, it makes sense to include line drag in all parachute computations. Figure 9.14 shows, from Tezduyar et al. (2010b), an example. All the suspension lines were defined as a single group, and $\mathbf{u}_W$ was updated at every nonlinear iteration.*

9.4 Starting Point for the FSI Computation

Having a good starting point for the FSI computation plays an important role in FSI modeling. This becomes especially important in FSI problems where the structure is light and

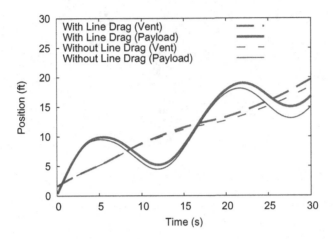

Figure 9.13 Vent and payload positions in the wind direction

Figure 9.14 Parachute FSI modeling with line drag. Parachute shape and line drag vectors. Parachute horizontal speed coupled with the suspension line orientation produces a larger drag on the upwind lines

therefore very sensitive to the fluid dynamics forces. The starting point is defined in terms of the parachute shape, descent speed, and a developed flow field. In the parachute computations reported in Tezduyar *et al.* (2008a,c) and those reported earlier by the T★AFSM, the starting parachute shape was determined by a stand-alone structural mechanics computation with a uniform parachute inflation pressure equal to the stagnation pressure corresponding to the starting descent speed. In Tezduyar et al. (2008a,c), the starting descent speed was 25 ft/s. The starting shape and descent speed were also the bases of the four-gore model used in the stand-alone fluid mechanics computation carried out as part of the HMGP (see Tezduyar et al., 2008a,c). The developed flow field was obtained with a stand-alone fluid mechanics computation based on the starting shape and descent speed, with the equivalent, locally-varying homogenized porosity coming from the HMGP.

In some later parachute FSI computations carried out (but not published) by the T⋆AFSM, the starting parachute shape and descent speed were determined by an alternating sequence of stand-alone structural and fluid mechanics computations. The first parachute shape, with area A^1, was determined by a stand-alone structural mechanics computation with a uniform pressure equal to the stagnation pressure corresponding to an estimated descent speed, U^1 (25 ft/s). This was followed by a stand-alone fluid mechanics computation based on A^1 and U^1. The equivalent, locally-varying homogenized porosity used in the fluid mechanics computation was the one calculated earlier in Tezduyar et al. (2008a,c). The fluid dynamics forces calculated were then "symmetrized" (see "symmetric FSI" technique in Section 9.5 for more details) and used in another stand-alone structural mechanics computation to determine a new parachute shape. Altogether, seven pairs of stand-alone fluid mechanics and structural mechanics computations were carried out. The descent speed was updated at every fluid mechanics computation with the recipe given in Tezduyar et al. (2010b), and the seven pairs of computations were more than enough to have a reasonably close match between the computed drag and the total weight (payload weight plus parachute weight), at which point the descent speed was 25.7 ft/s.

In the recipe given in Tezduyar et al. (2010b), the descent speed is updated by using Newton–Raphson iterations based on

$$W - F_\text{D} = 0, \qquad F_\text{D} = \frac{1}{2} C_\text{D} \rho U^2 A. \tag{9.9}$$

Here W and F_D are the total weight and drag force, and the area A is based on the diameter measured at the parachute skirt. The Newton–Raphson iterations are based on first writing

$$\frac{1}{2} C_\text{D} \rho \left(2 U^i A^i \Delta U^i + (U^i)^2 \left. \frac{\partial A}{\partial U} \right|_i \Delta U^i \right) = W - F_\text{D}^i, \tag{9.10}$$

where i is the iteration counter and A^i is the area corresponding to the parachute shape used in computing F_D^i. Assuming that C_D is not changing in the range of Reynolds numbers we are working with, we can write

$$\frac{1}{2} C_\text{D} \rho = \frac{F_\text{D}^i}{(U^i)^2 A^i}, \tag{9.11}$$

and based on that rewrite Equation (9.10) as

$$\frac{F_\text{D}^i}{(U^i)^2 A^i} \left(2 U^i A^i \Delta U^i + (U^i)^2 \left. \frac{\partial A}{\partial U} \right|_i \Delta U^i \right) = W - F_\text{D}^i, \tag{9.12}$$

and then as

$$\left(2 A^i + U^i \left. \frac{\partial A}{\partial U} \right|_i \right) \Delta U^i = \frac{U^i A^i \left(W - F_\text{D}^i \right)}{F_\text{D}^i}, \tag{9.13}$$

where

$$\left.\frac{\partial A}{\partial U}\right|_i = \frac{A^{i+1} - A^i}{U^i - U^{i-1}}. \tag{9.14}$$

We start with A^1 and U^1, use Equation (9.14) for $i \geq 2$, and set $\left.\frac{\partial A}{\partial U}\right|_1 = 0$.

REMARK 9.3 *The parachute shape, descent speed and the developed flow field obtained with the sequence of stand-alone computations described above were not used as the starting point in any published FSI computation. However the shape and descent speed were the bases of a new four-gore model used in the stand-alone fluid mechanics computation that generated, as part of an improved HMGP, the equivalent locally-varying homogenized porosity used in the FSI computations reported in Tezduyar et al. (2010b). The shape and descent speed were also the components of the starting point for the symmetric FSI step described in Tezduyar et al. (2010b), which is also described in Section 9.5.*

9.5 "Symmetric FSI" Technique

The "symmetric FSI" technique, introduced in Tezduyar *et al.* (2010b), is helpful in building a good starting point in parachute computations. In the symmetric FSI step, we project to the structure the circumferentially-averaged fluid interface stress, $\left(\mathbf{h}_{II}^h\right)_{AVE}$, which is symmetric with respect to the parachute axis. This helps us build a good starting point, which can be a rather lengthy process, without generating any unsymmetric parachute deformation or gliding. After the symmetric FSI period, we project to the structure $(1 - r_S) \mathbf{h}_{II}^h + r_S \left(\mathbf{h}_{II}^h\right)_{AVE}$, where r_S is gradually varied from 1.0 to 0.0. In the actual computations reported in Tezduyar *et al.* (2010b) and later, for expedited implementation, the symmetrization of the interface stress projected to the structure and the de-symmetrization with the parameter r_S are done in terms of only the pressure component of the interface stress, $-p_{II}^h \mathbf{n}$. This expedited implementation was motivated by the SSP concept.

A symmetric FSI computation would of course benefit from having a good starting point of its own. The shape and descent speed components of the starting point used in Tezduyar *et al.* (2010b) were obtained from the alternating sequence of stand-alone structural mechanics and fluid mechanics computations described in Section 9.4. The developed-flow component of the starting point was obtained with a stand-alone fluid mechanics computation based on the starting shape and descent speed. The symmetric FSI step reported in Tezduyar *et al.* (2010b) had a duration of 100 s. The de-symmetrization period was approximately 7 s, during which r_S varies from 1.0 to 0.0 in a cosine form.

In the FSI computations, starting with those reported in Tezduyar *et al.* (2010b), in addition to moving the reference frame vertically with a reference descent speed, the mesh is moved horizontally and vertically, with the average displacement rate for the structure. The horizontal motion of the mesh becomes particularly helpful when the parachute glides significantly. It is then more convenient to use the velocity form of the free-stream conditions also at the lateral boundaries. For that reason, the computational domain was expanded laterally in Tezduyar *et al.* (2010b), based on comparison with using slip conditions at those boundaries. The dimensions of the computational domain, in ft, were 1740×1740×1566 (the dimensions of the domain used in Tezduyar *et al.* (2008a,c) were 870×870×1566). This approach reduces mesh stretching also in the horizontal direction, even when the parachute glides significantly, and therefore reduces the need for remeshing. All computations reported in Tezduyar et al. (2010b) were completed without any remeshing. The interface-stress projection was based on the SSP technique. Other computational conditions, including information on the fluid and structural mechanics meshes, homogenized-porosity values, time-step size, and iteration numbers, can be found in Tezduyar et al. (2010b).

After the de-symmetrization during the simulation, the parachute starts gaining a gliding speed, with the magnitude reaching 10 ft/s and with a direction that is rather steady, unless the payload is swinging. The parachute shape also loses its axial symmetry. Figure 9.15 shows, from Tezduyar et al. (2010b), the parachute shape and flow field at instants during the symmetric FSI and FSI computations.

9.6 Multiscale SCFSI M2C Computations

The spatially multiscale Sequentially-Coupled FSI (SCFSI) techniques were introduced in Tezduyar et al. (2009a) as spatially multiscale for the fluid mechanics part, which is called SCFSI M1C, and then in Tezduyar et al. (2010b) as spatially multiscale for the structural mechanics part, which is called SCFSI M2C (see Section 8.7). In SCFSI M2C, the time-dependent flow field is first computed with the (fully) coupled FSI (CFSI) technique and a relatively coarser structural mechanics mesh, followed by a structural mechanics computation with a more refined mesh, with the time-dependent interface stresses coming from the previously carried out CFSI computation. With this technique, the FSI computational effort is reduced where it is not needed, and the accuracy of the structural mechanics computation is increased where we need accurate, detailed structural mechanics computations, such as computing the fabric stresses. We can do this because the coarse mesh is sufficient for the purpose of FSI computations, and using more refined meshes does not change the FSI results that much. However, mesh refinement does make a difference in detailed structural mechanics computation.

9.6.1 Structural Mechanics Solution for the Reefed Stage

In Tezduyar et al. (2010b) the SCFSI M2C technique was used for increasing the accuracy of the structural mechanics solution for the parachute reefed to approximately 13%. During the descent of a spacecraft, the parachute skirt is initially constricted to reduce forces on the parachute structure and the crew, and this is called the reefed stage. The skirt diameter is constrained using a reefing line, with length characterized by the reefing ratio: $r_{REEF} = D_{REEF}/D_o$, where D_{REEF} is the reefed skirt diameter and D_o is the parachute nominal diameter. Starting with the fully open parachute geometry, which is relatively easier to compute, an incremental shape determination approach based on gradually shortening the reefing line was used in Tezduyar et al. (2010b) to compute the parachute shape at reefed configurations. Because the objective was just to determine the parachute shape, the symmetric FSI technique was used.

The coarse structure mesh used in the CFSI computation consists of 31 122 nodes and 26 320 four-node quadrilateral membrane elements, 12 441 two-node cable elements, and one payload point mass. The membrane part of the structure forms the structure interface and has 29 600 nodes. More information on the computational conditions, including the homogenized-porosity values, fluid mechanics mesh, time-step size, iteration numbers, and computational

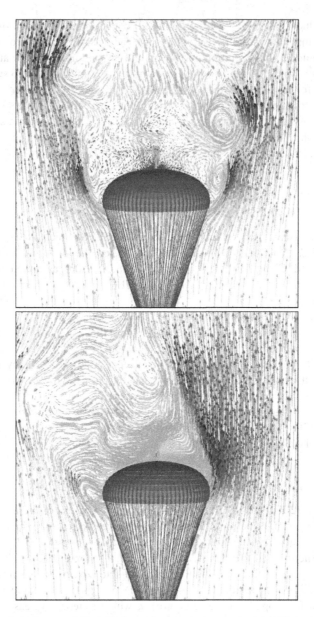

Figure 9.15 Parachute shape and flow field at an instant during symmetric FSI (top) and FSI (bottom)

steps followed, can be found in Tezduyar *et al.* (2010b). Figure 9.16 shows the structural mechanics solution for the parachute reefed to $\tau_{\text{REEF}} = 13\%$ (approximately).

In the SCFSI M2C computation, the interface stresses were extracted from the CFSI computation described above and were used in a structural mechanics computation with a more refined mesh. The interface stress projected to the structure consists of only the pressure

Figure 9.16 Structural mechanics solution for the parachute reefed to $\tau_{\text{REEF}} = 13\%$ (approximately), obtained with the CFSI computation and the coarse structure mesh

component of the interface stress, and the SSP technique is used for the projection. The refined structure mesh has 128 882 nodes and 119 040 four-node quadrilateral membrane elements, 23 001 two-node cable elements, and one payload point mass. The membrane part of the structure forms the structure interface and has 127 360 nodes. At this reefed configuration, the interface stresses obtained in the symmetric FSI computation do not have a significantly dynamic nature, and therefore the time-averaged values were used.

As a related technique, a "cable symmetrization" procedure to be applied to the canopy cables during the structural mechanics computation with the more refined mesh was proposed in Tezduyar *et al.* (2010b). In this procedure, it was proposed that for the cable nodes at each latitude, the tangential component of the displacement is set to zero, and the radial and axial components are set to the average values for that latitude. This can be done as frequently as every nonlinear iteration, or as few as just once. In the computation reported in Tezduyar *et al.* (2010b), it was done just once and that was during the starting phase of the computation. Also, in the computation reported in Tezduyar *et al.* (2010b), the actual symmetrization procedure used was a close approximation to the proposed one. Figure 9.17 shows the canopy cables before and after symmetrization. In addition to and following that symmetrization, the cable positions are fixed and the computation is continued until the membrane parts of the canopy structure settle. After that we release all the structural nodes (except for the payload) and compute until the solution settles. Figure 9.18 shows the structural mechanics solution obtained with the SCFSI M2C computation and the refined mesh. Figure 9.19 shows the structural mechanics solution obtained with the SCFSI M2C computation, which compares very well to what we see in the picture from a NASA drop test.

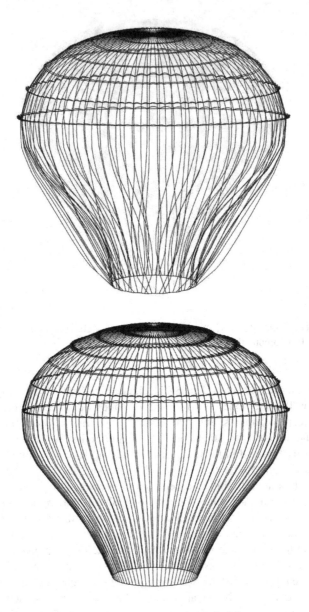

Figure 9.17 Structural mechanics solution for the parachute reefed to $\tau_{\text{REEF}} = 13\%$ (approximately). Canopy cables before (top) and after (bottom) symmetrization

9.6.2 Fabric Stress Computations

It was shown in Takizawa *et al.* (2011f) that the SCFSI M2C technique can be used for computing the fabric stresses more accurately by increasing the structural mesh refinement after the CFSI computation is carried out with a coarse mesh. It was also shown how the SCFSI M2C technique can be used for computing the fabric stress more accurately by adding the

Figure 9.18 Structural mechanics solution for the parachute reefed to $\tau_{\text{REEF}} = 13\%$ (approximately), obtained with the SCFSI M2C computation and the refined structure mesh

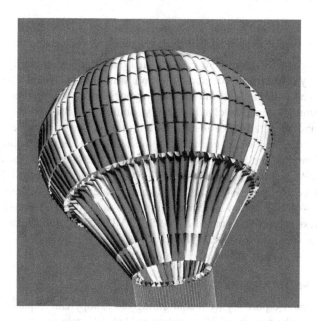

Figure 9.19 Structural mechanics solution for the parachute reefed to $\tau_{\text{REEF}} = 13\%$ (approximately). Obtained with the SCFSI M2C computation and the refined structure mesh

vent hoop after the FSI computation is carried out without it. The vent hoop is a reinforcement cable placed along the circumference of the vent. Again, we can do this because the structural model without the vent hoop is sufficient for the purpose of FSI computations, and including the vent hoop does not change the FSI results that much. However it makes a large difference in the fabric stresses near the vent.

In the tests carried out in Takizawa *et al.* (2011f) with the SCFSI M2C technique, the interface stresses are extracted from the FSI computation reported in Takizawa *et al.* (2011d) (for the case where the horizontal speed of the payload is instantaneously hiked to 20 ft/s to emulate the swinging motion). The stress projected to the structure consists of only the pressure component of the interface stress, and the SSP technique is used for the projection. Also, to expedite the tests, in Takizawa *et al.* (2011f) a time-averaged, circumferentially symmetric pressure was applied to the structure. The coarse structure mesh for the canopy has 29 200 nodes, 26 000 four-node membrane elements, and 10 920 two-node cable elements. The fine mesh has 115 680 nodes, 108 480 four-node membrane elements, and 21 640 two-node cable elements. Adding the vent hoop increases the number of cable elements by 80. Figures 9.20 and 9.21 show the coarse and fine meshes for one gore.

The cases with and without a vent hoop were computed in Takizawa *et al.* (2011f) using both meshes, resulting in a total of four test cases. Additional information on the computational conditions, including the time-step size, iteration numbers, and computational steps followed, can be found in Takizawa *et al.* (2011f). Figures 9.22 and 9.23 show the fabric (maximum principal) tension for the coarse and fine meshes with no vent hoop. Figures 9.24 and 9.25 show the fabric tension for the coarse and fine meshes with a vent hoop.

9.7 Single-Parachute Computations

All computations reported here were carried out in a parallel computing environment. In all cases, the fully-discretized, coupled fluid and structural mechanics and mesh-moving equations are solved with the quasi-direct coupling technique (see Section 6.1.2). In solving the linear equation systems involved at every nonlinear iteration, the GMRES search technique Saad and Schultz (1986) is used with a diagonal preconditioner. The meshes are partitioned to enhance the parallel efficiency of the computations. Mesh partitioning is based on the METIS Karypis and Kumar (1998) algorithm. The computations were carried out using SSTFSI-TIP1 technique (see Remarks 4.10 and 5.9), with the SUPG test function option WTSA (see Remark 4.4). The stabilization parameters used are those given by Equations (4.117)–(4.122) and (4.125), with the τ_{SUGN2} term dropped from Equation (4.121). The interface-stress projection is based on the SSP. All computations were carried out using properties of air at standard sea-level conditions.

9.7.1 Various Canopy Configurations

Adjusting the geometric porosity by reconfiguring the canopy can impact both stability and drag performance of the parachute. The baseline and two alternate canopy configurations were investigated in Takizawa *et al.* (2011d). We describe that investigation here. The two alternate configurations are "missing" the 5th and 11th sail, respectively, when numbered starting from the top including the rings. These three parachutes were called "PA" (all sails are in place), "PM5" (missing the 5th sail) and "PM11" (missing the 11th sail) in Takizawa *et al.* (2011d).

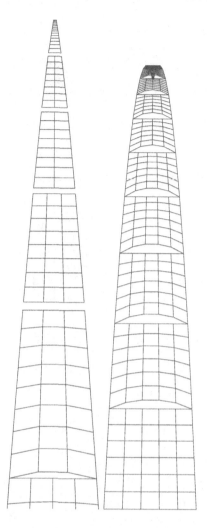

Figure 9.20 Coarse structure mesh for one gore

The payload weight is approximately 5570 lbs. The total weights, including the parachute for each of the three configurations: PA, PM5, and PM11, are approximately 5725 lbs, 5720 lbs and 5715 lbs, respectively.

The homogenized-porosity distribution used is based on the computation reported in Tezduyar *et al.* (2010b), which is given in Table 9.2 as the porosity coefficients for the 12 patches of the parachute. Figure 9.26 shows the porosity distribution for each of the three cases.

The geometry and material properties of PA are mostly the same as they are described in Section 2 of Tezduyar *et al.* (2008c), but it does not have a vent cap, and the riser is longer (approximately 100 ft). The dimensions of the computational domain are the same as those given in Section 9.5.

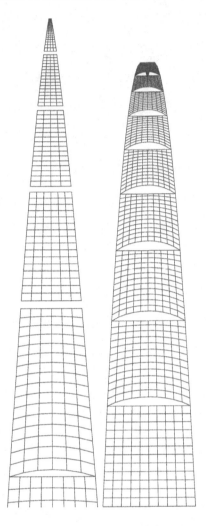

Figure 9.21 Fine structure mesh for one gore

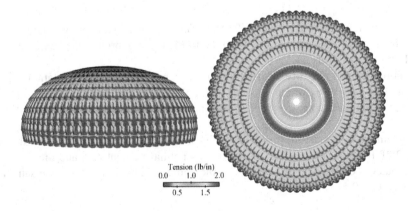

Figure 9.22 Fabric tension for the coarse mesh with no vent hoop

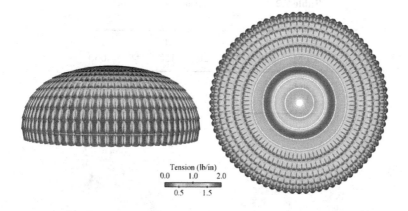

Figure 9.23 Fabric tension for the fine mesh with no vent hoop

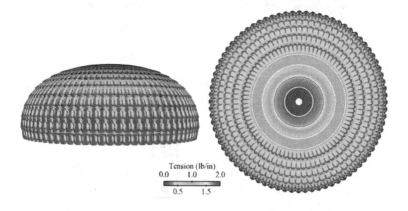

Figure 9.24 Fabric tension for the coarse mesh with a vent hoop

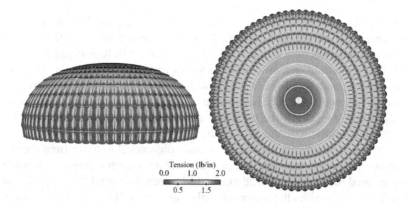

Figure 9.25 Fabric tension for the fine mesh with a vent hoop

Table 9.2 Porosity coefficients for the 12 patches of the parachute

Patch	1	2	3	4	5	6
CFM	314	278	201	157	59	66
Patch	7	8	9	10	11	12
CFM	62	79	107	145	150	149

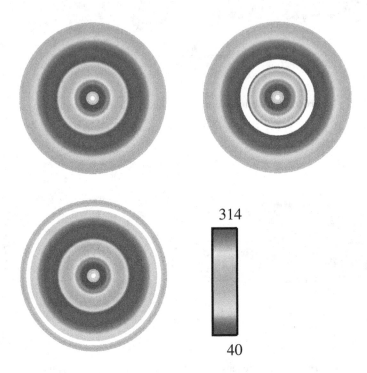

Figure 9.26 Porosity distribution (in CFM) at the fluid interface for PA, PM5, and PM11

The number of nodes and elements are given in Table 9.3. Figure 9.27 shows the parachute shape and structure mesh for each configuration. The time-step size is 0.0232 s. The number of nonlinear iterations per time step is 6, and the number of GMRES iterations per nonlinear iteration is 90 for the fluid and structural mechanics parts and 30 for the mesh moving part. We use Selective Scaling (see Section 6.1.2), with the scale for the structure part set to 10.

An important dynamical feature in parachute flight is the breathing in steady descent. The flow separates near the skirt of the parachute and a ring vortex forms. The vortex creates a large pressure differential near the skirt of the parachute at maximum inflation. The vortex then moves upward causing the maximum pressure differential to move from the skirt to the crown. Figure 9.28 shows the parachute and flow field for each configuration when the tilt angle, as measured between the vertical axis and a line connecting the payload and vent, is at a maximum. Table 9.4 provides the computational results for the three parachute configurations.

Table 9.3 Number of nodes and elements for each of the three parachute configurations. Here *nn* and *ne* are number of nodes and elements, respectively. The structural mechanics mesh consists of four-node quadrilateral membrane elements, two-node cable elements and one-node payload element. The structure interface mesh consists of four-node quadrilateral elements. The fluid volume mesh consists of four-node tetrahedral elements, while the fluid interface mesh consists of three-node triangular elements

			PA	PM5	PM11
Structure	Membr	nn	30 722	28 642	28 082
		ne	26 000	24 080	23 600
	Cable	ne	12 521	11 401	12 121
	Payload	ne	1	1	1
	Interface	nn	29 200	27 120	26 560
		ne	26 000	24 080	23 600
Fluid	Volume	nn	178 270	192 412	180 917
		ne	1 101 643	1 192 488	1 119 142
	Interface	nn	2140	2060	2060
		ne	4180	3860	3860

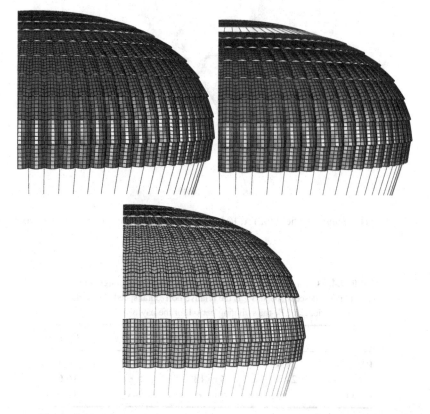

Figure 9.27 Parachute shape and structure mesh for PA, PM5, and PM11

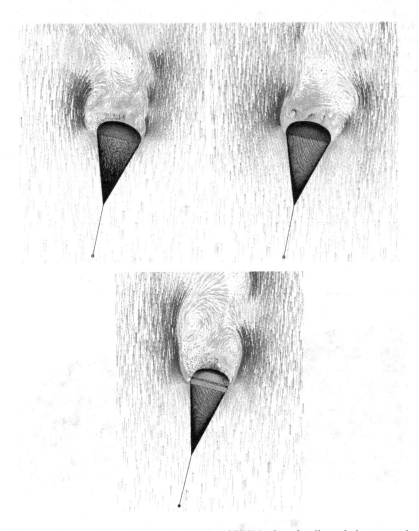

Figure 9.28 Parachute and flow field for PA, PM5 and PM11 when the tilt angle is at a maximum

Table 9.4 Computational results for the three parachute configurations. Here U is the descent speed, V_R is the relative horizontal speed, T_B is the breathing period, and T_S the swinging period for the payload

	U (ft/s)	V_R (ft/s)	T_B (s)	T_S (s)
PA	21.4	4 to 13	6.7	16.4
PM5	24.0	4 to 13	5.8	16.6
PM11	29.0	0 to 4	NA	17.0

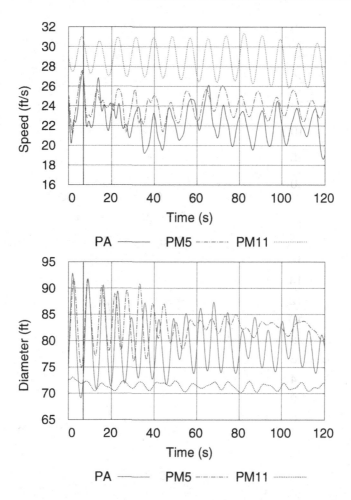

Figure 9.29 Payload descent speed and parachute skirt diameter for the three configurations. The thin vertical line at 7 s marks the end of the de-symmetrization

Figure 9.29 shows the payload descent speed and skirt diameter of the three parachutes. The descent speed is directly dependent on the diameter because drag production is dictated by the projected area of the canopy. It can be seen from the plots that the PA and PM5 have similar maximum diameters. Though the diameters are close, PM5 has a slightly higher average descent speed due to the decreased projected area with a missing sail. The similar maximum diameter is due to the fact that the pressurization of the lower sails for PA and PM5 are unaffected by the modification. The PM11 parachute drag performance is hindered largely by a lack of projected area due to the loss of pressurization of the bottom three sails. Figure 9.30 shows the horizontal-velocity magnitudes for the three configurations. In this case we define improved static stability as having lower gliding speed. PA and PM5 have larger gliding speeds while the glide speed of PM11 is very low. Thus the static stability of PM11 is much improved over that of PM5 and PA. This may have implications in parachute

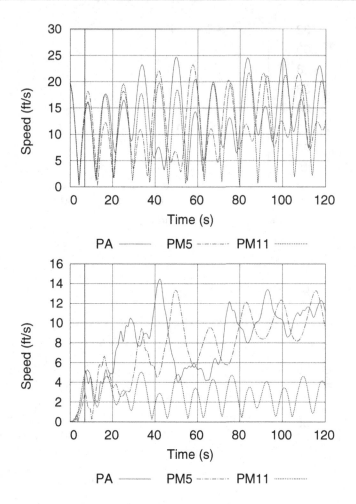

Figure 9.30 Horizontal-velocity magnitude for the payload (top) and vent (bottom) for the three parachute configurations. The thin vertical line at 7 s marks the end of the de-symmetrization

cluster applications. The horizontal-velocity magnitude for the PM11 payload exhibits sharp kinks near 0 ft/s, indicating that the horizontal velocity is reversing direction. This reversal happens when, in a swinging cycle, the tilt angle is maximum. On the other hand, PA and PM5 payloads are moving on an elliptical trajectory with respect to the vent, as clearly seen in Figure 9.31, and therefore do not exhibit sharp kinks. Overall, the study shows that the more stable configurations exhibit a loss of drag. However, it shows that geometric porosity can influence the descent characteristics. Therefore a parachute's performance might be tuned by readjusting the geometric porosity.

9.7.2 Various Suspension Line Length Ratios

Parachute performance could be improved by increasing the suspension line length ratio, which is defined as the ratio of the suspension line length to the nominal diameter of the

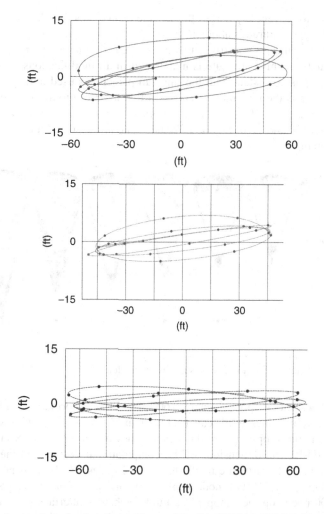

Figure 9.31 Payload trajectories (relative to the vent) for PA, PM5, and PM11. Lines are drawn from 60 to 120 s. Dots are placed every 2.3 s. We note that the y scale is stretched to twice that of the x scale

parachute, S_L/D_o. This was investigated in Takizawa et al. (2012e), and we describe that investigation here. Longer suspension lines may relieve the restoring forces of the suspension lines that restrict canopy diameter thus producing better performance by allowing for increased projected area. Modifications to the suspension line length are governed by two constraints: effective line length and riser wrap. First, it is desirable to maintain the effective line length, a parameter important for cluster efficiency, which is defined as the distance from the payload to the skirt. To maintain this constraint the riser is shortened a distance equivalent to the increase in suspension line length. The increased suspension line length also represents a small decrease in weight for the system as each individual suspension line has a lower factor of safety than the single riser. Second, a minimum riser length must be maintained due to the possibility of the riser wrapping around the vehicle during parachute deployment.

More specifically, to prevent any part of the capsule from contacting the suspension lines, the minimum riser length required is about 26 ft.

The baseline parachute is constructed with a suspension line length ratio of 1.15 and the effective line length of this parachute is maintained in all modifications. A suspension line length ratio of 1.44 is consistent with the ring-sail parachutes used during the Apollo lunar missions while 1.76 is the maximum suspension line length ratio achievable while meeting both effective line length and riser wrap constraints. The S_L/D_o values investigated are 1.0, 1.15, 1.30, 1.44, 1.60, and 1.76. Figure 9.32 shows a comparison of the parachute geometry for each configuration.

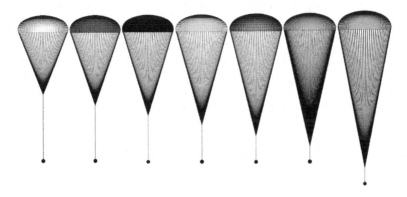

Figure 9.32 Parachute configurations with suspension line length ratios of 1.00, 1.15, 1.30, 1.44, 1.60, and 1.76

The payload weight is approximately 5 600 lbs. More information on the parachutes, including the HMGP values used, can be found in Tezduyar *et al.* (2008c, 2010b). The geometry and material properties are the same as they are described in Section 3.2 of Takizawa *et al.* (2011d). The dimensions of the computational domain are the same as those given in Section 9.5. The structure model has 30 722 nodes, 26 000 four-node quadrilateral membrane elements, 12 521 two-node cable elements and 1 one-node payload element. There are 29 200 nodes on the canopy. The fluid mechanics interface mesh has 2140 nodes and 4180 three-node triangular elements. The fluid mechanics mesh includes 178 270 nodes and 1 101 643 four-node tetrahedral elements. Selective Scaling (see Section 6.1.2) is used, with the scale for the structure part set to 10. The time-step size is 0.0232 s. The number of nonlinear iterations per time step is 6. The number GMRES iterations per nonlinear iteration is 90 for the fluid+structure block, and 30 for the mesh-moving block. The process begins with the parachute shape obtained from the shape determination reported in Tezduyar *et al.* (2010b). Modifications to the suspension line length is made during the symmetric FSI step. The payload and the parachute have no horizontal speed at the end of the symmetric FSI step, which does not match what is observed in the tests. To emulate the swinging motion observed in the drop tests, the horizontal speed of the payload is instantaneously hiked to 20 ft/s. Simultaneously, the de-symmetrization begins using a cosine form that spans over one breathing cycle (7 s).

Changing the suspension line length begins with a parachute shape obtained after the initial 40 s of the symmetric FSI computation reported in Tezduyar *et al.* (2010b). The suspension line and riser lengths are simultaneously increased or decreased in a linear fashion over a time period of 7 s of continued symmetric FSI computation. Symmetric FSI is then computed for an additional 53 s allowing the solution to settle, and we proceed with the payload velocity hike and de-symmetrization.

The drag coefficient of a parachute in "steady" descent (i.e., without events such as disreefing or offloading) is defined as the payload weight nondimensionalized with the parachutes nominal area and the instantaneous dynamic pressure based on payload vertical descent speed, $C_D = W/(S_o q)$. Here W is the payload weight and S_o is the nominal area with a constant value of approximately 10 500 ft^2. Instantaneous dynamic pressure is computed with the expression $q = \frac{1}{2}\rho U^2$. In these computations the suspension line drag (see Section 9.3) was not taken into account, because $S_L/D_o = 1.15$ case was computed before taking the line drag into account became standard in T★AFSM computations, and we wanted to compare the new cases to that earlier-computed case.

Increased S_L/D_o is expected to lead to an increase in projected area (S_p) and C_D. Table 9.5 provides a comparison of computational results in terms of the average values of S_p, C_D, and canopy horizontal velocity (V). The data was averaged over a time period ranging from

Table 9.5 Average values of the projected area (S_p), C_D, and canopy horizontal velocity (V) for different suspension line length ratios (S_L/D_o)

S_L/D_o	S_p (ft^2)	C_D	V (ft/s)
1.00	4897	0.85	7.7
1.15	5075	0.92	9.0
1.30	5241	0.97	9.4
1.44	5389	1.00	8.2
1.60	5473	1.03	8.1
1.76	5558	1.08	11.1

23 s after the end of the de-symmeterization period, a time when the full dynamics of the parachute have developed, to 120 s. For all cases, there is a significant increase in S_p and C_D as S_L/D_o is increased. Figure 9.33 shows the time-dependent behavior of S_p. There is not a large difference in the peaks of S_p from parachute to parachute. The valleys of S_p increase significantly with S_L/D_o, which in turn leads to a higher average S_p with decreased oscillations. These decreased oscillations could lower the peaks in the descent speed, which is limited by mission requirements. Figure 9.34 shows the time-dependent behavior of C_D. At lower S_L/D_o values, C_D appears to involve several modes. There are two dominant dynamics present in the descent speed: breathing and swinging. As S_L/D_o is increased, the descent speed is dominated by the swinging dynamics and the breathing dynamics is subdued due to the reduction in S_p oscillations.

The final aspect of this evaluation deals with the lateral stability of each parachute. Efficient parachutes, such as the ringsail parachutes employed in the recovery of the Orion spacecraft, are slightly unstable and develop a horizontal speed during steady descent as noted in Tezduyar et al. (2010b). Large instabilities may have implications of decreased cluster efficiency. The average V is presented in Table 9.5. There does not appear to be an appreciable trend in V for the cases investigated here. Increasing the S_L/D_o does not appear to have a significant impact on the canopy stability.

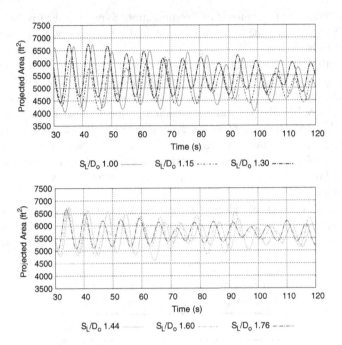

Figure 9.33 Projected area for different S_L/D_o

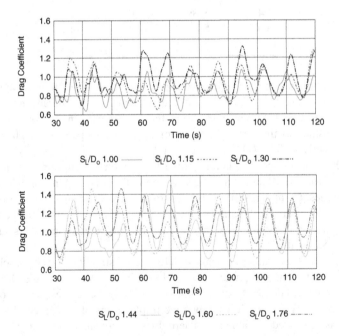

Figure 9.34 Drag coefficients for different S_L/D_o

Overall, the study shows that the increasing S_L/D_o leads to a significant improvement in C_D and decreased oscillations in S_p. This modification does not produce adverse stability characteristics. It is possible that further increases in S_L/D_o could yield even better performance. However, increasing S_L/D_o beyond 1.76 would incur a weight penalty as the overall length of the system would be increased.

9.8 Cluster Computations

A series of two-parachute cluster computations were carried out in Takizawa et al. (2011e) to determine how the parameters representing the payload models and starting-conditions affect long-term cluster dynamics. The parachute clusters reported in Takizawa et al. (2011e) were used with a 19 200 lb payload. Information on the parachutes can be found in Tezduyar et al. (2008c, 2010b) and Takizawa et al. (2011d). The parameters selected for testing were the payload-model configurations and initial coning angles (θ_{INIT}) and parachute diameters (D_{INIT}) (for readers not familiar with the term "coning angle," see Figure 9.53). We also investigated two scenarios to approximate the conditions immediately after parachute disreefing. This is explained in more detail in a later paragraph. A summary of the computations is shown in Table 9.6. In all cases, the θ_{INIT} is the same for both parachutes.

Table 9.6 Summary of all parachute cluster computations for different combinations of payload models and initial coning angles (θ_{INIT}) and parachute diameters (D_{INIT}). The values tabulated for θ_{INIT} apply to both parachutes, "P_1" and "P_2." The acronyms PAC, PLC, and PTE represent payload at the confluence, payload lower than the confluence, and payload as a truss element

Payload model	θ_{INIT} (°)	D_{INIT} (ft)	
		P_1	P_2
PAC	35	80	80
PLC	35	80	80
PTE	35	80	80
PTE	15	80	80
PTE	25	80	80
PTE	10	70	70
PTE	35	70	90

The first set of computations were carried out to investigate the effect of the payload model. In drop tests, the parachutes are connected to a rectangular pallet that is weighted to represent the mass and inertial properties of the crew capsule. The preliminary parachute cluster computations reported in Takizawa et al. (2011f) modeled the payload as a point mass located at the confluence of the risers. We will refer to this as the payload at the confluence (PAC) configuration. Two new computational payload models were created to see how they would influence parachute behavior. The payload lower than the confluence (PLC) configuration adds another cable element below the confluence and models the payload as a point mass at the location of the pallet center of gravity. The payload as a truss element (PTE) configuration further enhances the model by distributing the payload mass at 9 different

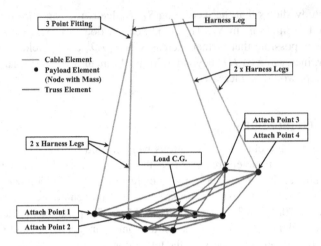

Figure 9.35 Payload as a truss element (PTE) configuration showing the cable, truss, and payload elements (figure from Moorman, 2010). The cable elements are the four longer ones and the truss elements are the remaining, shorter ones

points to match the mass, center of gravity, and six components of the inertia tensor of the pallet. This is accomplished by adding 5 cable elements and 26 truss elements below the confluence (see Moorman, 2010). Figure 9.35 shows the PTE configuration. In all of the payload comparison computations, $\theta_{INIT} = 35°$.

The second set of computations were carried out to investigate the effect of θ_{INIT}. Three values of θ_{INIT} were tested: 15°, 25°, and 35°. It should be noted that 35° is greater than the θ values seen in drop tests. The average θ during normal descent is around 15°, and the maximum θ does not usually exceed 25°. We used $\theta_{INIT} = 35°$ only to cause a large perturbation in order to analyze the dynamic response of the parachute cluster. All of the θ_{INIT} comparison computations used the PTE configuration.

In the third set of computations, two scenarios were computed to analyze how conditions immediately after disreefing could have an effect on long-term dynamics. In the first scenario, which we call "simulated disreef," $\theta_{INIT} = 10°$, and for both parachutes $D_{INIT} = 70$ ft. These values represent the approximate θ during final disreefing and the average minimum D during nominal descent. The second scenario represents an "asynchronous disreef" by using for one parachute $D_{INIT} = 70$ ft, and for the other $D_{INIT} = 90$ ft. These values represent the average minimum and maximum parachute diameters during nominal descent, respectively. Both scenarios used the PTE configuration.

9.8.1 Starting Conditions

First a starting condition was built in Takizawa et al. (2011e) for a single parachute. In this process, we begin with a parachute shape obtained with the symmetric FSI computation reported in Tezduyar et al. (2010b). We do another symmetric FSI computation with a horizontal inflow velocity of $24.0 \sin(\theta_{INIT})$ ft/s. This results in an angle of attack of θ_{INIT}, and we compute for three breathing cycles. We use the parachute shape and position corresponding to the time when the parachute skirt diameter is at its average value and assemble the cluster structural mechanics mesh with the parachutes at θ_{INIT}. After that we generate a fluid mechanics mesh. With the cluster mesh, holding the parachute shapes and positions fixed, we first do a fluid mechanics computation. The inflow velocity is 31.0 ft/s. Next, we do a fluid mechanics computation with a prescribed, time-dependent shape for both parachutes. The time-dependent shape comes from the single-parachute symmetric FSI computation carried out earlier at an angle of attack of θ_{INIT}. We use the solution from the fluid mechanics computation with prescribed parachute motion as the starting condition for the FSI computation.

9.8.2 Computational Conditions

Figure 9.36 shows, for a single parachute, the canopy structure mesh and the fluid mechanics interface mesh. The fluid mechanics mesh is cylindrical with a diameter of 1740 ft and a height of 1566 ft. It consists of four-node tetrahedral elements, while the fluid interface mesh consists of three-node triangular elements. The number of nodes and elements are given in Table 9.7.

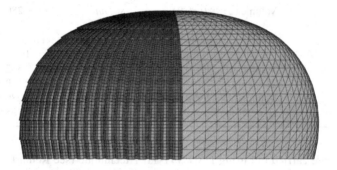

Figure 9.36 Canopy structure mesh (left) and fluid mechanics interface mesh (right) for a single parachute. The structure has 30 722 nodes, 26 000 four-node quadrilateral membrane elements, and 12 521 two-node cable elements. There are 29 200 nodes on the canopy. The fluid mechanics interface mesh has 2140 nodes and 4180 three-node triangular elements

All computations were carried out using air properties at standard sea-level conditions. The geometry and material properties are the same as they are described in Section 3.2 of Takizawa *et al.* (2011d). In addition to moving our reference frame vertically with a reference descent speed, as originally proposed in Tezduyar *et al.* (2010b), we move the mesh horizontally and vertically, with the average displacement rate for the structure.

All computations were carried out in a parallel computing environment. The meshes are partitioned to enhance the parallel efficiency of the computations. Mesh partitioning is based on the METIS Karypis and Kumar (1998) algorithm. In solving the linear equation systems involved at every nonlinear iteration, the GMRES search technique Saad and Schultz (1986) is used with a diagonal preconditioner.

The stand-alone fluid mechanics computations were performed in two parts. The first-part computations were carried out with the semi-discrete formulation given in Tezduyar (2003a). We computed 1000 time steps with a time-step size of 0.232 s and 7 nonlinear iterations per time step. The number of GMRES iterations per nonlinear iteration is 90. The second-part

Table 9.7 Number of nodes and elements for the two-parachute clusters before any payload modifications. Here nn and ne are the number of nodes and elements, respectively. The fluid mechanics volume mesh is tabulated for different combinations of θ_{INIT} and D_{INIT} values. The PLC configuration has 1 more structure node and 1 more cable element. The PTE configuration has 10 more structure nodes, 5 more cable elements, 26 more truss elements, and 8 more payload elements

Structure	Membrane	nn	61 443
		ne	52 000
	Cable	ne	25 042
	Payload	ne	1
	Interface	nn	58 400
		ne	52 000
Fluid	Interface	nn	4280
		ne	8360
	Volume (15°, 80/80 ft)	nn	197 288
		ne	1 210 349
	Volume (25°, 80/80 ft)	nn	280 601
		ne	1 739 739
	Volume (35°, 80/80 ft)	nn	289 679
		ne	1 797 003
	Volume (10°, 70/70 ft)	nn	352 861
		ne	2 199 472
	Volume (35°, 70/90 ft)	nn	289 221
		ne	1 795 542

computations were carried out using DSD/SST-TIP1 technique (see Remark 4.10), with the SUPG test function option WTSA (see Remark 4.4). The stabilization parameters used are those given by Equations (4.117)–(4.122) and (4.125), with the τ_{SUGN2} term dropped from Equation (4.121). The porosity model is HMGP-FG. We computed 600 time steps with a time-step size of 0.0232 s, 6 nonlinear iterations per time step, and 90 GMRES iterations per nonlinear iteration.

For the fluid mechanics computations with prescribed, time-dependent shapes, again the DSD/SST-TIP1 technique was used, with the same SUPG test function option and stabilization parameters as those described above. We computed roughly 300 time steps with a time-step size of 0.0232 s, 6 nonlinear iterations per time step, and 90 GMRES iterations per nonlinear iteration.

The FSI computations were carried out with the SSTFSI-TIP1 technique (see Remarks 4.10 and 5.9), again with the same SUPG test function option and stabilization parameters as those described above. The fully-discretized, coupled fluid and structural mechanics and mesh-moving equations are solved with the quasi-direct coupling technique (see Section 6.1.2). The time-step size is 0.0232 s, and the number of nonlinear iterations per time step is 6. The porosity model is HMGP-FG. The interface-stress projection is based on the SSP. We used Selective Scaling (see Section 6.1.2), with the scale for the structure part set to 100. The SENCT-FC contact algorithm (see Section 6.8) was used with $\epsilon_A^S = \epsilon_A^C = 1.45$ ft, which is approximately equal to the radial distance between the valley nodes and the outermost part of the sails at the parachute skirt. The number of GMRES iterations per nonlinear iteration is for most of the time steps 140 for the fluid+structure block, and 30 for the mesh-moving block. When the parachutes are close to each other, the number of GMRES iterations per nonlinear iteration for the fluid+structure block is increased as needed to control the residuals, especially those corresponding to the structural mechanics part. The maximum number of GMRES iterations used per nonlinear iteration for the fluid+structure block was 1400.

We computed each parachute cluster for a total of about 75 s, with remesh as needed to preserve mesh quality. The frequency of remeshing varies for each computation and usually depends on how often the parachutes collide, how much the cluster rotates about the vertical axis, and how much each parachute rotates about its own axis. Depending on the computation, remeshing was needed every 170 to 370 time steps.

9.8.3 Results

The critical measure of performance for the parachute system described in Takizawa et al. (2011e) is the descent speed of the payload. The maximum payload descent speed ultimately determines if the system meets mission requirements. Another common measure of performance is the drag coefficient, C_D. Figures 9.37–9.40 show the computational results for the parachute cluster computations in terms of U and C_D.

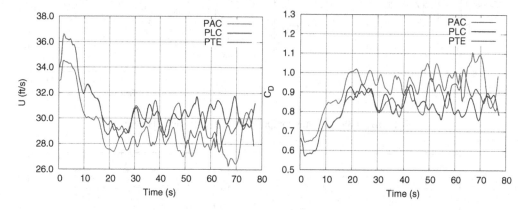

Figure 9.37 Cluster computations for different payload models and $\theta_{INIT} = 35°$

The geometry of parachute clusters usually forces individual parachutes to fly at angles of attack that are higher than the angle of attack at which they would fly as single parachutes. If the forced angle of attack in the cluster is not a stable one for the parachutes, they tend to collide with each other as they attempt to reach an angle of attack that is stable. Figures 9.41–9.43 show the contact between two parachutes from the asynchronous-disreef computation. Parachute clusters often experience reductions in drag due to this mutual interference between parachutes. The oscillatory motion of parachutes in the cluster and the frequency of collisions between parachutes can be used to characterize cluster stability. Figures 9.44–9.50 show the vent-separation distance ("L_{VS}") for all cluster computations. The horizontal black line on each plot shows the approximate vent-separation distance when the parachutes are in contact.

Tables 9.8–9.10 summarize the payload descent speeds and drag coefficients for all of the cluster computations.

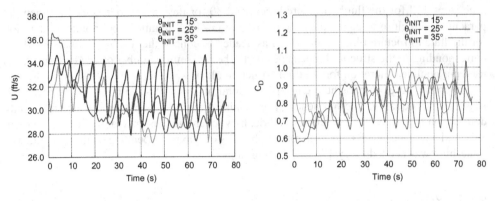

Figure 9.38 Cluster computations for PTE and different θ_{INIT} values

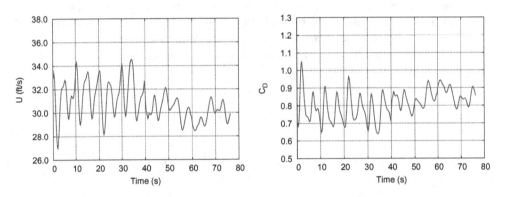

Figure 9.39 Cluster computations for simulated disreef

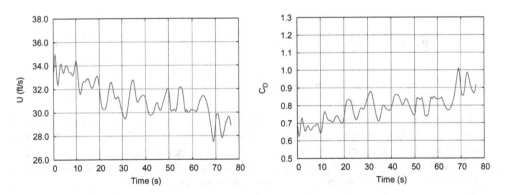

Figure 9.40 Cluster computations for asynchronous disreef

One of the goals of computational analysis is to assist parachute design engineers in determining which factors contribute to the payload descent speed oscillations seen in drop tests. For example, collisions between parachutes are usually associated with increased payload descent speed, but this is not always true. Previous analyses have also noted some

Figure 9.41 Parachutes at $t = 52.20$ s and $t = 53.36$ s during the asynchronous-disreef computation modeling the contact between parachutes

correlation between parachute coning angles and payload descent speed. However, the correlation between these parameters is not strong enough to conclude that coning angle is the only, or even the most important, factor. The payload descent speed is composed of several overlapping frequencies caused by various parachute dynamics. The overlapping frequencies make it very difficult to determine which individual parachute behaviors and parachute cluster behaviors are contributing to changes in payload descent speed. In order to address this complex problem, a technique to decompose the payload descent speed into components was developed in Takizawa *et al.* (2011e). This technique is described in Section 9.9.

9.9 Techniques for Dynamical Analysis and Model-Parameter Extraction

We describe the techniques and analyses from Takizawa *et al.* (2011e).

9.9.1 Contributors to Parachute Descent Speed

To better understand the descent speed fluctuations, we decompose the payload velocity into components based on geometric contributing factors. First we do that for a single parachute, and then for a cluster of parachutes.

Figure 9.42 Parachutes at $t = 54.52$ s and $t = 55.68$ s during the asynchronous-disreef computation modeling the contact between parachutes

9.9.1.1 Single Parachute

We represent the parachute in a spherical polar coordinate system. In that system, the payload is the origin, and the basis vectors are given in terms of the Cartesian basis vectors $\mathbf{e}_x$, $\mathbf{e}_y$, and $\mathbf{e}_z$ as follows:

$$\mathbf{g}_r = \sin\theta \cos\phi\, \mathbf{e}_x + \sin\theta \sin\phi\, \mathbf{e}_y + \cos\theta\, \mathbf{e}_z, \tag{9.15}$$

$$\mathbf{g}_\theta = \cos\theta \cos\phi\, \mathbf{e}_x + \cos\theta \sin\phi\, \mathbf{e}_y - \sin\theta\, \mathbf{e}_z, \tag{9.16}$$

$$\mathbf{g}_\phi = -\sin\phi\, \mathbf{e}_x + \cos\phi\, \mathbf{e}_y. \tag{9.17}$$

Here $\mathbf{g}_r$ and $\mathbf{g}_\theta$ represent the direction of the parachute axis and swinging, respectively, and $\mathbf{r} = r\mathbf{g}_r$ as shown in Figure 9.51. We separate the payload velocity $\mathbf{u}_p \equiv \frac{d\mathbf{x}_p}{dt}$ into its geometric ($\mathbf{u}_G$) and aerodynamic ($\mathbf{u}_A$) contributors as follows:

$$\mathbf{u}_p = \mathbf{u}_G + \mathbf{u}_A, \tag{9.18}$$

$$\mathbf{u}_G = \frac{d(\mathbf{x}_p - \mathbf{x}_A)}{dt}, \tag{9.19}$$

Figure 9.43 Parachutes at $t = 56.84$ s and $t = 58.00$ s during the asynchronous-disreef computation modeling the contact between parachutes

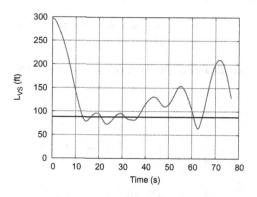

Figure 9.44 Vent-separation distance during the cluster computation with PAC and $\theta_{\text{INIT}} = 35°$

where $\mathbf{x}_A$ is a reference point, meant to be selected to give us a good way of differentiating between the factors contributing to the payload velocity. Here we use the canopy centroid as that reference point. We define the relative position vector $\mathbf{r} = (\mathbf{x}_A - \mathbf{x}_p)$, and obtain

$$\mathbf{u}_G = -\dot{\mathbf{r}}. \tag{9.20}$$

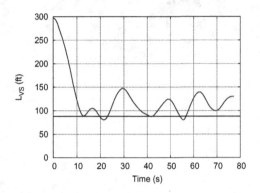

Figure 9.45 Vent-separation distance during the cluster computation with PLC and $\theta_{INIT} = 35°$

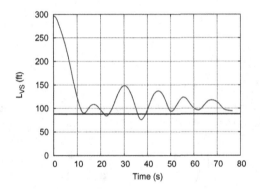

Figure 9.46 Vent-separation distance during the cluster computation with PTE and $\theta_{INIT} = 35°$

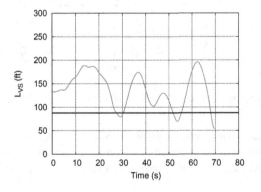

Figure 9.47 Vent-separation distance during the cluster computation with PTE and $\theta_{INIT} = 15°$

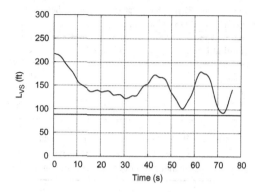

Figure 9.48 Vent-separation distance during the cluster computation with PTE and $\theta_{INIT} = 25°$

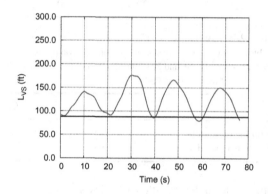

Figure 9.49 Vent-separation distance during the simulated-disreef cluster computation

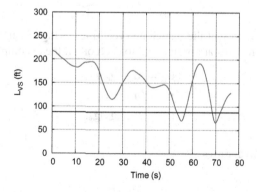

Figure 9.50 Vent-separation distance during the asynchronous-disreef cluster computation

Table 9.8 Average U and C_D for different payload models with $\theta_{INIT} = 35°$. Statistical analysis begins 20 s after the start of the computation

Payload model	U (ft/s)	C_D
PAC	28.1	0.97
PLC	30.1	0.85
PTE	29.5	0.88

Table 9.9 Average U and C_D for PTE and different values of θ_{INIT}. Statistical analysis begins 20 s after the start of the computation

θ_{INIT}	U (ft/s)	C_D
15°	29.9	0.86
25°	31.4	0.78
35°	29.5	0.88

Table 9.10 Average U and C_D for the disreef cases. Statistical analysis begins 5 s after the start of the computation for the simulated-disreef case, and 20 s after the start of the computation for the asynchronous-disreef case

	U (ft/s)	C_D
Simulated disreef	30.6	0.82
Asynchronous disreef	30.8	0.81

The geometric contribution can be rewritten as follows:

$$\mathbf{u}_G = -\frac{d(r\mathbf{g}_r)}{dt} = -\dot{r}\mathbf{g}_r - r\frac{d\mathbf{g}_r}{dt} \tag{9.21}$$

$$= \underbrace{-\dot{r}\mathbf{g}_r}_{\mathbf{u}_B} \underbrace{-r\mathbf{g}_\theta \dot{\theta}}_{\mathbf{u}_S} \underbrace{-r\sin\theta \mathbf{g}_\phi \dot{\phi}}_{\mathbf{u}_C}, \tag{9.22}$$

where $\mathbf{u}_B$, $\mathbf{u}_S$, and $\mathbf{u}_C$ represent the parachute breathing, swinging, and coning, respectively. We note that because the axes are orthogonal to each other, the decomposition becomes

$$\mathbf{u}_B = (\mathbf{u}_G \cdot \mathbf{g}_r)\mathbf{g}_r, \tag{9.23}$$

$$\mathbf{u}_S = (\mathbf{u}_G \cdot \mathbf{g}_\theta)\mathbf{g}_\theta, \tag{9.24}$$

$$\mathbf{u}_C = (\mathbf{u}_G \cdot \mathbf{g}_\phi)\mathbf{g}_\phi. \tag{9.25}$$

9.9.1.2 Cluster of Parachutes

We define $\bar{\mathbf{x}}_A$ as

$$\bar{\mathbf{x}}_A = \frac{1}{n_{para}} \sum_{k=1}^{n_{para}} (\mathbf{x}_A)_k, \tag{9.26}$$

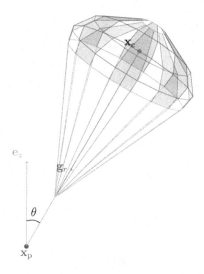

Figure 9.51 Parachute axis g_r. Swinging angle θ

where n_{para} is the number of parachutes. We again use a spherical polar coordinate systems with the payload being the origin and the basis vectors given as

$$\mathbf{g}_r = \sin\theta\cos\phi\,\mathbf{e}_x + \sin\theta\sin\phi\,\mathbf{e}_y + \cos\theta\,\mathbf{e}_z, \tag{9.27}$$

$$\mathbf{g}_\theta = \cos\theta\cos\phi\,\mathbf{e}_x + \cos\theta\sin\phi\,\mathbf{e}_y - \sin\theta\,\mathbf{e}_z, \tag{9.28}$$

$$\mathbf{g}_\phi = -\sin\phi\,\mathbf{e}_x + \cos\phi\,\mathbf{e}_y. \tag{9.29}$$

Here $\mathbf{g}_r$, $\mathbf{g}_\theta$, and $\mathbf{g}_\phi$ can be seen as the direction of the cluster axis, swinging, and coning, and $\bar{\mathbf{r}} = \bar{r}\,\mathbf{g}_r$. Figure 9.52 shows the relationship between the Cartesian axes and the basis vectors. We also define individual spherical coordinate systems for the parachutes of the cluster. In each of those coordinate systems, the payload is the origin, and the basis vectors are given in terms of the basis vectors: $\mathbf{g}_\theta$, $\mathbf{g}_\phi$ and $\mathbf{g}_r$:

$$(\mathbf{g}_r)_k = \sin\theta_k\cos\phi_k\,\mathbf{g}_\theta + \sin\theta_k\sin\phi_k\,\mathbf{g}_\phi + \cos\theta_k\,\mathbf{g}_r, \tag{9.30}$$

$$(\mathbf{g}_\theta)_k = \cos\theta_k\cos\phi_k\,\mathbf{g}_\theta + \cos\theta_k\sin\phi_k\,\mathbf{g}_\phi - \sin\theta_k\,\mathbf{g}_r, \tag{9.31}$$

$$(\mathbf{g}_\phi)_k = -\sin\phi_k\,\mathbf{g}_\theta + \cos\phi_k\,\mathbf{g}_\phi. \tag{9.32}$$

Here $(\mathbf{g}_r)_k$ is the axis direction for the kth parachute, θ_k is the coning angle, and $(\mathbf{x}_A)_k - \mathbf{x}_p \equiv \mathbf{r}_k = r_k(\mathbf{g}_r)_k$. Figure 9.53 shows the relationship between the cluster axes, the individual parachutes, and $(\mathbf{g}_r)_k$.

From Equation (9.26) and the definition of $\mathbf{r}_k$,

$$\mathbf{u}_p = \dot{\bar{\mathbf{x}}}_A - \dot{\bar{\mathbf{r}}}, \tag{9.33}$$

where

$$\bar{\mathbf{r}} = \frac{1}{n_{\text{para}}}\sum_{k=1}^{n_{\text{para}}}\mathbf{r}_k. \tag{9.34}$$

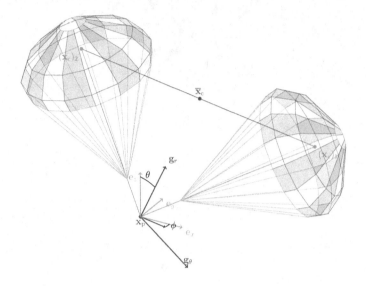

Figure 9.52 Direction of the cluster axis $\mathbf{g}_r$ and the cluster coning angle θ

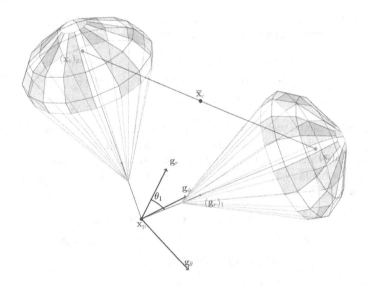

Figure 9.53 Direction of the parachute axis $(\mathbf{g}_r)_k$ and the individual parachute coning angle θ_k

We define

$$\bar{\mathbf{u}}_G = -\dot{\bar{\mathbf{r}}}, \tag{9.35}$$

therefore

$$\mathbf{u}_p = \bar{\mathbf{u}}_A + \bar{\mathbf{u}}_G. \tag{9.36}$$

First we consider the cluster system itself. As we did for a single parachute, we can decompose the geometric contribution as follows:

$$\bar{\mathbf{u}}_G = \mathbf{u}_B + \mathbf{u}_S + \mathbf{u}_C, \tag{9.37}$$

where $\mathbf{u}_B$, $\mathbf{u}_S$ and $\mathbf{u}_C$ represent the cluster system "breathing," swinging, and coning:

$$\mathbf{u}_B = (\bar{\mathbf{u}}_G \cdot \mathbf{g}_r)\mathbf{g}_r, \tag{9.38}$$
$$\mathbf{u}_S = (\bar{\mathbf{u}}_G \cdot \mathbf{g}_\theta)\mathbf{g}_\theta, \tag{9.39}$$
$$\mathbf{u}_C = (\bar{\mathbf{u}}_G \cdot \mathbf{g}_\phi)\mathbf{g}_\phi. \tag{9.40}$$

Furthermore, we decompose the velocity into contributions from individual parachutes, which are written in terms of their breathing, "swinging," and coning parts:

$$(\mathbf{u}_G)_k = (\mathbf{u}_B)_k + (\mathbf{u}_S)_k + (\mathbf{u}_C)_k. \tag{9.41}$$

Here

$$(\mathbf{u}_B)_k = ((\mathbf{u}_G)_k \cdot (\mathbf{g}_r)_k)(\mathbf{g}_r)_k, \tag{9.42}$$
$$(\mathbf{u}_S)_k = ((\mathbf{u}_G)_k \cdot (\mathbf{g}_\theta)_k)(\mathbf{g}_\theta)_k, \tag{9.43}$$
$$(\mathbf{u}_C)_k = ((\mathbf{u}_G)_k \cdot (\mathbf{g}_\phi)_k)(\mathbf{g}_\phi)_k. \tag{9.44}$$

By definition,

$$\bar{\mathbf{u}}_G = \frac{1}{n_{\text{para}}} \sum_{k=1}^{n_{\text{para}}} (\mathbf{u}_G)_k. \tag{9.45}$$

We also define the averages of the parts given by Equations (9.42)–(9.44):

$$\bar{\mathbf{u}}_B = \frac{1}{n_{\text{para}}} \sum_{k=1}^{n_{\text{para}}} (\mathbf{u}_B)_k, \tag{9.46}$$

$$\bar{\mathbf{u}}_S = \frac{1}{n_{\text{para}}} \sum_{k=1}^{n_{\text{para}}} (\mathbf{u}_S)_k, \tag{9.47}$$

$$\bar{\mathbf{u}}_C = \frac{1}{n_{\text{para}}} \sum_{k=1}^{n_{\text{para}}} (\mathbf{u}_C)_k. \tag{9.48}$$

The cluster "breathing" comes from $((\bar{\mathbf{u}}_B)_k \cdot \mathbf{g}_r)\mathbf{g}_r$ and $((\bar{\mathbf{u}}_S)_k \cdot \mathbf{g}_r)\mathbf{g}_r$. We note from Equation (9.32) that $\bar{\mathbf{u}}_C$ does not have a part in the direction of the cluster axis.

In summary,

$$\bar{\mathbf{u}}_G = \mathbf{u}_B + \mathbf{u}_S + \mathbf{u}_C = \bar{\mathbf{u}}_B + \bar{\mathbf{u}}_S + \bar{\mathbf{u}}_C. \tag{9.49}$$

We note that $\mathbf{u}_B$ and $\bar{\mathbf{u}}_B$, $\mathbf{u}_S$, and $\bar{\mathbf{u}}_S$, and $\mathbf{u}_C$ and $\bar{\mathbf{u}}_C$ represent things that are different from each other. We focus on the vertical components of all these velocities, which, for example for the cluster, can be expressed as follows:

$$\mathbf{u}_B \cdot \mathbf{e}_z = (((\bar{\mathbf{u}}_B)_k \cdot \mathbf{g}_r) + ((\bar{\mathbf{u}}_S)_k \cdot \mathbf{g}_r))(\mathbf{g}_r \cdot \mathbf{e}_z), \tag{9.50}$$
$$\mathbf{u}_S \cdot \mathbf{e}_z = (((\bar{\mathbf{u}}_B)_k \cdot \mathbf{g}_\theta) + ((\bar{\mathbf{u}}_S)_k \cdot \mathbf{g}_\theta) + ((\bar{\mathbf{u}}_C)_k \cdot \mathbf{g}_\theta))(\mathbf{g}_\theta \cdot \mathbf{e}_z), \tag{9.51}$$
$$\mathbf{u}_C \cdot \mathbf{e}_z = 0. \tag{9.52}$$

9.9.1.3 Results

Figures 9.54–9.60 show, for all cluster computations, the decomposition of the descent speed. For the cluster, the decomposition is in terms of $\bar{\mathbf{u}}_A$ and the breathing and swinging contributors of the descent speed. Top plots represent that. For the average of the individual parachute contributions, it is in terms of their breathing, swinging and coning parts. Bottom plots represent that.

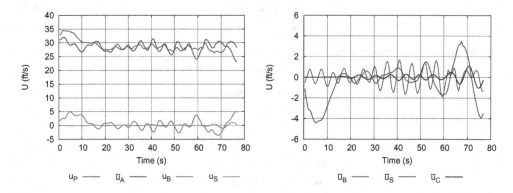

Figure 9.54 Decomposition of the descent speed for the cluster computation with PAC and $\theta_{INIT} = 35°$

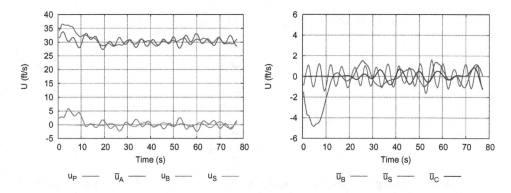

Figure 9.55 Decomposition of the descent speed for the cluster computation with PLC and $\theta_{INIT} = 35°$

From the top plots, we note that the contribution from $\mathbf{u}_B$ is always greater than the contribution from $\mathbf{u}_S$. Furthermore, cluster swinging does not appear to be a major contributor to payload descent speed oscillations. Most of the oscillations are come from the motion of the individual parachutes. Looking at the bottom plots, we observe that for all of the computations with $\theta_{INIT} = 35°$, the largest contribution comes from $\bar{\mathbf{u}}_S$. For the computations with

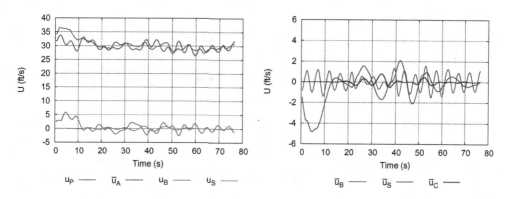

Figure 9.56 Decomposition of the descent speed for the cluster computation with PTE and $\theta_{INIT} = 35°$

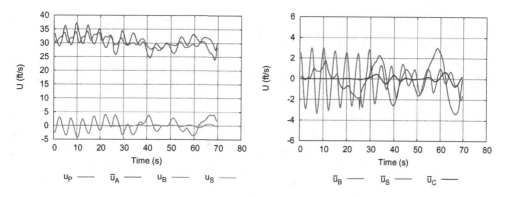

Figure 9.57 Decomposition of the descent speed for the cluster computation with PTE and $\theta_{INIT} = 15°$

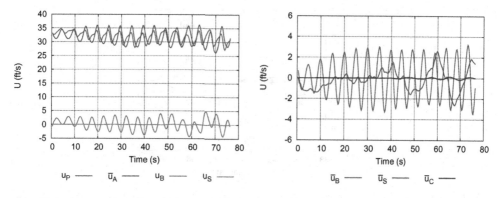

Figure 9.58 Decomposition of the descent speed for the cluster computation with PTE and $\theta_{INIT} = 25°$

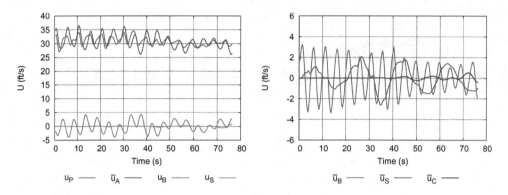

Figure 9.59 Decomposition of the descent speed for the simulated-disreef cluster computation

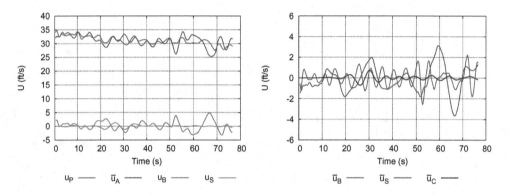

Figure 9.60 Decomposition of the descent speed for the asynchronous-disreef cluster computation

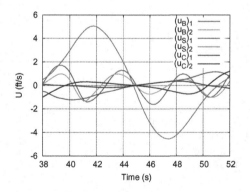

Figure 9.61 Individual-parachute contributions to descent speed for the cluster computation with PTE and $\theta_{\text{INIT}} = 35°$

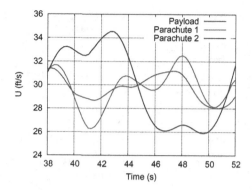

Figure 9.62 Payload and canopy-centroid descent speeds for the cluster computation with PTE and $\theta_{\text{INIT}} = 35°$

smaller θ_{INIT}, the $\bar{\mathbf{u}}_B$ contribution is dominant initially, and the $\bar{\mathbf{u}}_S$ contribution grows over time. The $\theta_{\text{INIT}} = 25°$ case seems to have fairly equal contributions from both $\bar{\mathbf{u}}_S$ and $\bar{\mathbf{u}}_B$. The contribution from the individual parachute coning, $\bar{\mathbf{u}}_C$, is the smallest.

We further analyze a particularly interesting result from the cluster computation with PTE and $\theta_{\text{INIT}} = 35°$ to better understand the parachute behavior that is causing the anomaly. Approximately 38 s into the computation, there is a large positive contribution from $(\mathbf{u}_S)_2$ followed by a large negative contribution (see Figure 9.61). In comparison, $(\mathbf{u}_S)_1$ remains fairly constant over the same time period. From the plots of payload and canopy-centroid descent speeds shown in Figure 9.62, we see that Parachute 2 initially has a much higher descent speed than Parachute 1. After about the 45 s mark, the descent speed for Parachute 2 becomes much lower than Parachute 1. The payload descent speed is relatively shielded because the fluctuations for Parachutes 1 and 2 are opposite each other and the average of the canopy speed fluctuations is small. We have determined that the large changes in $(\mathbf{u}_S)_2$ occur as Parachute 2 flies around Parachute 1 in a coning motion. Meanwhile, Parachute 1 descends in a relatively straight path and presents a larger projected area to the free stream. Therefore, Parachute 1 produces substantially more drag than Parachute 2 as shown in Figure 9.63. At the time of maximum drag differential, Parachute 1 produces nearly 70% of the total cluster drag as shown in Figure 9.64.

9.9.2 Added Mass

9.9.2.1 Relevance to Parachutes

The concept of added mass is applicable to parachutes during disreefing and unsteady full-open descent. When the parachute is disreefed, the sudden increase in drag area causes a rapid deceleration and increase in system inertia. During full-open descent, the parachute exhibits a periodic breathing motion. The frequency and damping of this motion is related to the added mass. Here we describe a technique developed in Takizawa et al. (2011e) for determining the added mass of a parachute from FSI computations.

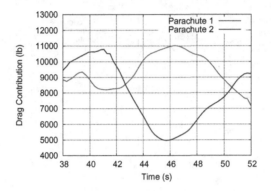

Figure 9.63 Drag contribution of each parachute for the cluster computation with PTE and $\theta_{INIT} = 35°$

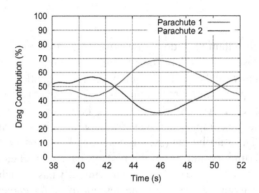

Figure 9.64 Drag contribution percentage of each parachute for the cluster computation with PTE and $\theta_{INIT} = 35°$

9.9.2.2 Determining Added Mass

The force at time t can be written in terms of coefficients C and m_A as

$$F(t) = CU_A^2(t) + m_A \dot{U}_A(t), \qquad (9.53)$$

where m_A is the added mass. Dividing both sides of Equation (9.53) by $U_A^2(t)$, we obtain

$$\frac{F(t)}{U_A^2(t)} = C + m_A \frac{\dot{U}_A(t)}{U_A^2(t)}. \qquad (9.54)$$

Plotting data points from parachute FSI computations in this form and fitting a trend line to the data can yield values for C and m_A.

REMARK 9.4 *While the parameter extraction here is based on Equation (9.54), another version was proposed in Takizawa et al. (2011e) as an alternative:*

$$\frac{1}{\rho A_E} \frac{F(t)}{U_A^2(t)} = C^* + m_A^* \frac{V_E}{A_E} \frac{\dot{U}_A(t)}{U_A^2(t)}, \qquad (9.55)$$

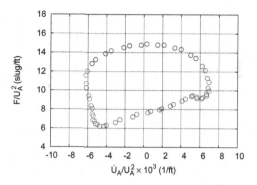

Figure 9.65 Symmetric FSI computation for a single parachute in full-open descent. Results are plotted in the form of Equation (9.54). Two breathing periods are shown

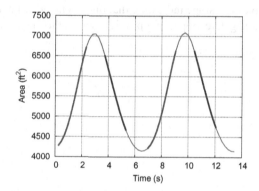

Figure 9.66 Parachute projected area for the time steps corresponding to the data points shown in Figure 9.65

where A_E and V_E are the estimated, time-dependent values for the parachute drag area and the volume of the added air mass, $C^* = C/(\rho A_E)$, and $m_A^* = m_A/(\rho V_E)$.

Figure 9.65 shows two breathing periods from a symmetric FSI computation for a single parachute in full-open descent. The parachute exhibits a periodic breathing motion resulting in a time-varying added mass. Figure 9.66 shows the projected area of the parachute for the time steps corresponding to the data points and the colors illustrate the chronological sequence of the data points during the breathing period.

Figure 9.67 shows the full results from the same symmetric FSI computation for approximately 150 s of full-open descent. Linear regression is used to fit a trend line to the part of the periodic motion when the parachute diameter is increasing. From the plot we find $m_A = 339$ slugs and $C = 7.7$ slugs/ft. Added mass during the inflation part of the breathing period is nearly constant at approximately 10 900 lbs. This is almost twice the payload

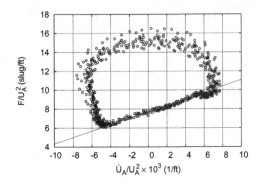

Figure 9.67 Full results for the single parachute symmetric FSI computation. The data points shown here cover approximately 150 s of full-open descent

weight of 5 570 lbs and approximately 10% larger than the enclosed air mass. Some parts of the breathing period have zero slope indicating that there is no added mass. Other parts of the breathing period have a negative slope. In a physical sense, this means that the fluid is doing work on the parachute.

10

Wind-Turbine Aerodynamics and FSI

Countries around the world are putting substantial effort into the development of wind energy technologies. The ambitious wind energy goals put pressure on the wind energy industry research and development to significantly enhance current wind generation capabilities in a short period of time and decrease the associated costs. This calls for transformative concepts and designs (e.g., floating offshore wind turbines) that must be created and analyzed with high-precision methods and tools. These include complex-geometry, 3D, time dependent, multi-physics predictive simulation methods and software that will play an increasingly important role as the demand for wind energy grows.

Currently most wind-turbine aerodynamics and aeroelasticity simulations are performed using low-fidelity methods, such as the Blade Element Momentum (BEM) theory for the rotor aerodynamics employed in conjunction with simplified structural models of the wind-turbine blades and tower (see, e.g., Jonkman and Buhl Jr., 2005; Jonkman et al., 2009). These methods are very fast to implement and execute. However, the cases involving unsteady flow, turbulence, 3D details of the wind-turbine blade and tower geometry, and other similarly-important features, are beyond their range of applicability.

To obtain high-fidelity predictive simulation results for wind turbines, 3D modeling is essential. However, simulation of wind turbines at full scale engenders a number of challenges: the flow is fully turbulent, requiring highly accurate methods and increased grid resolution. The presence of fluid boundary layers, where turbulence is created, complicates the situation further. Wind-turbine blades are long and slender structures, with complex distribution of material properties, for which the numerical approach must have good approximation properties and avoid locking. Wind-turbine simulations involve moving and stationary components, and the fluid–structure coupling must be accurate, efficient and robust. These explain the current, modest nature of the state-of-the-art in wind-turbine simulations.

FSI simulations at full scale are essential for accurate modeling of wind turbines. The motion and deformation of the wind-turbine blades depend on the wind speed and air flow, and the air flow patterns depend on the motion and deformation of the blades. In order to simulate the coupled problem, the equations governing the air flow and the blade motions and

---

Computational Fluid–Structure Interaction: Methods and Applications, First Edition.
Yuri Bazilevs, Kenji Takizawa and Tayfun E. Tezduyar.
© 2013 John Wiley & Sons, Ltd. Published 2013 by John Wiley & Sons, Ltd.

deformations need to be solved simultaneously. Without that the modeling cannot be realistic: unsteady blade deformation affects aerodynamic efficiency and noise generation, and response to wind gusts. Flutter analysis of large blades operating in offshore environments is of great importance and cannot be accomplished without FSI.

In recent years, several attempts were made to address the above mentioned challenges and to raise the fidelity and predictability levels of wind-turbine simulations. Stand alone aerodynamics simulations of wind-turbine configurations in 3D were reported in Sørensen *et al.* (2002), Pape and Lecanu (2004), Zahle *et al.* (2009), Bazilevs *et al.* (2011b), Takizawa *et al.* (2011a,b) and Li and K. J. Paik (2012), while stand alone structural analyses of rotor blades of complex geometry and material composition, but under assumed wind-load conditions or wind-load conditions coming from separate aerodynamic computations were reported in Guttierez *et al.* (2003), Kong *et al.* (2005), Hansen *et al.* (2006), Jensen *et al.* (2006), Kiendl *et al.* (2010) and Bazilevs *et al.* (2012a). In the recent work of Bazilevs *et al.* (2011c) it was shown that coupled FSI modeling and simulation of wind turbines is important for accurately predicting their mechanical behavior at full scale.

We feel that to address the above mentioned challenges one should employ a combination of numerical techniques, which are general, accurate, robust and efficient for the targeted class of problems. Such techniques are summarized in what follows and are described in greater detail in the body of this chapter.

IGA is adopted as the geometry modeling and simulation framework for wind turbines in some of the examples presented in this chapter. We use IGA based on NURBS, which are more efficient than standard finite elements for representing complex, smooth geometries, such as wind-turbine blades. The IGA was successfully employed for computation of turbulent flows (Bazilevs *et al.*, 2007a,c, 2010e; Akkerman *et al.*, 2008; Hsu *et al.*, 2010; Bazilevs and Akkerman, 2010), nonlinear structures (Elguedj *et al.*, 2008; Lipton *et al.*, 2010; Benson *et al.*, 2010a,b; Kiendl *et al.*, 2009, 2010), and FSI (Zhang *et al.*, 2007; Bazilevs *et al.*, 2006a, 2008; Isaksen *et al.*, 2008), and, in most cases, gave a clear advantage over standard low-order finite elements in terms of solution accuracy per degree-of-freedom. This is in part attributable to the higher-order smoothness of the basis functions employed. Flows about rotating components are naturally handled in an isogeometric framework because all conic sections, and in particular, circular and cylindrical shapes, are represented exactly (Bazilevs and Hughes, 2008).

The blade structure is governed by the isogeometric rotation-free shell formulation with the aid of the bending-strip method (Kiendl *et al.*, 2010). The method is appropriate for thin-shell structures comprised of multiple C^1- or higher-order continuous surface patches that are joined or merged with continuity no greater than C^0. The Kirchhoff–Love shell theory that relies on higher-order continuity of the basis functions is employed in the patch interior as in Kiendl *et al.* (2009). Although NURBS-based IGA is employed in this work, other discretizations such as T-splines (Bazilevs *et al.*, 2010a; Dörfel *et al.*, 2010) or subdivision surfaces (Cirak *et al.*, 2000, 2002; Cirak and Ortiz, 2001), are perfectly suited for the proposed structural modeling method.

In addition, an isogeometric representation of the analysis-suitable geometry may be used to construct tetrahedral and hexahedral meshes for computations using the FEM. In this chapter, we use such tetrahedral meshes for wind-turbine rotor computation using the ALE-VMS (see Section 4.6.1) and DSD/SST-VMST (ST-VMS) (see Section 4.6.3) methods. In application of the DSD/SST formulation to flows with moving mechanical components, the

Shear–Slip Mesh Update Method (SSMUM) (Tezduyar *et al.*, 1996; Behr and Tezduyar, 1999, 2001) has been very instrumental. The SSMUM was first introduced for computation of flow around two high-speed trains passing each other in a tunnel (see Tezduyar *et al.*, 1996). The challenge was to accurately and efficiently update the meshes used in computations based on the DSD/SST formulation and involving two objects in fast, linear relative motion. The idea behind the SSMUM was to restrict the mesh moving and remeshing to a thin layer of elements between the objects in relative motion. The mesh update at each time step can be accomplished by a "shear" deformation of the elements in this layer, followed by a "slip" in node connectivities. The slip in the node connectivities, to an extent, un-does the deformation of the elements and results in elements with better shapes than those that were shear-deformed. Because the remeshing consists of simply re-defining the node connectivities, both the projection errors and the mesh generation cost are minimized. A few years after the high-speed train computations, the SSMUM was implemented for objects in fast, rotational relative motion and applied to computation of flow past a rotating propeller (Behr and Tezduyar, 1999) and flow around a helicopter with its rotor in motion (Behr and Tezduyar, 2001).

A number of special techniques for wind-turbine aerodynamics and FSI simulation were developed recently. In Bazilevs *et al.* (2011b), a technique for describing the wind-turbine rotor geometry was developed based on NURBS and applied to the aerodynamics simulation of a 5MW wind-turbine rotor with its design specified in Jonkman *et al.* (2009). Follow-up computational studies of this rotor were presented in Takizawa *et al.* (2011a,b) and Hsu *et al.* (2011b). In Hsu *et al.* (2011a), the aerodynamics simulations of the National Renewable Energy Lab (NREL) Phase VI two-bladed rotor (see Hand *et al.*, 2001) were performed to validate the ALE-VMS computations against an extensive set of experimental data available for this test case. The structural mechanics formulation for wind-turbine blades, which is based on the Kirchhoff–Love thin shell theory and the bending strip method (see Kiendl *et al.*, 2009, 2010), was developed and applied in Bazilevs *et al.* (2011c) to fully-coupled FSI simulation of a 5MW wind-turbine rotor. We believe this was the first fully-coupled FSI simulation of a full-scale wind-turbine rotor. A special mesh moving procedure for FSI simulation of wind-turbine rotors was also proposed in Bazilevs *et al.* (2011c), where only the deflection part of the mesh motion is handled using the elasticity-based mesh moving method (Tezduyar *et al.*, 1992b, 1993), while the rotational part is handled exactly. A method and algorithm for pre-bending of wind-turbine blades to avoid the blade striking the tower during operation in high winds were recently proposed in Bazilevs *et al.* (2012a). In this chapter we provide a description of these techniques.

10.1 Aerodynamics Simulations of a 5MW Wind-Turbine Rotor

In this section we begin with a careful definition of the 5MW wind-turbine rotor geometry. We then present the NURBS-based and FEM-based simulations of the wind-turbine rotor. In this section, we only present pure aerodynamic simulations. Structural and FSI modeling and simulations will be presented in the later sections.

10.1.1 5MW Wind-Turbine Rotor Geometry Definition

As a first step we construct a template for the structural model of the rotor. Here, the structural model is limited to a surface (shell) representation of the wind-turbine blade, the hub, and their

Table 10.1 Wind-turbine rotor geometry definition

RNodes (m)	AeroTwst (°)	Chord (m)	AeroCent (-)	AeroOrig (-)	Airfoil
2.0000	0.000	3.542	0.2500	0.50	Cylinder
2.8667	0.000	3.542	0.2500	0.50	Cylinder
5.6000	0.000	3.854	0.2218	0.44	Cylinder
8.3333	0.000	4.167	0.1883	0.38	Cylinder
11.7500	13.308	4.557	0.1465	0.30	DU40
15.8500	11.480	4.652	0.1250	0.25	DU35
19.9500	10.162	4.458	0.1250	0.25	DU35
24.0500	9.011	4.249	0.1250	0.25	DU30
28.1500	7.795	4.007	0.1250	0.25	DU25
32.2500	6.544	3.748	0.1250	0.25	DU25
36.3500	5.361	3.502	0.1250	0.25	DU21
40.4500	4.188	3.256	0.1250	0.25	DU21
44.5500	3.125	3.010	0.1250	0.25	NACA64
48.6500	2.310	2.764	0.1250	0.25	NACA64
52.7500	1.526	2.518	0.1250	0.25	NACA64
56.1667	0.863	2.313	0.1250	0.25	NACA64
58.9000	0.370	2.086	0.1250	0.25	NACA64
61.6333	0.106	1.419	0.1250	0.25	NACA64
62.9000	0.000	0.700	0.1250	0.25	NACA64

attachment zone. The blade surface is assumed to be composed of a collection of airfoil shapes that are lofted in the blade-axis direction.

The geometry of the rotor blade is based on the NREL 5MW offshore baseline wind turbine described in Jonkman et al. (2009). The blade geometry data taken from the reference is summarized in Table 10.1. A 61 m blade is attached to a hub with radius of 2 m, which gives the total rotor radius of 63 m. The blade is composed of several airfoil types provided in the rightmost column of the table. The first portion of the blade is a perfect cylinder. Further away from the root the cylinder is smoothly blended into a series of DU (Delft University) airfoils. At the 44.55 m location away from the root the NACA64 profile is used to define the blade all the way to the tip (see Figure 10.1). The remaining parameters from Table 10.1 are defined in Figure 10.1: "RNodes" is the distance from the rotor center to the airfoil cross-section in the blade axis direction. "AeroTwst" is the twist angle for a given cross-section. The blades are twisted to enhance the aerodynamic performance. "Chord" is the chord length of the airfoil. "AeroOrig" is the location of the aerodynamic center. For most of the blade airfoil cross-sections, the aerodynamic center is taken at 25% of the chord length from the leading edge. To accommodate the cylindrical shape at the root, the aerodynamic center is gradually moved to 50% of the chord length. This is not reported in Jonkman et al. (2009), but mentioned in Kooijman et al. (2003).

REMARK 10.1 *There is some redundancy in the parameters given in Table 10.1. The variable "AeroCent" is used as an input to FAST (Jonkman and Buhl Jr., 2005), which is the aerodynamics modeling software that is typically used for wind-turbine rotor computations. FAST assumes that the blade-pitch axis passes through each airfoil section at 25% chord length, and defines AeroCent − 0.25 to be the fractional distance to the aerodynamic center*

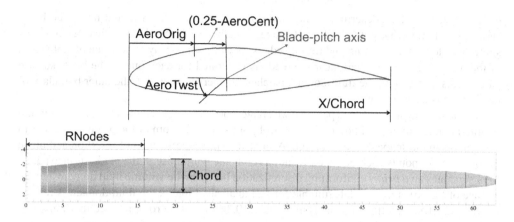

Figure 10.1 Illustration of quantities from Table 10.1

from the blade-pitch axis along the chordline, positive toward the trailing edge. Therefore, AeroOrig + (0.25 − AeroCent) gives the location of where the blade-pitch axis passes through each airfoil cross-section. Although for our purposes this added complexity is unnecessary, the same naming system is used for backward compatibility with the referenced reports.

For each blade cross-section, we use quadratic NURBS to represent the 2D airfoil shape. The weights of the NURBS functions are set to unity. The weights are adjusted near the root to represent the circular cross-sections of the blade exactly. The cross-sections are lofted in the blade axis direction, also using quadratic NURBS and unity weights. This geometry modeling procedure produces a smooth rotor-blade surface using a relatively small number of input parameters, which is an advantage of the isogeometric representation. Figure 10.2 shows a top view of the blade in which the twisting of the cross-sections is evident. Given

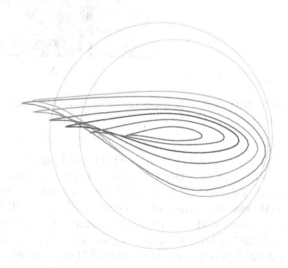

Figure 10.2 Top view of a subset of the airfoil cross-sections illustrating blade twisting

the rotor-blade surface description, the fluid-domain volume is constructed next. The blade surface is split into four patches of similar sizes, which we call the blade surface patches. The splitting is done at the leading and trailing edges, as well as half-way in between on both sides of the blade. The fluid domain near the blade is generated for each one of the blade surface patches. As a final step, the fluid-domain patches are merged such that the outer boundary of the domain is a perfect cylinder.

For each of the blade surface patches, we create a 60° pie-shaped domain using a minimum required number of control points. The control points at the bottom of the patch are moved to accommodate the shape of the rotor hub. As a next step, we perform knot insertion and move the new control points such that their locations coincide with those of the blade surface patch. This generates an *a priori* conforming discretization between the volume fluid domain and the surface of the structural model, suitable for FSI analysis. Finally, the fluid domain is refined in all parametric directions for analysis. Figure 10.3a shows the rotor surface mesh and one of the fluid-mesh subdomains adjacent to it. The remaining fluid subdomains are generated in the same manner.

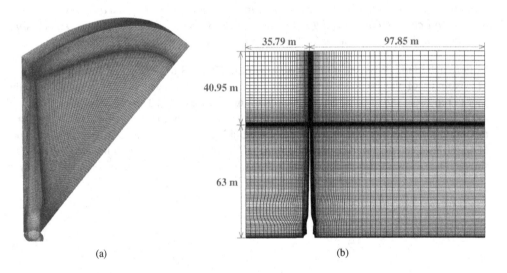

Figure 10.3 (a) Volume NURBS mesh of the computational domain. (b) A planar cut to illustrate mesh grading toward the rotor blade

The resultant fluid NURBS mesh may be embedded into a larger domain for the purposes of simulation. In this work we take this larger domain to also be a cylinder. For computational efficiency, only one-third of the domain is modeled. The fluid volume mesh, corresponding to one-third of the fluid domain, consists of 1 449 000 quadratic NURBS elements (and a similar number of control points). The fluid mesh cross-section that also shows the details of mesh refinement in the boundary layer is shown in Figure 10.3b. To carry out the simulations, rotational-periodicity conditions (see Sections 9.2 and 9.2.3) are imposed. Denoting by $\mathbf{u}_l^h$ and

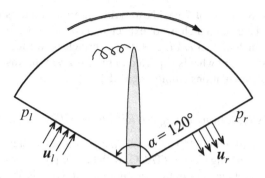

Figure 10.4 Rotational-periodicity conditions

$\mathbf{u}_r^h$ the discrete fluid velocities at the left and right boundary, respectively (see Figure 10.4), and by p_l^h and p_r^h the corresponding pressures, we set

$$p_l^h = p_r^h, \tag{10.1}$$
$$\mathbf{u}_l^h = \mathbf{R}(2/3\pi)\,\mathbf{u}_r^h, \tag{10.2}$$

where $\mathbf{R}(2/3\pi)$ is the rotation matrix evaluated at $\alpha = 2/3\pi$. That is, while the pressure degrees-of-freedom take on the same values, the fluid velocity degrees-of-freedom are related through a linear transformation corresponding to a rotation by $2/3\pi$ radians. Note that the transformation matrix is independent of the current domain position. Rotational-periodicity conditions are implemented through standard master-slave relationships. We note that rotational-periodicity conditions were employed earlier in Takizawa *et al.* (2011d,f, 2010c) for parachute simulations.

We compute the aerodynamics of the wind-turbine rotor with prescribed speed using a rotating mesh. The wind speed is uniform at 9 m/s and the rotor speed is 1.08 rad/s, giving a tip speed ratio of 7.55 (see Spera, 1994 for wind-turbine terminology). We use air properties at standard sea-level conditions. The Reynolds number (based on the cord length at $\frac{3}{4}R$ and the relative velocity there) is approximately 12 million. At the inflow boundary the velocity is set to the wind velocity, at the outflow boundary the stress vector is set to zero, and at the radial boundary the radial component of the velocity is set to zero. We start from a flow field where the velocity is equal to the inflow velocity everywhere in the domain except on the rotor surface, where the velocity matches the rotor velocity.

We carry out the computations at a constant time-step size of 4.67×10^{-4} s. Both NURBS and tetrahedral FEM simulations of this setup are performed.

The chosen wind velocity and rotor speed correspond to one of the cases given in Jonkman *et al.* (2009), where the aerodynamics simulations were performed using FAST (Jonkman and Buhl Jr., 2005). We note that FAST is based on lookup tables for airfoil cross-sections, which give planar, steady-state lift and drag data for a given wind speed and angle of attack. The effects of trailing-edge turbulence, hub, and tip are incorporated through empirical models. It was reported in Jonkman *et al.* (2009) that at these wind conditions and rotor speed, no blade pitching takes place and the rotor develops a favorable aerodynamic torque (i.e., torque in the direction of the rotation) of 2500 kN m. Although this value is used for comparison with our simulations, the exact match is not expected, as our computational modeling is very different than the one in Jonkman *et al.* (2009). Nevertheless, we feel that this value of the aerodynamic torque is close to what is expected in reality, given the vast experience of NREL with wind-turbine rotor simulations employing FAST.

10.1.2 ALE-VMS Simulations Using NURBS-based IGA

The computation is carried out on 240 cores on Ranger, a Sun Constellation Linux Cluster at the Texas Advanced Computing Center (TACC) with 62 976 processing cores. Near the blade surface, the size of the first element in the wall-normal direction is about 2 cm. The GMRES search technique Saad and Schultz (1986) is used with a block-diagonal preconditioner. Each nodal block consists of a 3×3 and 1×1 matrices, corresponding to the discrete momentum and continuity equations, respectively. The number of nonlinear iterations per time step is 4 and the number of GMRES iterations is 200 for the first nonlinear iteration, 300 for the second, and 400 for the third and fourth. Figure 10.5 shows the air speed at $t = 0.8$ s at

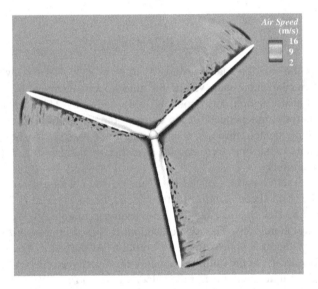

Figure 10.5 Air speed at $t = 0.8$ s

1 m behind the rotor plane. Note the fine-grained turbulent features at the trailing edge of the blade, which require enhanced mesh resolution for accurate representation. The fluid-traction vectors projected to the plane of rotation are shown in Figure 10.6. The traction vectors point in the direction of rotation and grow in magnitude toward the blade tip, creating favorable aerodynamic torque. However, at the blade tip the traction vectors rapidly decay to zero and even change sign, which introduces a small amount of inefficiency. The time history of the aerodynamic torque is shown in Figure 10.7, where the steady-state result from Jonkman *et al.* (2009) is also shown for reference. The figure shows that in less than 0.8 s the torque settles at a statistically-stationary value of 2670 kN m, which is within 6.4% of the reference. Given the significant differences in the computational modeling approaches, the two values

Wind-Turbine Aerodynamics and FSI

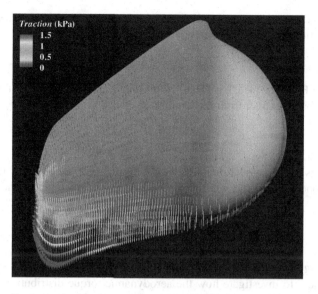

Figure 10.6 Fluid traction vectors at $t = 0.8$ s viewed from the back of the blade. The fluid-traction vectors, colored by magnitude, are projected to the rotor plane and illustrate the mechanism by which the favorable aerodynamic torque is created

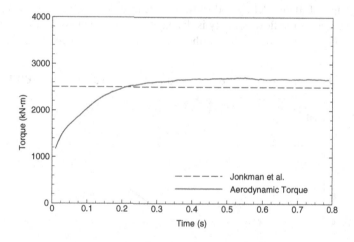

Figure 10.7 Time history of the aerodynamic torque. Statistically-stationary torque is attained in less than 0.8 s. The reference steady-state result from NREL is also shown for comparison

are remarkably close. This result is encouraging in that 3D time-dependent simulation with a manageable number of degrees-of-freedom and without any empiricism is able to predict important quantities of interest for wind-turbine rotors simulated at full scale. This result also gives us confidence that our procedures are accurate and may be applied to simulations cases where 3D, time-dependent modeling is indispensable (e.g., simulation of wind gusts or blade pitching).

Given the aerodynamic torque and the rotor speed, the power extracted from the wind with these wind conditions (based on our aerodynamic torque T_f) is

$$P = T_f \dot{\theta} \approx 2.88 \text{ MW}. \tag{10.3}$$

According to the Betz law (see, e.g., Hau, 2006), the maximum power that a horizontal-axis wind turbine is able to extract from the wind is

$$P_{\max} = \frac{16}{27} \frac{\rho A \|\mathbf{u}_{in}\|^3}{2} \approx 3.23 \text{ MW}, \tag{10.4}$$

where $A = \pi R^2$ is the cross-sectional area swept by the rotor, and $\|\mathbf{u}_{in}\|$ is the inflow speed. From this we conclude that the wind-turbine aerodynamic efficiency at the simulated wind conditions is

$$\frac{P}{P_{\max}} \approx 89\%, \tag{10.5}$$

which is quite high even for modern wind-turbine designs. The blade is segmented into 18 spanwise "patches" to investigate how the aerodynamic torque distribution varies along the blade span. The patch-wise torque distribution is shown in Figure 10.8. The torque is nearly zero in the cylindrical section of the blade. A favorable aerodynamic torque is created on Patch 4 and its magnitude continues to increase until Patch 15. The torque magnitude decreases rapidly after Patch 15, however, the torque remains favorable all the way to the last patch.

The importance of 3D modeling and simulation is further illustrated in Figure 10.9, where the axial component of the flow velocity is displayed at a blade cross-section located at 56 m above the rotor center. The magnitude of the axial velocity component exceeds 15 m/s in the

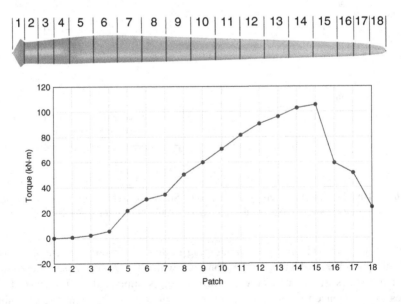

Figure 10.8 Patches along the blade (top) and the aerodynamic torque contribution from each patch (bottom) at $t = 0.8$ s

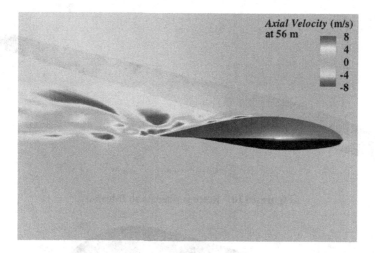

Figure 10.9 Axial flow velocity over the blade cross-section at 56 m at $t = 0.8$ s. The level of axial flow in the boundary layer is significant, which illustrates the importance of 3D modeling

boundary layer, showing that 3D effects are important, especially in the regions of the blade with the largest contribution to the aerodynamic torque.

10.1.3 Computations with the DSD/SST Formulation Using Finite Elements

We describe from Takizawa et al. (2011a,b) the computations with the DSD/SST formulation and linear finite elements. To generate the triangular mesh on the rotor surface, we started with a quadrilateral surface mesh generated by interpolating the NURBS geometry of the rotor at each knot intersection. We subdivided each quadrilateral element into triangles and then made minor modifications to improve the mesh quality near the hub. We use three different meshes: Mesh-2, Mesh-3 and Mesh-4, with the surface mesh refined along the blade 2, 3 and 4 times, respectively, compared to the finite element mesh used in Bazilevs et al. (2011b). The number of nodes and elements for each blade surface mesh is shown in Table 10.2, and Figure 10.10 shows the surface mesh for Mesh-4. For computational efficiency, rotational-periodicity (Takizawa et al., 2011d,f) is utilized so that the domain includes only one of three blades, as shown in Figure 10.11. The inflow, outflow and radial boundaries lie $0.5R$, $2R$, and

Table 10.2 Summary of the meshes, where nn and ne are the number of nodes and elements

	Surface		Volume	
	nn	ne	nn	ne
Mesh-2	5748	11 452	155 494	898 640
Mesh-3	7552	15 060	205 855	1 195 452
Mesh-4	9268	18 492	253 340	1 475 175

Figure 10.10 Rotor surface mesh (Mesh-4)

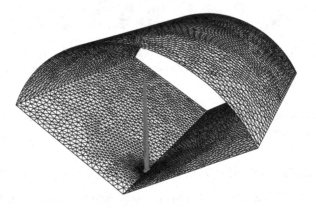

Figure 10.11 Rotationally-periodic domain with wind-turbine blade shown in blue

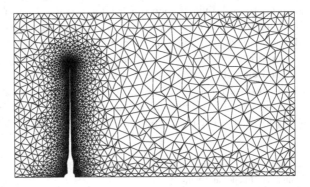

Figure 10.12 Cut plane of the fluid volume mesh along rotor axis (Mesh-4)

$1.43R$ from the hub center, respectively. This can be more easily seen in Figure 10.12, where the inflow, outflow, and radial boundaries are the left, right and top edges, respectively, of the cut plane along the rotation axis. Each periodic boundary contains 1430 nodes and 2697 triangles. Near the rotor surface, we have 22 layers of refined mesh with first-layer thickness of

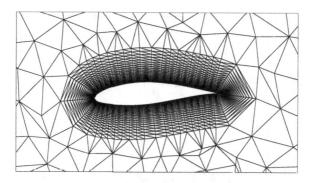

Figure 10.13 Boundary layer mesh at $\frac{3}{4}R$

1 cm and a progression factor of 1.1. The boundary layer mesh at $\frac{3}{4}R$ is shown in Figure 10.13. The number of nodes and elements for each volume mesh is shown in Table 10.2.

We compute the problem with the DSD/SST-SUPS and the conservative form of DSD/SST-VMST. The SUPS version is used without the LSIC stabilization (i.e., $\nu_{LSIC} = 0$), while for the VMST version ν_{LSIC} is defined according to Equation (4.126), and is referred to as "TGI."

In solving the linear equation systems involved at every nonlinear iteration, the GMRES search technique (Saad and Schultz, 1986) is used with a diagonal preconditioner. The computation is carried out in a parallel computing environment. The mesh is partitioned to enhance the parallel efficiency of the computations. Mesh partitioning is based on the METIS algorithm (Karypis and Kumar, 1998). The time-step size is 4.67×10^{-4} s. The number of nonlinear iterations per time step is 3 with 30, 60, and 500 GMRES iterations for the first, second, and third nonlinear iterations, respectively.

Prior to the computations reported here, we performed a series of brief computations with the DSD/SST-SUPS technique, starting from a lower Reynolds number and gradually reaching the actual Reynolds number. This solution is used as the initial condition also for the computations with the DSD/SST-VMST technique. The purpose is to generate a divergence-free and reasonable flow field at this Reynolds number. We note that it was especially difficult with the VMST option to start from nonphysical conditions, such as setting all nodes except those on the blade to the inflow velocity.

Figures 10.14–10.16 show the time history of the aerodynamic torque and the torque contribution from each patch for a single blade at $t = 1.0$ s. The patches are defined as shown in Figure 10.8. Figure 10.17 shows the pressure coefficient at $t = 1.0$ s for Patch 16 (at $0.90R$), which is a representative section of the blade. For most of the patches, the angle of attack and Reynolds number do not vary much from one patch to another. For example, the angle of attack and Reynolds number are $7.4°$ and 9.9×10^6 at $0.65R$ for Patch 12 (at $0.65R$) and $7.6°$ and 9.6×10^6 for Patch 16 (at $0.90R$).

Mesh refinement studies for both the SUPS and VMST versions indicate good convergence in the quantities of interest such as the aerodynamic torque and pressure coefficient. The VMST version on the finest mesh gives more or less the same value of the aerodynamic torque as the ALE-VMS simulation using NURBS, which is taken as a reference solution for this study. The results from the SUPS version are also very good, however the torque is slightly underpredicted with respect to the VMST and NURBS-based ALE-VMS simulations.

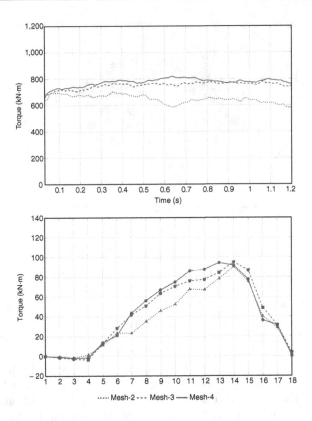

Figure 10.14 The aerodynamic torque generated by a single blade. Comparison between different meshes with the DSD/SST-SUPS technique. Time history (top). The torque contribution from each patch at $t = 1.0$ s (bottom)

Figure 10.17 indicates that smoother (i.e., more stable) pressure solution is obtained with the VMST version. We note that the main reason behind the higher VMST torque is the wider low-pressure region on the upper surface of the NACA64 geometry, as can be seen in the figure. The lower pressure indicates that the flow is attached; thus, the VMST version, for the level of mesh refinement presented here, is able to better represent the turbulent boundary layer solution than the SUPS version.

10.2 NREL Phase VI Wind-Turbine Rotor: Validation and the Role of Weakly-Enforced Essential Boundary Conditions

The proposed method is applied to predict the aerodynamics of the Unsteady Aerodynamics Experiment (UAE) Phase VI two-bladed wind-turbine rotor (Hand *et al.*, 2001) from NREL. In this experiment, a two-bladed twisted and tapered 10.058 m diameter wind turbine, which has a rated power of 19.8 kW, was tested in the NASA Ames 80 ft × 120 ft wind tunnel in 2000 (see Figure 10.18). This is one of the most comprehensive, accurate, and reliable experiments carried out on a full-scale wind turbine. This test case was also studied by many computational

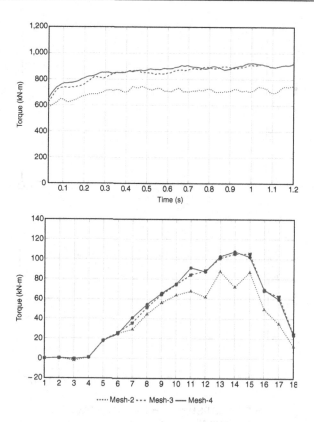

Figure 10.15 The aerodynamic torque generated by a single blade. Comparison between different meshes with the DSD/SST-VMST (TGI) technique. Time history (top). The torque contribution from each patch at $t = 1.0$ s (bottom)

researchers (Sørensen et al., 2002; Laino et al., 2002; Pape and Lecanu, 2004; Tongchitpakdee et al., 2005; Schmitz and Chattot, 2006; Zahle et al., 2009) for the purposes of validating their simulations and improving their ability to predict wind-turbine aerodynamic performance.

The Phase VI rotor geometry makes use of a single NREL S809 airfoil (Hand et al., 2001). Selected blade cross-section geometry data are summarized in Table 10.3. The detailed documentation of the rotor configuration and its technical specifications are available in Hand et al. (2001). Two cases from the experiment were selected for the validation study. The first case has a wind speed of 5 m/s and the second case 25 m/s. For both cases, we have upwind configuration, 0° yaw angle, 0° cone angle, blade tip pitch angle of 3°, and rotational speed of 72 rpm. The influence of the hub and tower on the rotor aerodynamics was neglected, which is a fair approximation for an upwind turbine (see, e.g., Sørensen et al., 2002). The two cases we consider here present very different flow conditions. For the 5 m/s case the flow is fully attached for the entire blade. On the contrary, the stall occurs in most part of the blade for the 25 m/s case and the simulation is considered to be more challenging (Li and K. J. Paik, 2012).

The mesh resolution and computational domain are shown in Figure 10.19. The rotor radius R is 5.029 m and the blade is assumed to be rigid. At the inflow boundary the wind speed is set to either 5 m/s or 25 m/s. At the outflow boundary the traction vector is set to zero. At the radial

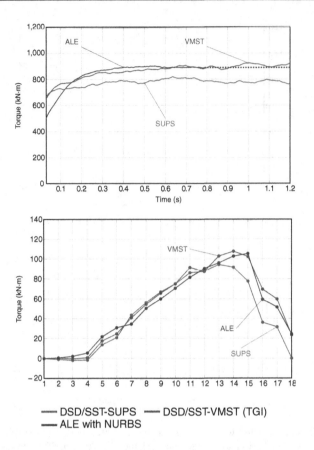

Figure 10.16 The aerodynamic torque generated by a single blade. Computed with different techniques using Mesh-4. Time history (top). The torque contribution from each patch at $t = 1.0$ s (bottom). We note that the curve labeled "ALE with NURBS" represents the aerodynamic torque data from the NURBS computation presented in the previous section

boundary the radial component of the velocity is set to zero. The air density and viscosity are 1.23 kg/m^3 and 1.78×10^{-5} kg/m s, respectively. The mesh is comprised of 1 508 983 nodes and 8 494 182 linear tetrahedral elements. Figure 10.20 shows a 2D blade cross-section at 0.8R to illustrate the type of mesh near the boundary. Near the blade surface at 0.8R, the size of the first element in the wall-normal direction is about 0.008 m. No special boundary layer meshing was used in this study, in part to test the ability of the ALE-VMS method to deal with coarse boundary layer meshes.

The computations were carried out in a parallel computing environment on a Dell Cluster at the Texas Advanced Computing Center (TACC), TACC. The linear system is solved using a block-diagonal preconditioned GMRES method (Saad and Schultz, 1986; Shakib et al., 1989). The time-step size is 0.0001 s. The number of nonlinear iterations per time step is 3, with 50 GMRES iterations for the first and second nonlinear iterations, and 50–80 GMRES iterations for the third nonlinear iteration, depending on the nonlinear convergence.

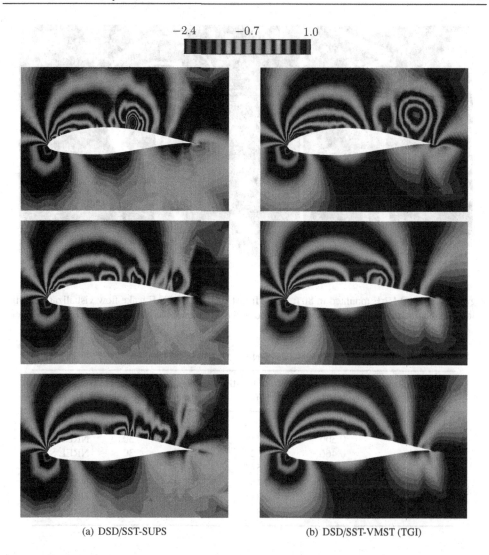

(a) DSD/SST-SUPS (b) DSD/SST-VMST (TGI)

Figure 10.17 Pressure coefficient at $t = 1.0$ s for Patch 16 (at $0.90R$). Top: Mesh-2. Middle: Mesh-3. Bottom: Mesh-4

The time history of aerodynamic (low-speed shaft) torque is shown in Figure 10.21. Good agreement in the aerodynamic torque is found between the weak-boundary-condition simulations and experimental data for both flow conditions. However, the results for the strongly-enforced boundary condition are not at all accurate.

Pressure, air speed contours, and streamlines at $0.8R$ for the 5 m/s and 25 m/s cases are shown in Figures 10.22 and 10.23, respectively. Figure 10.22a shows the weak-boundary-condition prediction of the air flow for the 5 m/s case. The flow is fully attached, and the torque is correctly predicted. However, the strong-boundary-condition simulation predicts flow separation at the trailing edge (see Figure 10.22b). The blade stalls and, as a result, the torque

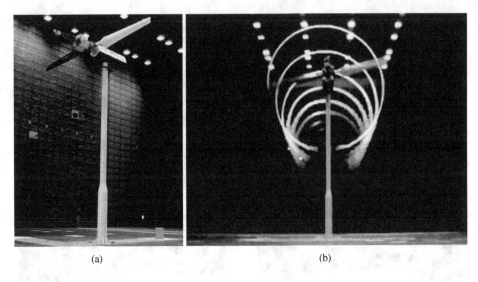

Figure 10.18 (a) UAE mounted in 80 ft × 120 ft test section. (b) UAE wake flow visualization in the field at the National Wind Technology Center and in the wind tunnel at NASA Ames Research Center

Table 10.3 Selected cross-sectional data for NREL UAE Phase VI blade

Radial distance r (m)	Span station ($r/5.029$ m)	Chord length (m)	Twist (degrees)	Twist axis (% chord)	Airfoil (-)
0.508	0.100	0.218	0.0	50	Cylinder
1.510	0.300	0.711	14.292	30	NREL S809
2.343	0.466	0.627	4.715	30	NREL S809
3.185	0.633	0.542	1.115	30	NREL S809
4.023	0.800	0.457	−0.381	30	NREL S809
4.780	0.950	0.381	−1.469	30	NREL S809
5.029	1.000	0.355	−1.815	30	NREL S809

is underpredicted by 126% (see Figure 10.21a). For the 25 m/s case, small differences are found in the pressure contours and air flow patterns between the weak- and strong-boundary-condition computations. This is due to the fact that the flow is already separated at the edges, the entire airfoil is stalled, and the boundary layer resolution is not so important for these flow conditions. In this case, the weak boundary condition again correctly predicts the torque, while the strong boundary condition underpredicts the torque, but only by 11% (see Figure 10.21b).

These results are not surprising. For strongly-enforced boundary conditions, the coarse boundary layer discretization gives rise to artificially "thick" boundary layers, which retard the flow and lead to nonphysical aerodynamics, such as premature flow separation. In the case of weakly-enforced boundary conditions, the flow is allowed to slip on the solid surface without forming these undesired thick boundary layers. Of course, with sufficient boundary layer mesh refinement, both approaches will capture the boundary layer, and the strongly-enforced

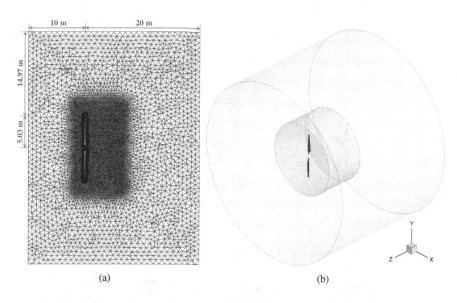

Figure 10.19 A 2D cut of the mesh (left) and the computational domain (right)

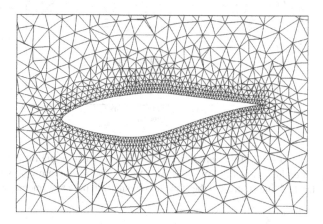

Figure 10.20 A 2D blade cross-section at 0.8R illustrating the boundary layer mesh used in the computations, which is relatively coarse. The size of the first element in the wall-normal direction is about 0.008 m

boundary condition will also produce the correct result (see Bazilevs *et al.*, 2007c). Figure 10.24 shows the pressure coefficient at $0.466R$, $0.633R$, and $0.8R$ for the 5 m/s and 25 m/s cases. The predicted values (using the weak boundary condition) are plotted against the experimental data. Very good agreement is likewise achieved for both attached and separated flow conditions at different radial locations.

Figure 10.25 shows the flow visualization (isosurfaces of air speed) for the 5 m/s case. The tip vortex generated by the blade is carried downstream of the rotor with little decay. The figure also shows the pressure contours on the rotor surface.

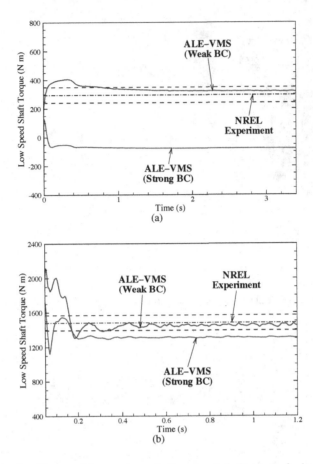

Figure 10.21 The time history of the aerodynamic (low-speed shaft) torque for both weak and strong boundary condition simulations at (a) 5 m/s and (b) 25 m/s. The results are compared to the NREL experimental data. Dashed lines represent the standard deviation in the experimental data

10.3 Structural Mechanics of Wind-Turbine Blades

10.3.1 The Bending-Strip Method

We employ the composite Kirchhoff–Love shell formulation in Section 1.2.9.1 to model the structural mechanics of wind-turbine blades. Composite materials are typically used in the manufacturing of modern wind-turbine blades. Due to the smoothness requirements on the basis functions, NURBS-based IGA is used to discretize the variational equations of the Kirchhoff–Love shell. However, the expression of the internal virtual work given by Equation (1.191) is only meaningful when the shell midsurface is described using a smooth geometrical mapping. In the case when the regularity of the mapping reduces to the C^0 level, the terms involving the curvature tensors, which rely on the second derivatives of the geometrical mapping, lead to non-integrable singularities, and the formulation may not be used as is. However, for complex structures, the geometry definition often requires that the

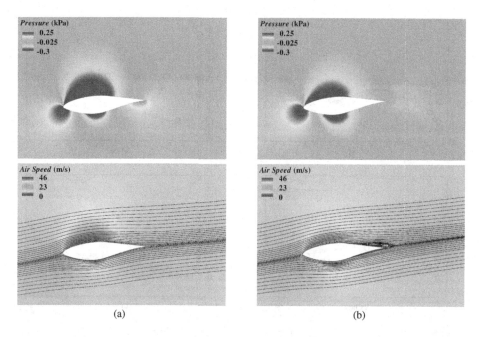

Figure 10.22 Pressure, air speed contours, and streamlines at $0.8R$ for the 5 m/s case. (a) Weakly-enforced boundary condition. (b) Strongly-enforced boundary condition

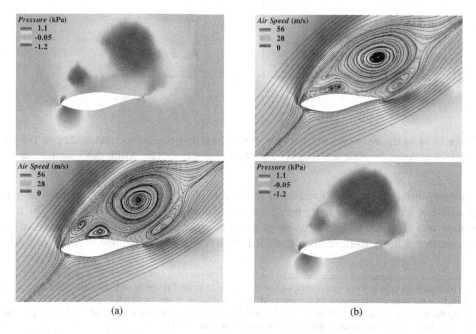

Figure 10.23 Pressure, air speed contours, and streamlines at $0.8R$ for the 25 m/s case. (a) Weakly-enforced boundary condition. (b) Strongly-enforced boundary condition

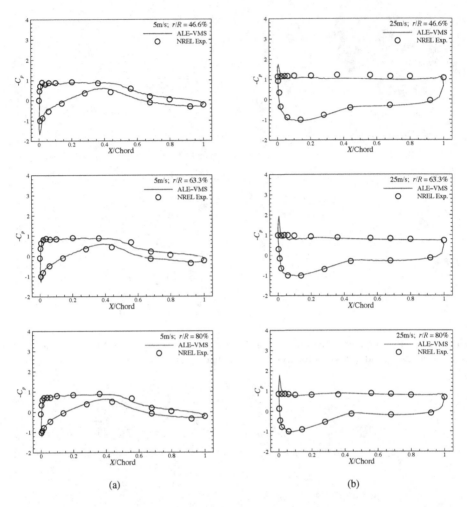

Figure 10.24 Pressure coefficients at $0.466R$, $0.633R$, and $0.8R$ for (a) 5 m/s and (b) 25 m/s. The predicted values (using the weak boundary condition) are plotted against the NREL experimental data

continuity of the geometrical mapping is reduced to the C^0 level (think of a trailing edge of an airfoil or an I-beam, the latter being a non-manifold surface). In Kiendl *et al.* (2010), a method was proposed to handle complex multi-patch shell structures in the context of the rotation-free Kirchhoff–Love theory, called the "bending-strip method.[1]" The main idea

---

[1] The method in its current form was developed and implemented at the University of California, San Diego, when J. Kiendl, at the time a PhD student in the group of K.-U. Bletzinger at the Technical University of Munich, was visiting the research group of Y. Bazilevs. The method has similarities with the concept of "continuity patches," introduced by K.-U. Bletzinger and collaborators in Bletzinger *et al.* (1991).

Wind-Turbine Aerodynamics and FSI

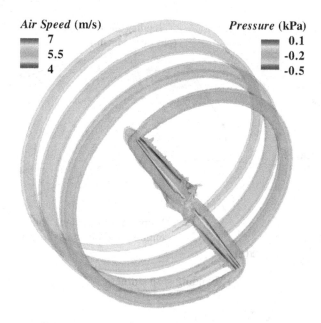

Figure 10.25 Isosurfaces of air speed at an instant for the 5 m/s case. The tip vortex generated by the blade is carried downstream of the rotor with little decay. The figure also shows the pressure contours on the rotor surface

behind the method, illustrated in Figure 10.26, consists of the following. It is assumed that the shell structure is comprised of smooth subdomains, such as NURBS patches, that are joined with C^0-continuity. In addition, thin strips of fictitious material, also modeled as surface NURBS patches, are placed at patch intersections. The triples of control points at the patch interface, consisting of a shared control point and one on each side, are extracted and used as a control mesh for the bending strips. The parametric domain of each bending strip consists of one quadratic element in the direction transverse to the interface and, for simplicity and computational efficiency, of as many linear elements as necessary to accommodate all the control points along the length of the strip. The material is assumed to have zero mass, zero membrane stiffness, and *nonzero bending stiffness only in the direction transverse to the interface*. The transverse direction may be obtained using the local basis construction given by Equations (1.168) and (1.169), however, other options may be explored.

Recall that Γ_0^s and Γ_t^s denote the structure midsurface in the reference and deformed configurations, respectively. The structure midsurface is composed of surface patches joined in a C^0-continuous fashion. Let Γ_0^b denote the bending-strip domain, which is a collection of the bending-strip patch subdomains. Let $\mathcal{S}_y^h$ and $\mathcal{V}_y^h$ denote the discrete trial and test function spaces for the structural mechanics problem. We take the variational equation of the Kirchhoff–Love shell given by Equation (1.192) in Section 1.2.9.1 as the starting point, and formulate the bending-strip method as: find the shell midsurface displacement $\mathbf{y}^h \in \mathcal{S}_y^h$, such

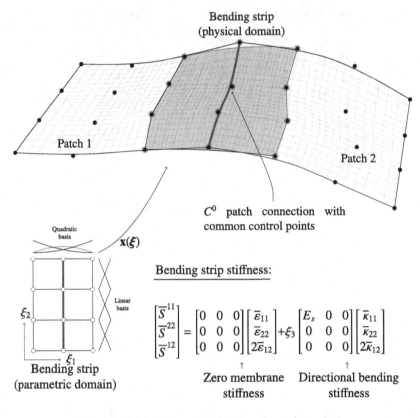

Figure 10.26 Schematic of the bending-strip method

that $\forall \, \mathbf{w}_2^h \in \mathcal{V}_y^h$:

$$\int_{\Gamma_0^s} \mathbf{w}_2^h \cdot \overline{\rho}_0 h_{\text{th}} \left(\frac{d^2 \mathbf{y}^h}{dt^2} - \mathbf{f}^h \right) d\Gamma$$

$$+ \int_{\Gamma_0^s} \delta \overline{\boldsymbol{\varepsilon}}^h \cdot \left(\mathbf{K}_{\text{exte}} \overline{\boldsymbol{\varepsilon}}^h + \mathbf{K}_{\text{coup}} \overline{\boldsymbol{\kappa}}^h \right) d\Gamma$$

$$+ \int_{\Gamma_0^s} \delta \overline{\boldsymbol{\kappa}}^h \cdot \left(\mathbf{K}_{\text{coup}} \overline{\boldsymbol{\varepsilon}}^h + \mathbf{K}_{\text{bend}} \overline{\boldsymbol{\kappa}}^h \right) d\Gamma$$

$$+ \int_{\Gamma_0^b} \delta \overline{\boldsymbol{\kappa}}^h \cdot \mathbf{K}_{\text{bstr}} \overline{\boldsymbol{\kappa}}^h \, d\Gamma - \int_{(\Gamma_t^s)_h} \mathbf{w}_2^h \cdot \mathbf{h}^h \, d\Gamma = 0. \quad (10.6)$$

In the above formulation, the superscript h denotes the discrete nature of the quantities involved. The term $\int_{\Gamma_0^b} \delta \overline{\boldsymbol{\kappa}}^h \cdot \mathbf{K}_{\text{bstr}} \overline{\boldsymbol{\kappa}}^h \, d\Gamma$, which is not present in Equation (1.192), is penalty-like, and represents the contribution of the bending strips to the structural formulation. Here,

$\mathbf{K}_{bstr}$ is the bending stiffness of the strips:

$$\mathbf{K}_{bstr} = \frac{h_{th}^3}{12}\overline{\mathbb{C}}_{bstr},\qquad(10.7)$$

where

$$\overline{\mathbb{C}}_{bstr} = \begin{bmatrix} E_s & 0 & 0 \\ 0 & 0 & 0 \\ 0 & 0 & 0 \end{bmatrix},\qquad(10.8)$$

and E_s is the scalar bending-strip stiffness, typically chosen as a multiple of the local Young's modulus of the shell. This design of the material constitutive matrix ensures that the bending strips add no extra stiffness to the structure. They only penalize the change in the angle during the deformation between the triples of control points at the patch interface. The stiffness E_s must be high enough so that the change in angle is within an acceptable tolerance. However, if E_s is chosen too high, the global stiffness matrix becomes badly conditioned, which may lead to divergence in the computations.

The following illustrates the use of the bending-strip method applied to the construction of the structural model of the wind-turbine blade. A symmetric fiberglass/epoxy composite with $[\pm 45/0/90_2/0_3]_s$ lay-up, which enhances flap-wise and edge-wise stiffness, is considered for the rotor-blade material. The 0° fiber points in the direction of a tangent vector to the airfoil cross-section curve. The orthotropic elastic moduli for each ply are given in Table 10.4. For simplicity, the entire blade is assumed to have the same lay-up. The resulting $\mathbf{K}_{exte}$, $\mathbf{K}_{coup}$, and $\mathbf{K}_{bend}$ matrices from Equations (1.184)–(1.186) are

$$\mathbf{K}_{exte} = h_{th}\begin{bmatrix} 26.315 & 4.221 & 0 \\ 4.221 & 18.581 & 0 \\ 0 & 0 & 5.571 \end{bmatrix}\times 10^9 \text{ (N/m)},\qquad(10.9)$$

$$\mathbf{K}_{coup} = \mathbf{0},\qquad(10.10)$$

$$\mathbf{K}_{bend} = h_{th}^3\begin{bmatrix} 1.727 & 0.545 & 0.053 \\ 0.545 & 1.627 & 0.053 \\ 0.053 & 0.053 & 0.658 \end{bmatrix}\times 10^9 \text{ (N m)}.\qquad(10.11)$$

Table 10.4 Material properties of a unidirectional E-glass/epoxy composite

E_1 (GPa)	E_2 (GPa)	G_{12} (GPa)	ν_{12}	$\bar{\rho}_0$ (g/cm^3)
39	8.6	3.8	0.28	2.1

The total laminate thickness distribution is shown in Figure 10.27a. The blade shell model together with the bending strips covering the regions of C^0-continuity is shown in Figure 10.27b. The FSI simulations with this blade model will be presented in the sequel. Further examples of calculations using the bending-strip method, which include verification and validation, and the possibility of using the approach for simple coupling of solids and shells, may be found in Kiendl *et al.* (2010).

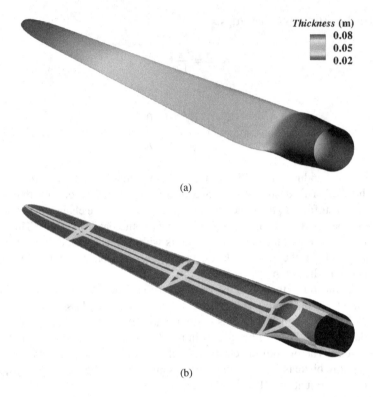

Figure 10.27 NREL 5MW offshore baseline wind-turbine rotor-blade model. (a) Blade thickness. (b) Bending strips

REMARK 10.2 *Because of the structure of the bending-strip term in Equation (10.6), the method may be interpreted as a physically-motivated penalty formulation.*

REMARK 10.3 *In IGA, the possibility to employ smooth surface descriptions directly in analysis has led to the development of new shell element formulations. Besides the references cited in this section, the reader is referred to Cirak et al. (2000, 2002), Cirak and Ortiz (2001) and Benson et al. (2010b, 2011) for relevant work on shells. We would also like to note that references Cirak et al. (2000, 2002) and Cirak and Ortiz (2001) predate the development of IGA.*

10.3.2 Time Integration of the Structural Mechanics Equations

In the case of wind-turbine rotors, the structural motions are dominated by the rotation of the blades around the hub axis. In Bazilevs *et al.* (2011c), the authors proposed to take advantage of this fact and modify a class of standard time integration techniques to exactly account for the rotational part of the structural motion.

For this, as a first step, it is useful to decompose the structural displacement **y** into its rotation and deflection components as

$$\mathbf{y} = \mathbf{y}_\theta + \mathbf{y}_d. \tag{10.12}$$

Wind-Turbine Aerodynamics and FSI

The rotational component of the displacement may be computed as

$$\mathbf{y}_\theta = (\mathbf{R}(\theta) - \mathbf{I})(\mathbf{X} - \mathbf{X}_0), \tag{10.13}$$

where $\mathbf{X}$ are the coordinates of the structure reference configuration, $\mathbf{X}_0$ is a fixed point, θ is the time-dependent angle of rotation, $\mathbf{R}(\theta)$ is the rotation matrix, and $\mathbf{I}$ is the identity matrix. We specialize to the case of rotation about the x_3-axis, which gives

$$\mathbf{R}(\theta) = \begin{bmatrix} \cos\theta & -\sin\theta & 0 \\ \sin\theta & \cos\theta & 0 \\ 0 & 0 & 1 \end{bmatrix}. \tag{10.14}$$

The total structural velocity and acceleration may be computed as

$$\frac{d\mathbf{y}}{dt} = \dot{\mathbf{y}} = \dot{\mathbf{y}}_\theta + \dot{\mathbf{y}}_d = \dot{\mathbf{R}}(\theta)(\mathbf{X} - \mathbf{X}_0) + \dot{\mathbf{y}}_d, \tag{10.15}$$

$$\frac{d^2\mathbf{y}}{dt^2} = \ddot{\mathbf{y}} = \ddot{\mathbf{y}}_\theta + \ddot{\mathbf{y}}_d = \ddot{\mathbf{R}}(\theta)(\mathbf{X} - \mathbf{X}_0) + \ddot{\mathbf{y}}_d, \tag{10.16}$$

where

$$\dot{\mathbf{R}}(\theta) = \begin{bmatrix} -\sin\theta & -\cos\theta & 0 \\ \cos\theta & -\sin\theta & 0 \\ 0 & 0 & 0 \end{bmatrix} \dot{\theta}, \tag{10.17}$$

$$\ddot{\mathbf{R}}(\theta) = \begin{bmatrix} -\cos\theta & \sin\theta & 0 \\ -\sin\theta & -\cos\theta & 0 \\ 0 & 0 & 0 \end{bmatrix} \dot{\theta}^2 + \begin{bmatrix} -\sin\theta & -\cos\theta & 0 \\ \cos\theta & -\sin\theta & 0 \\ 0 & 0 & 0 \end{bmatrix} \ddot{\theta}. \tag{10.18}$$

We repeat this decomposition at the discrete level, where we operate directly on the nodal or control-point displacement degrees-of-freedom. For this, we let $\mathbf{Y}$, $\dot{\mathbf{Y}}$, and $\ddot{\mathbf{Y}}$ be the vectors of nodal or control-point displacements, velocities, and accelerations, respectively. We set

$$\mathbf{Y} = \mathbf{Y}_\theta + \mathbf{Y}_d, \tag{10.19}$$

$$\dot{\mathbf{Y}} = \dot{\mathbf{Y}}_\theta + \dot{\mathbf{Y}}_d, \tag{10.20}$$

$$\ddot{\mathbf{Y}} = \ddot{\mathbf{Y}}_\theta + \ddot{\mathbf{Y}}_d, \tag{10.21}$$

where $\mathbf{Y}_\theta$, $\dot{\mathbf{Y}}_\theta$, and $\ddot{\mathbf{Y}}_\theta$ are given by

$$\mathbf{Y}_\theta = (\mathbf{R}(\theta) - \mathbf{I})(\mathbf{X} - \mathbf{X}_0), \tag{10.22}$$

$$\dot{\mathbf{Y}}_\theta = \dot{\mathbf{R}}(\theta)(\mathbf{X} - \mathbf{X}_0), \tag{10.23}$$

$$\ddot{\mathbf{Y}}_\theta = \ddot{\mathbf{R}}(\theta)(\mathbf{X} - \mathbf{X}_0). \tag{10.24}$$

The above Equations (10.22)–(10.24) present an exact relationship between the nodal or control point displacements, velocities, and accelerations corresponding to the rotational motion. To relate the deflection degrees-of-freedom between time levels t_n and t_{n+1}, we make use of the standard Newmark formulas (see e.g., Hughes, 2000):

$$(\dot{\mathbf{Y}}_d)_{n+1} = (\dot{\mathbf{Y}}_d)_n + \Delta t\left((1-\gamma)(\ddot{\mathbf{Y}}_d)_n + \gamma(\ddot{\mathbf{Y}}_d)_{n+1}\right), \tag{10.25}$$

$$(\mathbf{Y}_d)_{n+1} = (\mathbf{Y}_d)_n + \Delta t(\dot{\mathbf{Y}}_d)_n + \frac{\Delta t^2}{2}\left((1-2\beta)(\ddot{\mathbf{Y}}_d)_n + 2\beta(\ddot{\mathbf{Y}}_d)_{n+1}\right), \tag{10.26}$$

where γ and β are the time integration parameters chosen to maintain second-order accuracy and unconditional stability of the method.

Combining exact rotations given by Equations (10.22)–(10.24) and time-discrete deflections given by Equations (10.25) and (10.26), we obtain the following modified Newmark formulas for the total discrete solution:

$$\dot{\mathbf{Y}}_{n+1} = \left(\dot{\mathbf{R}}_{n+1} - \left(\dot{\mathbf{R}}_n + \Delta t\left((1-\gamma)\ddot{\mathbf{R}}_n + \gamma\ddot{\mathbf{R}}_{n+1}\right)\right)\right)(\mathbf{X} - \mathbf{X}_0)$$
$$+ \dot{\mathbf{Y}}_n + \Delta t\left((1-\gamma)\ddot{\mathbf{Y}}_n + \gamma\ddot{\mathbf{Y}}_{n+1}\right), \quad (10.27)$$

$$\mathbf{Y}_{n+1} = \left(\mathbf{R}_{n+1} - \left(\mathbf{R}_n + \Delta t\dot{\mathbf{R}}_n + \frac{\Delta t^2}{2}\left((1-2\beta)\ddot{\mathbf{R}}_n + 2\beta\ddot{\mathbf{R}}_{n+1}\right)\right)\right)(\mathbf{X} - \mathbf{X}_0)$$
$$+ \mathbf{Y}_n + \Delta t\dot{\mathbf{Y}}_n + \frac{\Delta t^2}{2}\left((1-2\beta)\ddot{\mathbf{Y}}_n + 2\beta\ddot{\mathbf{Y}}_{n+1}\right). \quad (10.28)$$

We employ Equations (10.27) and (10.28), in conjunction with the generalized-α method, for the time discretization of the structure.

REMARK 10.4 *In the case of no rotation, for which $\mathbf{R}$ is an identity, Equations (10.27) and (10.28) reduce to the standard Newmark formulas. In the case of no deflection, the rigid-body rotation is likewise recovered.*

10.4 FSI Coupling and Aerodynamics Mesh Update

In this section we briefly summarize our FSI coupling procedures for wind-turbine simulations. The fluid and structural equations are integrated in time using the generalized-α method. In the case of the structure, the modified Newmark formulas given by Equations (10.27) and (10.28) are employed to enhance the accuracy of the time integration procedures in the presence of large rotation. Within each time step, the coupled equations are solved using an inexact Newton approach. For every Newton iteration the following steps are performed. 1. We obtain the fluid solution increment holding the structure, and mesh fixed. 2. We update the fluid solution, compute the aerodynamic force on the structure, and compute the structural solution increment. The aerodynamic force at control points or nodes is computed using the conservative definition (see, e.g., Melbø and Kvamsdal, 2003; van Brummelen et al., 2012 for the importance of using the conservative definitions of fluxes near essential boundaries and in coupled problems). 3. We update the structural solution and use elastic mesh moving to update the fluid-domain velocity and position. We note that only the deflection part of the mesh motion is computed using linear elastostatics, while the rotation part is computed exactly. This three-step iteration is repeated until convergence to an appropriately coupled discrete solution is achieved. This block-iterative coupling (see Section 6.1.1) is stable because the wind-turbine blades are relatively heavy structures.

We conclude this section with the discussion of a special technique we devised in Bazilevs et al. (2011c) to update the kinematics (position and velocity) of the fluid mesh. Typically, one employs the equations of linear elastostatics subject to dynamic boundary conditions coming from the structural displacement to update the position and velocity of the fluid mesh (see Section 4.7). In the case of wind turbines, which are dominated by rotation, this may not be a preferred procedure due to the fact that the linear elastostatics operator does not vanish on large rotational motions. This, in turn, may lead to the loss of the fluid-mesh quality if one plans to simulate the FSI problem for many revolutions of the wind-turbine rotor. As a result, for the present application, we modify our mesh moving strategy as follows. We take advantage of the fact that the structural displacement vector is already decomposed into the rotation and deflection parts. As a result, as the increment of the structural displacement is computed, we extract the deflection part, apply the elasticity-based mesh-moving method (see Section 4.7) to computation of just the deflection part of the mesh displacement, rotate the (deformed) mesh from the previous time level to the current time, and add the mesh deflection increment to obtain its current position. For a precise mathematical formulation of this procedure see (Bazilevs et al., 2011c).

REMARK 10.5 *For a variety of other mesh update strategies the reader is referred to Section 5.4 and Johnson and Tezduyar (1994), Tezduyar (2001b) and Takizawa et al. (2012d).*

10.5 FSI Simulations of a 5MW Wind-Turbine Rotor

The wind-turbine rotor is simulated at prescribed steady inlet wind velocity of 11.4 m/s and rotor angular velocity of 12.1 rpm. This setup corresponds to one of the cases reported in Jonkman et al. (2009). The dimensions of the problem domain and the NURBS mesh employed are the same as in Section 10.1. The time-step size is 0.0003 s. The structural model of the wind-turbine blade is described in Section 10.3. A rigid rotor under the same wind and rotor speed conditions is simulated for comparison.

Rotational-periodicity conditions for the fluid are imposed in order to reduce the computational cost (see Figure 10.4). However, because the rotor blades are subject to gravity forces, a fully rotationally-periodic structural solution is not expected in this case. Nevertheless, we feel that the use of rotational-periodicity conditions for the fluid problem is justified due to the fact that the periodic boundaries are located sufficiently far away from the structure and are not expected to influence the structural response.

Isosurfaces of air speed at an instant during the simulation are shown in Figure 10.28. Note that, for visualization purposes the rotationally-periodic 120° domain was merged into a full 360° domain. Fine-grained turbulent structures are generated at the trailing edge of the blade along its entire length. The vortex forming at the tip of the blades is advected downstream of the rotor with little decay. There is also high-intensity turbulence in the blade aerodynamic zone, which is a segment of the blade where the cylindrical root rapidly transitions to a thin airfoil shape. This suggests that the blade trailing edge in this location is subjected to high-frequency loads that are fatiguing the blade.

Figure 10.29 shows the isocontours of relative wind speed at a 30 m radial cut at different instants during the simulation. For every snapshot the blade is rotated to the reference configuration to better illustrate the deflection part of the motion. The blade deflection is quite significant. On the pressure side, the boundary layer is attached to the blade for the entire cord length. On the suction side, the flow detaches near the trailing edge and transitions to turbulence. The aerodynamic torque (for a single blade) is plotted in Figure 10.30 for both rigid- and flexible-blade simulations. Both cases compare favorably to the data reported in Jonkman et al. (2009) for this setup obtained using FAST Jonkman and Buhl Jr. (2005). Note that the aerodynamic torque for the flexible blade exhibits low-magnitude, high-frequency oscillations, while the rigid-blade torque is smooth. To better understand this behavior, we examine the twisting motion of the wind-turbine blade about its axis. Time histories of the twist angle

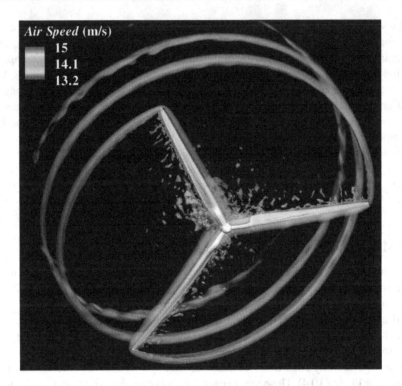

Figure 10.28 Isosurfaces of air speed at an instant during the simulation, showing that the flow exhibits complex behavior. The tip vortex is advected downstream of the rotor with very little decay

at four different cross-sections are shown in Figure 10.31. The twist angle increases with distance from the root and reaches almost 2° near the tip in the early stages of the simulation. However, starting at $t = 1.2$ s, when the blade tip reaches its lowest vertical position, the magnitude of the twist angle is reduced significantly. The reversal of the gravity vector with respect to the lift direction clearly affects the edge-wise bending and twisting behavior of the blade. The blade twist angle undergoes high frequency oscillations, which are in part driven by the trailing-edge vortex shedding and turbulence. Local oscillations in the twist angle lead to the temporal fluctuations in the aerodynamic torque.

10.6 Pre-Bending of the Wind-Turbine Blades

The rotor blades of a wind turbine need to be designed such that they do not strike the tower as the rotor turns in strong winds. This may be accomplished with blade pre-bending. In this case, the blades are manufactured to flex toward the wind when the rotor is mounted on the tower. Once the blades are exposed to the wind, and the rotor starts turning, the blades are straightened to achieve their design shape. This situation is graphically illustrated in Figure 10.32. Besides tower clearance, pre-bending of the blades engenders additional benefits. For example, the blades need not be quite as rigid because the amount of allowable deflection is greater. This makes it possible to use less material overall, and fewer processed materials, resulting in

Wind-Turbine Aerodynamics and FSI

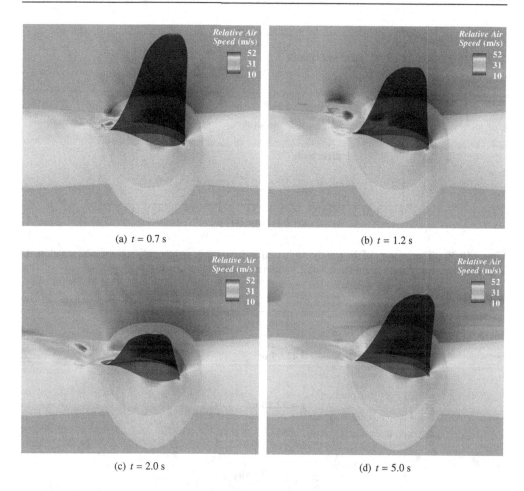

Figure 10.29 Isocontours of relative wind speed at a 30 m radial cut at different instants superposed on the moving blade. The air flow is fully attached on the pressure side of the blade and separates on the suction side. The flow separation point varies as the blade moves under the action of the wind, inertial, and gravitational forces

lighter and more economical blades. Pre-bending of the blades also results in a more compact nacelle design. During operation, the pre-bent blades straighten to their designed configuration, which is typically optimized for best possible aerodynamic performance.

Given the above advantages, it is important that one is able to determine the correct pre-bent shape given the blade structural and aerodynamic design, and the wind-turbine operating conditions (i.e., wind and rotor speeds). In Bazilevs *et al.* (2012a), we proposed a method that makes use of stand alone computational fluid and structural mechanics procedures to obtain a pre-bent shape of the wind-turbine blades. The main idea consists of performing an aerodynamics simulation of a rigidly-rotating rotor to obtain the aerodynamic load acting on the blade. Given the aerodynamic and inertial loads acting in the design configuration, a stress-free pre-bent blade configuration is found using a simple iterative procedure that

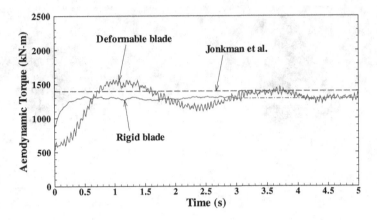

Figure 10.30 Time history of the aerodynamic torque. Both rigid- and flexible-rotor results are plotted. The reference steady-state result from NREL is also shown for comparison

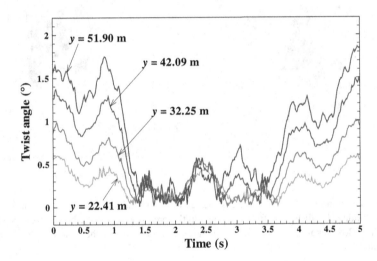

Figure 10.31 Time histories of the twist angle at four cross-sections along the blade axis

requires a sequence of structural mechanics simulations. Note that in the proposed approach the aerodynamic and structural computations are decoupled, which avoids the challenges involved in solving the coupled FSI problem. In this section, we summarize the method and show the supporting computations. We follow the developments of Bazilevs *et al.* (2012a).

10.6.1 Problem Statement and the Pre-Bending Algorithm

We begin with the statement of virtual work for the structure from Equation (10.6), where only the stress terms are left on the left-hand-side: find the displacement of the shell midsurface $\mathbf{y}^h \in \mathcal{S}_y^h$, such that for $\forall \, \mathbf{w}_2^h \in \mathcal{V}_y^h$:

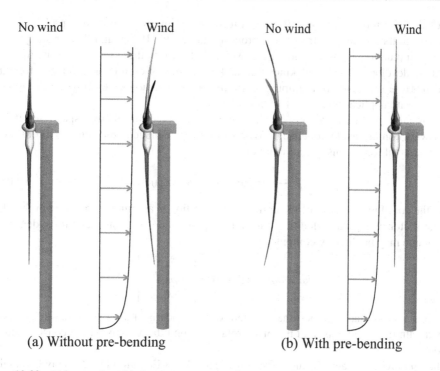

(a) Without pre-bending   (b) With pre-bending

Figure 10.32 Using pre-bent blades to ensure tower clearance and rotor operation in its design configuration

$$\int_{\Gamma_0^s} \delta \overline{\boldsymbol{\varepsilon}}^h \cdot \left(\mathbf{K}_{\text{exte}} \overline{\boldsymbol{\varepsilon}}^h + \mathbf{K}_{\text{coup}} \overline{\boldsymbol{\kappa}}^h \right) d\Gamma$$

$$+ \int_{\Gamma_0^s} \delta \overline{\boldsymbol{\kappa}}^h \cdot \left(\mathbf{K}_{\text{coup}} \overline{\boldsymbol{\varepsilon}}^h + \mathbf{K}_{\text{bend}} \overline{\boldsymbol{\kappa}}^h \right) d\Gamma$$

$$+ \int_{\Gamma_0^b} \delta \overline{\boldsymbol{\kappa}}^h \cdot \mathbf{K}_{\text{bstr}} \overline{\boldsymbol{\kappa}}^h \, d\Gamma = - \int_{\Gamma_t^s} \mathbf{w}_2^h \cdot \overline{\rho} h_{\text{th}} \left(\frac{d^2 \mathbf{y}^h}{dt^2} - \mathbf{f}^h \right) d\Gamma + \int_{(\Gamma_t^s)_h} \mathbf{w}_2^h \cdot \mathbf{h}^h \, d\Gamma, \quad (10.29)$$

where $\overline{\rho}$ is the shell density averaged through the thickness in the current configuration, which is related to its undeformed configuration counterpart as

$$\overline{\rho}_0 = \overline{J} \, \overline{\rho}. \tag{10.30}$$

In Equation (10.30) it is assumed that the shell thickness remains unchanged through the deformation, and $\overline{J}$, the surface transformation Jacobian, is given by

$$\overline{J} = \frac{\|\mathbf{g}_1 \times \mathbf{g}_2\|}{\|\mathbf{G}_1 \times \mathbf{G}_2\|}, \tag{10.31}$$

where $\mathbf{g}$'s and $\mathbf{G}$'s are the surface basis vectors defined in Equations (1.163) and (1.164), respectively.

Although the virtual work equations hold true, the problem setup is unusual in that the stress-free reference configuration Γ_0^s is unknown and the final configuration Γ_t^s is given. The formulation given by Equation (10.29) is a form of the inverse deformation problem, whose general formulation and treatment was proposed in Shield (1967), and further analyzed and studied computationally in Govindjee and Mihalic (1996). In these references, the focus was placed on developing the right kinematic and stress measures for the inverse deformation problem. Here we develop a simple algorithm for the solution of the inverse deformation equations with application to wind-turbine blades.

We assume that the rotor spins around its axis at a constant angular speed and that the inflow wind conditions do not change. With this setup, the blade is subjected to a constant-in-time centripetal force density (per unit volume) given by

$$\bar{\rho}\frac{d^2 y^h}{dt^2} = \bar{\rho}\boldsymbol{\omega} \times (\boldsymbol{\omega} \times (\mathbf{x} - \mathbf{x}_0)), \tag{10.32}$$

where the coordinate system of the current configuration is assumed to rotate with the blade, $\boldsymbol{\omega}$ is the vector of angular velocities, and $\mathbf{x}_0$ is a fixed point. The centripetal force density per unit volume may be directly computed as

$$\bar{\rho}\boldsymbol{\omega} \times (\boldsymbol{\omega} \times (\mathbf{x} - \mathbf{x}_0)) = \begin{bmatrix} -\bar{\rho} x_1 \dot{\theta}^2 \\ -\bar{\rho} x_2 \dot{\theta}^2 \\ 0 \end{bmatrix}, \tag{10.33}$$

where we assume that the coordinate system of the blade is chosen such that the x_2-axis is aligned with the blade axis, and the blade rotates around the x_3-axis with a constant angular speed $\dot{\theta}$.

The time-averaged aerodynamics traction vector $\mathbf{h}^h$ in Equation (10.29) may be obtained from a separate aerodynamics computation of a rigidly-spinning rotor using the methods described in the earlier sections (see Section 8.1.4 for computation of time-averaged stress vectors in rotational coordinate frames).

In Bazilevs *et al.* (2012a), we proposed the following two-stage iterative approach to solve Equation (10.29) for the shell midsurface displacement, which, in turn, gives the stress-free reference configuration.

Initialization: We initialize the unknown reference configuration to coincide with the current configuration, that is,

$$\Gamma_0^s = \Gamma_t^s, \tag{10.34}$$

which implies

$$\mathbf{y}^h = \mathbf{0}. \tag{10.35}$$

Step 1: Given the reference configuration Γ_0^s, we solve the standard nonlinear structural problem: find the structural displacement $\mathbf{y}^h \in \mathcal{S}_y^h$ relative to Γ_0^s, such that $\forall\, \mathbf{w}_2^h \in \mathcal{V}_y^h$:

$$\int_{\Gamma_0^s} \delta\bar{\boldsymbol{\varepsilon}}^h \cdot \left(\mathbf{K}_{\text{exte}}\bar{\boldsymbol{\varepsilon}}^h + \mathbf{K}_{\text{coup}}\bar{\boldsymbol{\kappa}}^h\right) d\Gamma$$

$$+ \int_{\Gamma_0^s} \delta\bar{\boldsymbol{\kappa}}^h \cdot \left(\mathbf{K}_{\text{coup}}\bar{\boldsymbol{\varepsilon}}^h + \mathbf{K}_{\text{bend}}\bar{\boldsymbol{\kappa}}^h\right) d\Gamma$$

$$+ \int_{\Gamma_0^s} \delta\bar{\boldsymbol{\kappa}}^h \cdot \mathbf{K}_{\text{bstr}}\bar{\boldsymbol{\kappa}}^h\, d\Gamma = -\int_{\Gamma_t^s} \mathbf{w}_2^h \cdot (\bar{\rho} h_{\text{th}} \boldsymbol{\omega} \times (\boldsymbol{\omega} \times (\mathbf{x} - \mathbf{x}_0)))\, d\Gamma + \int_{(\Gamma_t^s)_h} \mathbf{w}_2^h \cdot \mathbf{h}^h\, d\Gamma. \tag{10.36}$$

Standard Newton–Raphson iteration is employed in this work to compute the solution of the nonlinear structural problem given by Equation (10.36).

Step 2: Given $\mathbf{y}^h$ from Step 1, we update the reference configuration as

$$\Gamma_0^s = \{\mathbf{X} \mid \mathbf{X} = \mathbf{x} - \mathbf{y}^h,\ \forall \mathbf{x} \in \Gamma_t^s\}, \tag{10.37}$$

and return to Step 1 using $\mathbf{y}^h$ as the initial data.

Steps 1–2 are repeated until convergence, that is, until $\mathbf{y}^h$ satisfies Equation (10.36).

The above algorithm is based on the idea of computing negative increments of the displacement, or increments of the displacement away from the current configuration, until the reference configuration is found. The mathematical justification for this approach may be found in the appendix of Bazilevs *et al.* (2012a). In what follows, we will illustrate the good performance of the proposed algorithm on a full-scale wind-turbine blade subject to realistic wind and inertial loads.

10.6.2 Pre-Bending Results for the NREL 5MW Wind-Turbine Blade

The same blade design and wind conditions as in the previous section are taken for the pre-bending computations presented here. Figure 10.33 shows the tip displacement convergence of the iterative pre-bending algorithm. After a few (5–6) iterations of the two-step pre-bending algorithm the tip exhibits no further visible displacements, and the computation is stopped after a total of 15 iterations. Figure 10.34 shows the initial and the final stress-free blade shapes. As expected, the blade bends into the wind. The tip deflection is predicted to be 5.61 m.

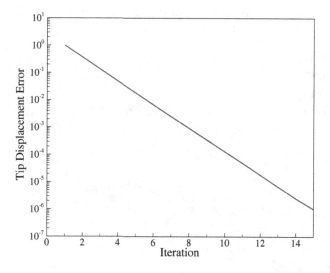

Figure 10.33 Blade tip displacement convergence as a function of the iteration number. The error is normalized by the magnitude of the tip displacement

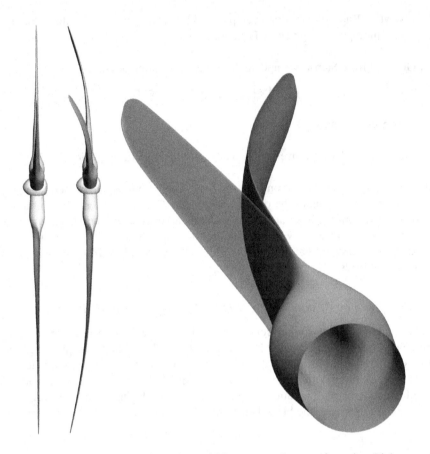

Figure 10.34 Left: rotor design configuration. Middle: rotor pre-bent configuration. Right: rotor-blade design and pre-bent configurations superposed

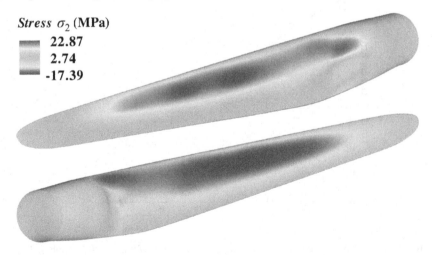

Figure 10.35 Normal stress distribution in the direction of the matrix material for the 0° ply number 14. Top: pressure side. Bottom: suction side

We examine the stress distribution in the composite plies of the blade skin. For each ply we compute the Cauchy stress tensor components with respect to the local Cartesian basis that is aligned with the material axes. The first basis vector points in the direction of the fiber and the second in the direction of the matrix, which is orthogonal to the fiber direction (see Equations (1.168) and (1.169)). The maximum values of the tensile (σ^t), compressive (σ^c), and in-plane shear stresses are computed for each ply. The highest ratio of the predicted Cauchy stress and the composite strength, that is, $\sigma_2^t/\sigma_2^{t,u}$, occurs for the tensile stress in the direction of the matrix material. Although the ratio does not exceed 0.6, which means the predicted stress is below the composite failure strength, we feel this value is somewhat high. In the rest of the stress components the ratios are significantly lower. Figure 10.35 shows the distribution of σ_2 in the 0° ply number 14, which has the highest ratio of $\sigma_2^t/\sigma_2^{t,u}$. The pressure side of the blade is in tension, while the suction side of the blade is in compression, as expected. However, the level of the tensile stress is not very far from the tensile failure strength, which suggests that stronger matrix material may be desirable for this blade design.

References

Akin, J.E., Tezduyar, T., Ungor, M., and Mittal, S. (2003). Stabilization parameters and Smagorinsky turbulence model. *Journal of Applied Mechanics*, 70:2–9.

Akin, J.E. and Tezduyar, T.E. (2004). Calculation of the advective limit of the SUPG stabilization parameter for linear and higher-order elements. *Computer Methods in Applied Mechanics and Engineering*, 193:1909–1922.

Akin, J.E., Tezduyar, T.E., and Ungor, M. (2007). Computation of flow problems with the mixed interface-tracking/interface-capturing technique (MITICT). *Computers & Fluids*, 36:2–11.

Akkerman, I., Bazilevs, Y., Benson, D.J., Farthing, M.W., and Kees, C.E. (2012). Free-surface flow and fluid–object interaction modeling with emphasis on ship hydrodynamics. *Journal of Applied Mechanics*, 79:010905.

Akkerman, I., Bazilevs, Y., Calo, V.M., Hughes, T.J.R., and Hulshoff, S. (2008). The role of continuity in residual-based variational multiscale modeling of turbulence. *Computational Mechanics*, 41:371–378.

Akkerman, I., Bazilevs, Y., Kees, C.E., and Farthing, M.W. (2011). Isogeometric analysis of free-surface flow. *Journal of Computational Physics*, 230:4137–4152.

Arnold, D.N., Brezzi, F., Cockburn, B., and Marini, L.D. (2002). Unified analysis of Discontinuous Galerkin methods for elliptic problems. *SIAM Journal of Numerical Analysis*, 39:1749–1779.

Bajaj, C., Schaefer, S., Warren, J., and Xu, G. (2002). A subdivision scheme for hexahedral meshes. *Visual Computer*, 18:343–356.

Bazilevs, Y. and Akkerman, I. (2010). Large eddy simulation of turbulent Taylor–Couette flow using isogeometric analysis and the residual–based variational multiscale method. *Journal of Computational Physics*, 229:3402–3414.

Bazilevs, Y., Calo, V.M., Cottrell, J.A., Hughes, T.J.R., Reali, A., and Scovazzi, G. (2007a). Variational multiscale residual-based turbulence modeling for large eddy simulation of incompressible flows. *Computer Methods in Applied Mechanics and Engineering*, 197:173–201.

Bazilevs, Y., Calo, V.M., Hughes, T.J.R., and Zhang, Y. (2008). Isogeometric fluid–structure interaction: theory, algorithms, and computations. *Computational Mechanics*, 43:3–37.

Bazilevs, Y., Calo, V.M., Cottrell, J.A., Evans, J.A., Hughes, T.J.R., Lipton, S., Scott, M.A., and Sederberg, T.W. (2010a). Isogeometric analysis using T-splines. *Computer Methods in Applied Mechanics and Engineering*, 199:229–263.

Computational Fluid–Structure Interaction: Methods and Applications, First Edition.
Yuri Bazilevs, Kenji Takizawa and Tayfun E. Tezduyar.
© 2013 John Wiley & Sons, Ltd. Published 2013 by John Wiley & Sons, Ltd.

Bazilevs, Y., Calo, V.M., Tezduyar, T.E., and Hughes, T.J.R. (2007b). YZβ discontinuity-capturing for advection-dominated processes with application to arterial drug delivery. *International Journal for Numerical Methods in Fluids*, 54:593–608.

Bazilevs, Y., Calo, V.M., Zhang, Y., and Hughes, T.J.R. (2006a). Isogeometric fluid–structure interaction analysis with applications to arterial blood flow. *Computational Mechanics*, 38:310–322.

Bazilevs, Y., da Veiga, L.B., Cottrell, J.A., Hughes, T.J.R., and Sangalli, G. (2006b). Isogeometric analysis: Approximation, stability and error estimates for h-refined meshes. *Mathematical Models and Methods in Applied Sciences*, 16:1031–1090.

Bazilevs, Y., del Alamo, J.C., and Humphrey, J.D. (2010b). From imaging to prediction: Emerging non-invasive methods in pediatric cardiology. *Progress in Pediatric Cardiology*, 30:81–89.

Bazilevs, Y., Gohean, J.R., Hughes, T.J.R., Moser, R.D., and Zhang, Y. (2009a). Patient-specific isogeometric fluid–structure interaction analysis of thoracic aortic blood flow due to implantation of the Jarvik 2000 left ventricular assist device. *Computer Methods in Applied Mechanics and Engineering*, 198:3534–3550.

Bazilevs, Y., Hsu, M.-C., Akkerman, I., Wright, S., Takizawa, K., Henicke, B., Spielman, T., and Tezduyar, T.E. (2011b). 3D simulation of wind turbine rotors at full scale. Part I: Geometry modeling and aerodynamics. *International Journal for Numerical Methods in Fluids*, 65:207–235.

Bazilevs, Y., Hsu, M.-C., Benson, D., Sankaran, S., and Marsden, A. (2009b). Computational fluid–structure interaction: Methods and application to a total cavopulmonary connection. *Computational Mechanics*, 45:77–89.

Bazilevs, Y., Hsu, M.-C., Kiendl, J., and Benson, D.J. (2012a). A computational procedure for pre-bending of wind turbine blades. *International Journal for Numerical Methods in Engineering*, 89:323–336.

Bazilevs, Y., Hsu, M.-C., Kiendl, J., Wüchner, R., and Bletzinger, K.-U. (2011c). 3D simulation of wind turbine rotors at full scale. Part II: Fluid–structure interaction modeling with composite blades. *International Journal for Numerical Methods in Fluids*, 65: 236–253.

Bazilevs, Y., Hsu, M.-C., Takizawa, K., and Tezduyar, T.E. (2012b). ALE-VMS and ST-VMS methods for computer modeling of wind-turbine rotor aerodynamics and fluid–structure interaction. *Mathematical Models and Methods in Applied Sciences*, 22(supp02): 1230002.

Bazilevs, Y., Hsu, M.-C., Zhang, Y., Wang, W., Kvamsdal, T., Hentschel, S., and Isaksen, J. (2010c). Computational fluid–structure interaction: Methods and application to cerebral aneurysms. *Biomechanics and Modeling in Mechanobiology*, 9:481–498.

Bazilevs, Y., Hsu, M.-C., Zhang, Y., Wang, W., Liang, X., Kvamsdal, T., Brekken, R., and Isaksen, J. (2010d). A fully-coupled fluid–structure interaction simulation of cerebral aneurysms. *Computational Mechanics*, 46:3–16.

Bazilevs, Y. and Hughes, T.J.R. (2007). Weak imposition of Dirichlet boundary conditions in fluid mechanics. *Computers and Fluids*, 36:12–26.

Bazilevs, Y. and Hughes, T.J.R. (2008). NURBS-based isogeometric analysis for the computation of flows about rotating components. *Computational Mechanics*, 43:143–150.

References

Bazilevs, Y., Michler, C., Calo, V.M., and Hughes, T.J.R. (2007c). Weak Dirichlet boundary conditions for wall-bounded turbulent flows. *Computer Methods in Applied Mechanics and Engineering*, 196:4853–4862.

Bazilevs, Y., Michler, C., Calo, V.M., and Hughes, T.J.R. (2010e). Isogeometric variational multiscale modeling of wall-bounded turbulent flows with weakly enforced boundary conditions on unstretched meshes. *Computer Methods in Applied Mechanics and Engineering*, 199:780–790.

Behr, M. and Tezduyar, T. (1999). The Shear-Slip Mesh Update Method. *Computer Methods in Applied Mechanics and Engineering*, 174:261–274.

Behr, M. and Tezduyar, T. (2001). Shear-slip mesh update in 3D computation of complex flow problems with rotating mechanical components. *Computer Methods in Applied Mechanics and Engineering*, 190:3189–3200.

Belytschko, T., Liu, W.K., and Moran, B. (2000). *Nonlinear Finite Elements for Continua and Structures*. Wiley.

Belytschko, T., Lu, Y., and Gu, L. (1994). Element-free galerkin methods. *International Journal for Numerical Methods in Engineering*, 37:229–256.

Benson, D.J., Bazilevs, Y., De Luycker, E., Hsu, M.C., Scott, M., Hughes, T.J.R., and Belytschko, T. (2010a). A generalized finite element formulation for arbitrary basis functions: from isogeometric analysis to XFEM. *International Journal for Numerical Methods in Engineering*, 83:765–785.

Benson, D.J., Bazilevs, Y., Hsu, M.C., and Hughes, T.J.R. (2010b). Isogeometric shell analysis: The Reissner–Mindlin shell. *Computer Methods in Applied Mechanics and Engineering*, 199:276–289.

Benson, D.J., Bazilevs, Y., Hsu, M.C., and Hughes, T.J.R. (2011). A large deformation, rotation-free, isogeometric shell. *Computer Methods in Applied Mechanics and Engineering*, 200:1367–1378.

Betsch, P., Gruttmann, F., and Stein, E. (1996). A 4-node finite shell element for the implementation of general hyperelastic 3d-elasticity at finite strains. *Computer Methods in Applied Mechanics and Engineering*, 130:57–79.

Bischoff, M., Wall, W.A., Bletzinger, K.U., and Ramm, E. (2004). Models and finite elements for thin-walled structures. In Stein, E., de Borst, R., and Hughes, T.J.R., editors, *Encyclopedia of Computational Mechanics, Vol. 2, Solids, Structures and Coupled Problems*, chapter 3. Wiley.

Bletzinger, K.U., Kimmich, S., and Ramm, E. (1991). Efficient modeling in shape optimal design. *Computing Systems in Engineering*, 2:483–495.

Bove, E., de Leval, M., Migliavacca, F., Guadagni, G., and Dubini, G. (2003). Computational fluid dynamics in the evaluation of hemodynamic performance of cavopulmonary connections after the Norwood procedure for hypoplastic left heart syndrome. *Journal of Thoracic and Cardiovascular Surgery*, 126:1040–1047.

Brenner, S.C. and Scott, L.R. (2002). *The Mathematical Theory of Finite Element Methods, 2nd ed.* Springer.

Brooks, A.N. and Hughes, T.J.R. (1982). Streamline upwind/Petrov-Galerkin formulations for convection dominated flows with particular emphasis on the incompressible Navier-Stokes equations. *Computer Methods in Applied Mechanics and Engineering*, 32:199–259.

Calo, V., Brasher, N., Bazilevs, Y., and Hughes, T. (2008). Multiphysics model for blood flow and drug transport with application to patient-specific coronary artery flow. *Computational Mechanics*, 43:161–177.

Catabriga, L., Coutinho, A.L.G.A., and Tezduyar, T.E. (2005). Compressible flow SUPG parameters computed from element matrices. *Communications in Numerical Methods in Engineering*, 21:465–476.

Catabriga, L., Coutinho, A.L.G.A., and Tezduyar, T.E. (2006). Compressible flow SUPG parameters computed from degree-of-freedom submatrices. *Computational Mechanics*, 38:334–343.

Chung, J. and Hulbert, G.M. (1993). A time integration algorithm for structural dynamics withimproved numerical dissipation: The generalized-α method. *Journal of Applied Mechanics*, 60:371–75.

Cirak, F. and Ortiz, M. (2001). Fully C^1-conforming subdivision elements for finite deformation thin shell analysis. *International Journal for Numerical Methods in Engineering*, 51:813–833.

Cirak, F., Ortiz, M., and Schröder, P. (2000). Subdivision surfaces: a new paradigm for thin shell analysis. *International Journal for Numerical Methods in Engineering*, 47:2039–2072.

Cirak, F., Scott, M.J., Antonsson, E.K., Ortiz, M., and Schröder, P. (2002). Integrated modeling, finite-element analysis, and engineering design for thin-shell structures using subdivision. *Computer-Aided Design*, 34:137–148.

Codina, R., Principe, J., Guasch, O., and Badia, S. (2007). Time dependent subscales in the stabilized finite element approximation of incompressible flow problems. *Computer Methods in Applied Mechanics and Engineering*, 196:2413–2430.

Corsini, A., Iossa, C., Rispoli, F., and Tezduyar, T.E. (2010). A DRD finite element formulation for computing turbulent reacting flows in gas turbine combustors. *Computational Mechanics*, 46:159–167.

Corsini, A., Rispoli, F., Santoriello, A., and Tezduyar, T.E. (2006). Improved discontinuity-capturing finite element techniques for reaction effects in turbulence computation. *Computational Mechanics*, 38:356–364.

Corsini, A., Rispoli, F., and Tezduyar, T.E. (2011). Stabilized finite element computation of NOx emission in aero-engine combustors. *International Journal for Numerical Methods in Fluids*, 65:254–270.

Cottrell, J.A., Hughes, T.J.R., and Bazilevs, Y. (2009). *Isogeometric Analysis: Toward Integration of CAD and FEA*. Wiley, Chichester.

Cottrell, J.A., Hughes, T.J.R., and Reali, A. (2007). Studies of refinement and continuity in isogeometric structural analysis. *Computer Methods in Applied Mechanics and Engineering*, 196:4160–4183.

Cox, M. (1971). The numerical evaluation of B-splines. Technical report, National Physics Laboratory DNAC 4.

Cruchaga, M.A., Celentano, D.J., and Tezduyar, T.E. (2007). A numerical model based on the Mixed Interface-Tracking/Interface-Capturing Technique (MITICT) for flows with fluid–solid and fluid–fluid interfaces. *International Journal for Numerical Methods in Fluids*, 54:1021–1030.

da Veiga, L.B., Cho, D., and Sangalli, G. (2012). Anisotropic NURBS approximation in isogeometric analysis. *Computer Methods in Applied Mechanics and Engineering*, 209–212:1–11.

Dörfel, M.R., Jüttler, B., and Simeon, B. (2010). Adaptive isogeometric analysis by local h-refinement with T-splines. *Computer Methods in Applied Mechanics and Engineering*, 199:264–275.

de Boor, C. (1972). On calculation with B-splines. *Journal of Approximation Theory*, 6:50–62.

de Leval, M.R., Dubini, G., Migliavacca, F., Jalali, H., camporini, G., Redington, A., and Pietrabissa, R. (1996). Use of computational fluid dynamics in the design of surgical procedures: application to the study of competitive flows in cavo-pulmonary connections. *Journal of Thoracic and Cardiovascular Surgery*, 111(3):502–13.

Dettmer, W.G. and Peric, D. (2006). A computational framework for fluid-structure interaction: Finite element formulation and applications. *Computer Methods in Applied Mechanics and Engineering*, 195:5754–5779.

Dettmer, W.G. and Peric, D. (2008). On the coupling between fluid flow and mesh motion in the modelling of fluid–structure interaction. *Computational Mechanics*, 43:81–90.

Dubini, G., de Leval, M.R., Pietrabissa, R., Montevecchi, F.M., and Fumero, R. (1996). A numerical fluid mechanical study of repaired congenital heart defects: Application to the total cavopulmonary connection. *Journal of Biomechanics*, 29(1):111–121.

Elguedj, T., Bazilevs, Y., Calo, V.M., and Hughes, T.J.R. (2008). B-bar and F-bar projection methods for nearly incompressible linear and nonlinear elasticity and plasticity using higher-order nurbs elements. *Computer Methods in Applied Mechanics and Engineering*, 197:2732–2762.

Ensley, A., Ramuzat, A., Healy, T., Chatzimavroudis, G., Lucas, C., Sharma, S., Pettigrew, R., and Yoganathan, A. (2000). Fluid mechanic assessment of the total cavopulmonary connection using magnetic resonance phase velocity mapping and digital particle image velocimetry. *Annals of Biomedical Engineering*, 28:1172–1183.

Ern, A. and Guermond, J.L. (2004). *Theory and Practice of Finite Elements*. Springer.

Farhat, C., Geuzaine, P., and Grandmont, C. (2001). The discrete geometric conservation law and the nonlinear stability of ALE schemes for the solution of flow problems on moving grids. *Journal of Computational Physics*, 174(2):669–694.

Farin, G.E. (1995). *NURBS Curves and Surfaces: From Projective Geometry to Practical Use*. A. K. Peters, Ltd., Natick, MA.

Fernandez, M.A. and Moubachir, M. (2005). A Newton method using exact Jacobians for solving fluid–structure coupling. *Computers and Structures*, 83:127–142.

Fontan, F. and Baudet, E. (1971). Surgical repair of tricuspid atresia. *Thorax*, 26:240–248.

Formaggia, L., Gerbeau, J.F., Nobile, F., and Quarteroni, A. (2001). On the coupling of 3D and 1D Navier-Stokes equations for flow problems in compliant vessels. *Computer Methods in Applied Mechanics and Engineering*, 191:561–582.

Frank, O. (1899). Die grundform des arteriellen pulses. *Zeitung fur Biologie*, 37:483–586.

Fujisawa, T., Inaba, M., and Yagawa, G. (2003). Parallel computing of high-speed compressible flows using a node-based finite element method. *International Journal for Numerical Methods in Fluids*, 58:481–511.

Gerbeau, J.-F., Vidrascu, M., and Frey, P. (2005). Fluid–structure interaction in blood flows on geometries based on medical imaging. *Computers and Structures*, 83:155–165.

Giardini, A., Balducci, A., and Specchia, S.t. (2008). Effect of sildenafil on haemodynamic response to exercise capacity in fontan patients. *European Heart Journal*, 29:1681–1687.

Glagov, S., Zarins, C., Giddens, D.P., and Ku, D.N. (1988). Hemodynamics and atherosclerosis: insights and perspectives gained from studies of human arteries. *Archives of Pathology and Laboratory Medicine*, 112:1018–1031.

Gohean, J.R. (2007). A closed-loop multi-scale model of the cardiovascular system for evaluation of ventricular devices. Master's thesis, University of Texas, Austin.

Govindjee, S. and Mihalic, P.A. (1996). Computational methods for inverse finite elastostatics. *Computer Methods in Applied Mechanics and Engineering*, 136:47–57.

Green, A.E. and Naghdi, P.M. (1976). A derivation of equations for wave propagation in water of variable depth. *Journal of Fluid Mechanics*, 78:237–26.

Gresho, P.M. and Sani, R.L. (2000). *Incompressible Flow and the Finite Element Method*. Wiley, New York, NY.

Guler, I., Behr, M., and Tezduyar, T. (1999). Parallel finite element computation of free-surface flows. *Computational Mechanics*, 23:117–123.

Guttierez, F., Primi, S., Taucer, F., Caperan, P., Tirelli, D., Mieres, J., Calvo, I., Rodriguez, J., Vallano, F., Galiotis, G., and Mouzakis, D. (2003). A wind turbine tower design based on fibre-reinforced composites. Technical report, Joint Research Centre - Ispra, European Laboratory for Structural Assessment (ELSA), Institute For Protection and Security of the Citizen (IPSC), European Commission.

Hand, M.M., Simms, D.A., Fingersh, L.J., Jager, D.W., Cotrell, J.R., Schreck, S., and Larwood, S.M. (2001). Unsteady aerodynamics experiment phase VI: Wind tunnel test configurations and available data campaigns. Technical Report NREL/TP-500-29955, National Renewable Energy Laboratory, Golden, CO.

Hansen, M.O.L., Sørensen, J.N., N. Sørensen, S.V., and Madsen, H.A. (2006). State of the art in wind turbine aerodynamics and aeroelasticity. *Progress in Aerospace Sciences*, 42:285–330.

Hau, E. (2006). *Wind Turbines: Fundamentals, Technologies, Application, Economics. 2nd Edition*. Springer, Berlin.

Hetzer, R., Jurmann, M.J., Potapov, E.V., Hennig, E., Stiller, B., Muller, J.H., and Weng, Y. (2002). Heart assist systems: current status. *Hertz*, 20:407.

Hilber, H.M., Hughes, T.J.R., and Taylor, R.L. (1977). Improved numerical dissipation for time integration algorithms in structural dynamics. *Earthquake Engineering and Structural Dynamics*, 5:283–292.

Hjortdal, V.E., Emmertsen, K., Stenbog, E., Frund, T., Rahbek Schmidt, M., Kromann, O., Sorensen, K., and Pedersen, E.M. (2003). Effects of exercise and respiration on blood flow in total cavopulmonary connection: A real-time magnetic resonance flow study. *Circulation*, 108:1227–1231.

Hoerner, S.F. (1993). *Fluid Dynamic Drag*. Hoerner Fluid Dynamics.

Holzapfel, G.A. (2000). *Nonlinear Solid Mechanics, a Continuum Approach for Engineering*. Wiley, Chichester.

Holzapfel, G. and Ogden, R. (2010). Constitutive modelling of arteries. *Proceedings of The Royal Society A*, 466:1551–1596.

Hsu, M.-C., Akkerman, I., and Bazilevs, Y. (2012). Wind turbine aerodynamics using ALE–VMS: Validation and the role of weakly enforced boundary conditions. *Computational Mechanics*, 50:499–511.

Hsu, M.-C., Akkerman, I., and Bazilevs, Y. (2011). High-performance computing of wind turbine aerodynamics using isogeometric analysis. *Computers and Fluids*, 49:93–100.

Hsu, M.-C. and Bazilevs, Y. (2011). Blood vessel tissue prestress modeling for vascular fluid–structure interaction simulations. *Finite Elements in Analysis and Design*, 47: 593–599.

Hsu, M.-C., Bazilevs, Y., Calo, V.M., Tezduyar, T.E., and Hughes, T.J.R. (2010). Improving stability of stabilized and multiscale formulations in flow simulations at small time steps. *Computer Methods in Applied Mechanics and Engineering*, 199:828–840.

Huang, H., Virmani, R., Younis, H., Burke, A.P., Kamm, R.D., and Lee, R.T. (2001). The impact of calcification on the biomechanical stability of atherosclerotic plaques. *Circulation*, 103:1051–1056.

Hughes, T.J.R. (1995). Multiscale phenomena: Green's functions, the Dirichlet-to-Neumann formulation, subgrid scale models, bubbles, and the origins of stabilized methods. *Computer Methods in Applied Mechanics and Engineering*, 127:387–401.

Hughes, T.J.R. (2000). *The Finite Element Method: Linear Static and Dynamic Finite Element Analysis*. Dover Publications, Mineola, NY.

Hughes, T.J.R., Cottrell, J.A., and Bazilevs, Y. (2005). Isogeometric analysis: CAD, finite elements, NURBS, exact geometry, and mesh refinement. *Computer Methods in Applied Mechanics and Engineering*, 194:4135–4195.

Hughes, T.J.R., Feijóo, G.R., Mazzei, L., and Quincy, J.B. (1998). The variational multiscale method–A paradigm for computational mechanics. *Computer Methods in Applied Mechanics and Engineering*, 166:3–24.

Hughes, T.J.R., Franca, L.P., and Balestra, M. (1986a). A new finite element formulation for computational fluid dynamics: V. Circumventing the Babuška–Brezzi condition: A stable Petrov–Galerkin formulation of the Stokes problem accommodating equal-order interpolations. *Computer Methods in Applied Mechanics and Engineering*, 59:85–99.

Hughes, T.J.R. and Hulbert, G.M. (1988). Space–time finite element methods for elastodynamics: formulations and error estimates. *Computer Methods in Applied Mechanics and Engineering*, 66:339–363.

Hughes, T.J.R., Liu, W.K., and Zimmermann, T.K. (1981). Lagrangian–Eulerian finite element formulation for incompressible viscous flows. *Computer Methods in Applied Mechanics and Engineering*, 29:329–349.

Hughes, T.J.R., Mallet, M., and Mizukami, A. (1986b). A new finite element formulation for computational fluid dynamics: II. Beyond SUPG. *Computer Methods in Applied Mechanics and Engineering*, 54:341–355.

Hughes, T.J.R., Mazzei, L., and Jansen, K.E. (2000). Large-eddy simulation and the variational multiscale method. *Computing and Visualization in Science*, 3:47–59.

Hughes, T.J.R. and Oberai, A.A. (2003). Calculation of shear stress in Fourier–Galerkin formulations of turbulent channel flows: projection, the Dirichlet filter and conservation. *Journal of Computational Physics*, 188:281–295.

Hughes, T.J.R., Reali, A., and Sangalli, G. (2010). Efficient quadrature for NURBS-based isogeometric analysis. *Computer Methods in Applied Mechanics and Engineering*, 199: 301–313.

Hughes, T.J.R. and Sangalli, G. (2007). Variational multiscale analysis: the fine-scale Green's function, projection, optimization, localization, and stabilized methods. *SIAM Journal of Numerical Analysis*, 45:539–557.

Hughes, T.J.R., Scovazzi, G., and Franca, L.P. (2004). Multiscale and stabilized methods. In Stein, E., de Borst, R., and Hughes, T.J.R., editors, *Encyclopedia of Computational Mechanics, Vol. 3, Fluids*, chapter 2. Wiley.

Hughes, T.J.R. and Tezduyar, T.E. (1984). Finite element methods for first-order hyperbolic systems with particular emphasis on the compressible Euler equations. *Computer Methods in Applied Mechanics and Engineering*, 45:217–284.

Humphrey, J. (2002). *Cardiovascular Solid Mechanics*. Springer-Verlag.

Idelsohn, S.R., Marti, J., Limache, A., and Onate, E. (2008a). Unified Lagrangian formulation for elastic solids and incompressible fluids: Application to fluid–structure interaction problems via the PFEM. *Computer Methods in Applied Mechanics and Engineering*, 197:1762–1776.

Idelsohn, S.R., Marti, J., Souto-Iglesias, A., and Onate, E. (2008b). Interaction between an elastic structure and free-surface flows: experimental versus numerical comparisons using the PFEM. *Computational Mechanics*, 43:125–132.

Irons, B. (1966). Engineering application of numerical integration in stiffness method. *American Institute of Aeronautics and Astronautics*, 14:2035–2037.

Isaksen, J.G., Bazilevs, Y., Kvamsdal, T., Zhang, Y., Kaspersen, J.H., Waterloo, K., Romner, B., and Ingebrigtsen, T. (2008). Determination of wall tension in cerebral artery aneurysms by numerical simulation. *Stroke*, 39:3172–3178.

Jansen, K.E., Whiting, C.H., and Hulbert, G.M. (2000). A generalized-α method for integrating the filtered Navier-Stokes equations with a stabilized finite element method. *Computer Methods in Applied Mechanics and Engineering*, 190:305–319.

Jensen, F.M., Falzon, B.G., Ankersen, J., and Stang, H. (2006). Structural testing and numerical simulation of a 34 m composite wind turbine blade. *Composite Structures*, 76:52–61.

Johan, Z., Hughes, T.J.R., and Shakib, F. (1991). A globally convergent matrix-free algorithm for implicit time-marching schemes arising in finite element analysis in fluids. *Computer Methods in Applied Mechanics and Engineering*, 87:281–304.

Johan, Z., Mathur, K.K., Johnsson, S.L., and Hughes, T.J.R. (1995). A case study in parallel computation: Viscous flow around an Onera M6 wing. *International Journal for Numerical Methods in Fluids*, 21:877–884.

Johnson, A.A. and Tezduyar, T.E. (1994). Mesh update strategies in parallel finite element computations of flow problems with moving boundaries and interfaces. *Computer Methods in Applied Mechanics and Engineering*, 119:73–94.

Johnson, A.A. and Tezduyar, T.E. (1996). Simulation of multiple spheres falling in a liquid-filled tube. *Computer Methods in Applied Mechanics and Engineering*, 134:351–373.

Johnson, A.A. and Tezduyar, T.E. (1997). Parallel computation of incompressible flows with complex geometries. *International Journal for Numerical Methods in Fluids*, 24:1321–1340.

Johnson, A.A. and Tezduyar, T.E. (1999). Advanced mesh generation and update methods for 3D flow simulations. *Computational Mechanics*, 23:130–143.

Johnson, C. (1987). *Numerical solution of partial differential equations by the finite element method*. Cambridge University Press, Sweden.

Jonkman, J., Butterfield, S., Musial, W., and Scott, G. (2009). Definition of a 5-MW reference wind turbine for offshore system development. Technical Report NREL/TP-500-38060, National Renewable Energy Laboratory.

Jonkman, J.M. and Buhl Jr., M.L. (2005). FAST user's guide. Technical Report NREL/EL-500-38230, National Renewable Energy Laboratory, Golden, CO.

Kalro, V. and Tezduyar, T. (1998). A parallel finite element methodology for 3D computation of fluid–structure interactions in airdrop systems. In *Proceedings of the 4th Japan-US Symposium on Finite Element Methods in Large-Scale Computational Fluid Dynamics*, Tokyo, Japan.

Kalro, V. and Tezduyar, T.E. (2000). A parallel 3D computational method for fluid–structure interactions in parachute systems. *Computer Methods in Applied Mechanics and Engineering*, 190:321–332.

Kar, B., III, R.M.D., Frazier, O.H., Gregoric, I., Harting, M.T., Wadia, Y., Myers, T., Moser, R., and Freund, J. (2005). The effect of LVAD aortic outflow-graft placement on hemodynamics and flow. *Journal of the Texas Heart Institute*, 32:294–298.

Karypis, G. and Kumar, V. (1998). A fast and high quality multilevel scheme for partitioning irregular graphs. *SIAM Journal of Scientific Computing*, 20:359–392.

Khunatorn, Y., Mahalingam, S., DeGroff, C., and Shandas, R. (2002). Influence of connection geometry and SVC-IVC flow rate ratio on flow structures within the total cavopulmonary connection: A numerical study. *Journal of Biomechanical Engineering-Transactions of the ASME*, 124:364–377.

Khurram, R.A. and Masud, A. (2006). A multiscale/stabilized formulation of the incompressible Navier–Stokes equations for moving boundary flows and fluid–structure interaction. *Computational Mechanics*, 38:403–416.

Kiendl, J., Bazilevs, Y., Hsu, M.-C., Wüchner, R., and Bletzinger, K.-U. (2010). The bending strip method for isogeometric analysis of Kirchhoff–Love shell structures comprised of multiple patches. *Computer Methods in Applied Mechanics and Engineering*, 199:2403–2416.

Kiendl, J., Bletzinger, K.U., Linhard, J., and Wüchner, R. (2009). Isogeometric shell analysis with Kirchhoff–Love elements. *Computer Methods in Applied Mechanics and Engineering*, 198:3902–3914.

Kilner, P.J., Yang, G.Z., Mohiaddin, R.H., Firmin, D.N., and Longmore, D.B. (1993). Helical and retrograde secondary flow patterns in the aortic arch studied by three-directional magnetic resonance velocity mapping. *Circulation*, 88:2235–2247.

Kong, C., Bang, J., and Sugiyama, Y. (2005). Structural investigation of composite wind turbine blade considering various load cases and fatigue life. *Energy*, 30:2101–2114.

Kooijman, H.J.T., Lindenburg, C., Winkelaarand, D., and van der Hooft, E.L. (2003). DOWEC 6 MW pre-design: Aero-elastic modelling of the DOWEC 6 MW pre-design in PHATAS. Technical Report DOWEC-F1W2-HJK-01-046/9.

Laino, D.J., Hansen, A.C., and Minnema, J.E. (2002). Validation of the AeroDyn subroutines using NREL Unsteady Aerodynamics Experiment data. *Wind Energy*, 5:227–244.

Launder, B.E. and Spalding, D.B. (1974). The numerical computation of turbulent flows. *Computer Methods in Applied Mechanics and Engineering*, 3:269–289.

Le Tallec, P. and Mouro, J. (2001). Fluid structure interaction with large structural displacements. *Computer Methods in Applied Mechanics and Engineering*, 190:3039–3068.

Levesque, M.J., Liepsch, D., Moravec, S., and Nerem, R. (1986). Correlation of endothelial cell shape and wall shear stress in a stenosed dog aorta. *Arteriosclerosis*, 6:220–229.

Levesque, M.J. and Nerem, R. (1985). The elongation and orientation of cultured endothelial cells in response to shear stress. *Journal of Biomechanical Engineering*, 107:341–347.

Li, X., Zheng, J., Sederberg, T., Hughes, T., and Scott, M.A. (2012). On linear independence of T-splines. *Computer-Aided Geometric Design*. Accepted for publication.

Li, Y. and K. J. Paik T. Xing, P.M.C. (2012). Dynamic overset CFD simulations of wind turbine aerodynamics. *Renewable Energy*, 37:285–298.

Limache, A., Sanchez, P., DalcŠn, L., and Idelsohn, S. (2008). Objectivity tests for Navier–Stokes simulations: The revealing of non-physical solutions produced by Laplace formulations. *Computer Methods in Applied Mechanics and Engineering*, 197:4180–4192.

Lipton, S., Evans, J.A., Bazilevs, Y., Elguedj, T., and Hughes, T.J.R. (2010). Robustness of isogeometric structural discretizations under severe mesh distortion. *Computer Methods in Applied Mechanics and Engineering*, 199:357–373.

Liu, S.Q., Zhong, L., and Goldman, J. (2002). Control of the shape of a thrombus-neointima-like structure by blood shear stress. *Journal of Biomechanical Engineering*, 124:30.

Lohner, R., Cebral, J.R., Yang, C., Baum, J.D., Mestreau, E.L., and Soto, O. (2006). Extending the range of applicability of the loose coupling approach for FSI simulations. In Bungartz, H.-J. and Schafer, M., editors, *Fluid–Structure Interaction*, volume 53 of *Lecture Notes in Computational Science and Engineering*, pages 82–100. Springer.

Manguoglu, M., Sameh, A.H., Saied, F., Tezduyar, T.E., and Sathe, S. (2009). Preconditioning techniques for nonsymmetric linear systems in computation of incompressible flows. *Journal of Applied Mechanics*, 76:021204.

Manguoglu, M., Sameh, A.H., Tezduyar, T.E., and Sathe, S. (2008). A nested iterative scheme for computation of incompressible flows in long domains. *Computational Mechanics*, 43:73–80.

Manguoglu, M., Takizawa, K., Sameh, A.H., and Tezduyar, T.E. (2010). Solution of linear systems in arterial fluid mechanics computations with boundary layer mesh refinement. *Computational Mechanics*, 46:83–89.

Manguoglu, M., Takizawa, K., Sameh, A.H., and Tezduyar, T.E. (2011a). Nested and parallel sparse algorithms for arterial fluid mechanics computations with boundary layer mesh refinement. *International Journal for Numerical Methods in Fluids*, 65:135–149.

Manguoglu, M., Takizawa, K., Sameh, A.H., and Tezduyar, T.E. (2011b). A parallel sparse algorithm targeting arterial fluid mechanics computations. *Computational Mechanics*, 48:377–384.

Marsden, A.L., Bernstein, A.D., Reddy, V.M., Shadden, S., Spilker, R., Chan, F.P., Taylor, C.A., and Feinstein, J.A. (2008). Evaluation of a novel Y-shaped extracardiac fontan baffle using computational fluid dynamics. *Journal of Thoracic and Cardiovascular Surgery*. To appear.

Marsden, A.L., Vignon-Clementel, I.E., Chan, F., Feinstein, J.A., and Taylor, C.A. (2007). Effects of exercise and respiration on hemodynamic efficiency in CFD simulations of the total cavopulmonary connection. *Annals of Biomedical Engineering*, 35:250–263.

Masud, A. and Hughes, T.J.R. (1997). A space–time Galerkin/least-squares finite element formulation of the Navier-Stokes equations for moving domain problems. *Computer Methods in Applied Mechanics and Engineering*, 146:91–126.

McPhail, T. and Warren, J. (2008). An interactive editor for deforming volumetric data. In *International Conference on Biomedical Engineering 2008*, pages 137–144, Singapore.

Melbø, H. and Kvamsdal, T. (2003). Goal oriented error estimators for Stokes equations based on variationally consistent postprocessing. *Computer Methods in Applied Mechanics and Engineering*, 192:613–633.

Michler, C., van Brummelen, E.H., Hulshoff, S.J., and de Borst, R. (2003). The relevance of conservation for stability and accuracy of numerical methods for fluid–structure interaction. *Computer Methods in Applied Mechanics and Engineering*, 192:4195–4215.

Michler, C., van Brummelen, E.H., Hulshoff, S.J., and de Borst, R. (2004). A monolithic approach to fluid–structure interaction. *Computers & Fluids*, 33:839–848.

Migliavacca, F., Dubini, G., Bove, E.L., and de Leval, M.R. (2003). Computational fluid dynamics simulations in realistic 3-D geometries of the total cavopulmonary anastomosis: the influence of the inferior caval anastomosis. *Journal of Biomechanical Engineering*, 125:805–813.

Migliavacca, F., Dubini, G., Pietrabissa, R., and de Leval, M.R. (1997). Computational transient simulations with varying degree and shape of pulmonic stenosis in models of the bidirectional cavopulmonary anastomosis. *Medical Engineering and Physics*, 19:394–403.

Mittal, S. and Tezduyar, T.E. (1994). Massively parallel finite element computation of incompressible flows involving fluid-body interactions. *Computer Methods in Applied Mechanics and Engineering*, 112:253–282.

Moes, N., Dolbow, J., and Belytschko, T. (1999). A finite element method for crack growth without remeshing. *International Journal for Numerical Methods in Engineering*, 46:131–150.

Moghadam, M.E., Bazilevs, Y., Hsia, T.-Y., Vignon-Clementel, I.E., Marsden, A.L., and of Congenital Hearts Alliance (MOCHA), M. (2011). A comparison of outlet boundary treatments for prevention of backflow divergence with relevance to blood flow simulations. *Computational Mechanics*, 48:277–291.

Moorman, C.J. (2010). *Fluid–Structure Interaction Modeling of the Orion Spacecraft Parachutes*. PhD thesis, Rice University.

Morand, H.J.-P. and Ohayon, R. (1995). *Fluid-Structure Interaction: Applied Numerical Methods*. Wiley.

Okano, M. and Yoshida, Y. (1994). Junction complexes of endothelial cells in atherosclerosis-prone and atherosclerosis-resistant regions on flow dividers of brachiocephalic bifurcations in the rabbit aorta. *Biorheology*, 31:155–161.

Olufsen, M.S. (1998). *Modeling of the arterial system with reference to an anesthesia simulator*. PhD thesis, Roskilde University.

Onate, E., Valls, A., and Garcia, J. (2006). FIC/FEM formulation with matrix stabilizing terms for incompressible flows at low and high Reynolds numbers. *Computational Mechanics*, 38:440–455.

Pape, A.L. and Lecanu, J. (2004). 3D Navier–Stokes computations of a stall-regulated wind turbine. *Wind Energy*, 7:309–324.

Pedersen, E.M., Stenbog, E.V., Frund, T., Houlind, K., Kromann, O., Sorensen, K.E., Emmertsen, K., and Hjortdal, V.E. (2002). Flow during exercise in the total cavopulmonary connection measured by magnetic resonance velocity mapping. *Heart*, 87:554–558.

Peters, J. and Reif, U. (2008). *Subdivision Surfaces*. Springer-Verlag.

Petrossian, E., Reddy, V.M., and Collins, K.K. (2006). The extracardiac conduit Fontan operation using minimal approach extracorporeal circulation: Early and midterm outcomes. *Journal of Thoracic and Cardiovascular Surgery*, 132(5):1054–1063.

Piegl, L. and Tiller, W. (1997). *The NURBS Book (Monographs in Visual Communication)*, 2nd ed. Springer-Verlag, New York.

Reddy, J.N. (2004). *Mechanics of Laminated Composite Plates and Shells: Theory and Analysis, 2nd ed.* CRC Press, Boca Raton, FL.

Rispoli, F., Corsini, A., and Tezduyar, T.E. (2007). Finite element computation of turbulent flows with the discontinuity-capturing directional dissipation (DCDD). *Computers & Fluids*, 36:121–126.

Rogers, D.F. (2001). *An Introduction to NURBS With Historical Perspective.* Academic Press, San Diego, CA.

Saad, Y. and Schultz, M. (1986). GMRES: A generalized minimal residual algorithm for solving nonsymmetric linear systems. *SIAM Journal of Scientific and Statistical Computing*, 7:856–869.

Sahni, O., Muller, J., Jansen, K., Shephard, M., and Taylor, C. (2006). Efficient anisotropic adaptive discretization of the cardiovascular system. *Computer Methods in Applied Mechanics and Engineering*, 195:5634–5655.

Sameh, A. and Sarin, V. (1999). Hybrid parallel linear solvers. *International Journal of Computational Fluid Dynamics*, 12:213–223.

Sameh, A. and Sarin, V. (2002). Parallel algorithms for indefinite linear systems. *Parallel Computing*, 28:285–299.

Sathe, S. and Tezduyar, T.E. (2008). Modeling of fluid–structure interactions with the space–time finite elements: Contact problems. *Computational Mechanics*, 43:51–60.

Schmitz, S. and Chattot, J.J. (2006). Characterization of three-dimensional effects for the rotating and parked NREL Phase VI wind turbine. *Journal of Solar Energy Engineering*, 128:445–454.

Scott, M., Li, X., Sederberg, T., and Hughes, T. (2012). Local refinement of analysis-suitable T-splines. *Computer Methods in Applied Mechanics and Engineering*. Accepted for publication.

Sederberg, T.W., Cardon, D., Finnigan, G., North, N., Zheng, J., and Lyche, T (2004). T-spline simplification and local refinement. *ACM Transactions on Graphics*, 23(3):276–283.

Sederberg, T., Zheng, J., Bakenov, A., and Nasri, A. (2003). T-splines and T-NURCCS. *ACM Transactions on Graphics*, 22(3):477–484.

Shaaban, A.M. and Duerinckx, A.J. (2000). Wall shear stress and early atherosclerosis: A review. *American Journal of Roentgenology*, 174:1657–1665.

Shachar, G., Fuhrman, B., Wang, Y., Lucas Jr, R., and Lock, J. (1982). Rest and exercise hemodynamics after the fontan procedure. *Circulation*, 65:1043–1048.

Shakib, F., Hughes, T.J.R., and Johan, Z. (1989). A multi-element group preconditioined GMRES algorithm for nonsymmetric systems arising in finite element analysis. *Computer Methods in Applied Mechanics and Engineering*, 75:415–456.

Shakib, F., Hughes, T.J.R., and Johan, Z. (1991). A new finite element formulation for computational fluid dynamics: X. The compressible euler and navier-stokes equations. *Computer Methods in Applied Mechanics and Engineering*, 89:141–219.

Shield, R.T. (1967). Inverse deformation results in finite elasticity. *ZAMP*, 18:381–389.

Simo, J.C. and Hughes, T.J.R. (1998). *Computational Inelasticity.* Springer-Verlag, New York.

Sørensen, N.N., Michelsen, J.A., and Schreck, S. (2002). Navier–Stokes predictions of the NREL Phase VI rotor in the NASA Ames 80 ft × 120 ft wind tunnel. *Wind Energy*, 5:151–169.

Spera, D.A. (1994). Introduction to modern wind turbines. In Spera, D.A., editor, *Wind Turbine Technology: Fundamental Concepts of Wind Turbine Engineering*, pages 47–72. ASME Press.

Stein, K., Benney, R., Kalro, V., Tezduyar, T.E., Leonard, J., and Accorsi, M. (2000). Parachute fluid–structure interactions: 3-D Computation. *Computer Methods in Applied Mechanics and Engineering*, 190:373–386.

Stein, K., Benney, R., Tezduyar, T., and Potvin, J. (2001a). Fluid–structure interactions of a cross parachute: Numerical simulation. *Computer Methods in Applied Mechanics and Engineering*, 191:673–687.

Stein, K. and Tezduyar, T. (2002). Advanced mesh update techniques for problems involving large displacements. In *Proceedings of the Fifth World Congress on Computational Mechanics*, On-line publication: http://wccm.tuwien.ac.at/, Paper-ID: 81489, Vienna, Austria.

Stein, K., Tezduyar, T., and Benney, R. (2003a). Computational methods for modeling parachute systems. *Computing in Science and Engineering*, 5:39–46.

Stein, K., Tezduyar, T., and Benney, R. (2003b). Mesh moving techniques for fluid–structure interactions with large displacements. *Journal of Applied Mechanics*, 70:58–63.

Stein, K., Tezduyar, T., Kumar, V., Sathe, S., Benney, R., Thornburg, E., Kyle, C., and Nonoshita, T. (2003c). Aerodynamic interactions between parachute canopies. *Journal of Applied Mechanics*, 70:50–57.

Stein, K., Tezduyar, T.E., and Benney, R. (2004). Automatic mesh update with the solid-extension mesh moving technique. *Computer Methods in Applied Mechanics and Engineering*, 193:2019–2032.

Stein, K.R., Benney, R.J., Kalro, V., Johnson, A.A., and Tezduyar, T.E. (1997). Parallel computation of parachute fluid–structure interactions. In *Proceedings of AIAA 14th Aerodynamic Decelerator Systems Technology Conference*, AIAA Paper 97-1505, San Francisco, California.

Stein, K.R., Benney, R.J., Tezduyar, T.E., Leonard, J.W., and Accorsi, M.L. (2001b). Fluid–structure interactions of a round parachute: Modeling and simulation techniques. *Journal of Aircraft*, 38:800–808.

Stuparu, M. (2002). Human heart valves. hyperelastic material modeling. In *Proceedings of the X-th Conference on Mechanical Vibrations*, Timisoara, Romania.

(TACC), T.A.C.C. (2011). Available at: http://www.tacc.utexas.edu. Accessed October 6.

Takizawa, K., Bazilevs, Y., and Tezduyar, T.E. (2012a). Space–time and ALE-VMS techniques for patient-specific cardiovascular fluid–structure interaction modeling. *Archives of Computational Methods in Engineering*, 19:171–225.

Takizawa, K., Brummer, T., Tezduyar, T.E., and Chen, P.R. (2012b). A comparative study based on patient-specific fluid–structure interaction modeling of cerebral aneurysms. *Journal of Applied Mechanics*, 79:010908.

Takizawa, K., Christopher, J., Tezduyar, T.E., and Sathe, S. (2010a). Space–time finite element computation of arterial fluid–structure interactions with patient-specific data. *International Journal for Numerical Methods in Biomedical Engineering*, 26:101–116.

Takizawa, K., Henicke, B., Montes, D., Tezduyar, T.E., Hsu, M.-C., and Bazilevs, Y. (2011a). Numerical-performance studies for the stabilized space–time computation of wind-turbine rotor aerodynamics. *Computational Mechanics*, 48:647–657.

Takizawa, K., Henicke, B., Puntel, A., Kostov, N., and Tezduyar, T.E. (July 2012c). Space–time techniques for computational aerodynamics modeling of flapping wings of an actual locust. *Computational Mechanics*, published online, DOI: 10.1007/s00466-012-0759-x.

Takizawa, K., Henicke, B., Puntel, A., Spielman, T., and Tezduyar, T.E. (2012d). Space–time computational techniques for the aerodynamics of flapping wings. *Journal of Applied Mechanics*, 79:010903.

Takizawa, K., Henicke, B., Tezduyar, T.E., Hsu, M.-C., and Bazilevs, Y. (2011b). Stabilized space–time computation of wind-turbine rotor aerodynamics. *Computational Mechanics*, 48:333–344.

Takizawa, K., Moorman, C., Wright, S., Christopher, J., and Tezduyar, T.E. (2010b). Wall shear stress calculations in space–time finite element computation of arterial fluid–structure interactions. *Computational Mechanics*, 46:31–41.

Takizawa, K., Moorman, C., Wright, S., Purdue, J., McPhail, T., Chen, P.R., Warren, J., and Tezduyar, T.E. (2011c). Patient-specific arterial fluid–structure interaction modeling of cerebral aneurysms. *International Journal for Numerical Methods in Fluids*, 65:308–323.

Takizawa, K., Moorman, C., Wright, S., Spielman, T., and Tezduyar, T.E. (2011d). Fluid–structure interaction modeling and performance analysis of the Orion spacecraft parachutes. *International Journal for Numerical Methods in Fluids*, 65:271–285.

Takizawa, K., Moorman, C., Wright, S., and Tezduyar, T.E. (2010c). Computer modeling and analysis of the Orion spacecraft parachutes. In Bungartz, H.-J., Mehl, M., and Schafer, M., editors, *Fluid–Structure Interaction II – Modelling, Simulation, Optimization*, volume 73 of *Lecture Notes in Computational Science and Engineering*, pages 53–81. Springer.

Takizawa, K., Spielman, T., Moorman, C., and Tezduyar, T.E. (2012e). Fluid–structure interaction modeling of spacecraft parachutes for simulation-based design. *Journal of Applied Mechanics*, 79:010907.

Takizawa, K., Spielman, T., and Tezduyar, T.E. (2011e). Space–time FSI modeling and dynamical analysis of spacecraft parachutes and parachute clusters. *Computational Mechanics*, 48:345–364.

Takizawa, K., Tanizawa, K., Yabe, T., and Tezduyar, T.E. (2007a). Ship hydrodynamics computations with the CIP method based on adaptive Soroban grids. *International Journal for Numerical Methods in Fluids*, 54:1011–1019.

Takizawa, K. and Tezduyar, T.E. (2011). Multiscale space–time fluid–structure interaction techniques. *Computational Mechanics*, 48:247–267.

Takizawa, K. and Tezduyar, T.E. (2012a). Computational methods for parachute fluid–structure interactions. *Archives of Computational Methods in Engineering*, 19:125–169.

Takizawa, K. and Tezduyar, T.E. (2012b). Space–time fluid–structure interaction methods. *Mathematical Models and Methods in Applied Sciences*, 22(supp02):1230001.

Takizawa, K., Wright, S., Moorman, C., and Tezduyar, T.E. (2011f). Fluid–structure interaction modeling of parachute clusters. *International Journal for Numerical Methods in Fluids*, 65:286–307.

Takizawa, K., Yabe, T., Tsugawa, Y., Tezduyar, T.E., and Mizoe, H. (2007b). Computation of free-surface flows and fluid–object interactions with the CIP method based on adaptive meshless Soroban grids. *Computational Mechanics*, 40:167–183.

Taylor, C.A., Hughes, T.J.R., and Zarins, C.K. (1998a). Finite element modeling of blood flow in arteries. *Computer Methods in Applied Mechanics and Engineering*, 158:155–196.

Taylor, C.A., Hughes, T.J.R., and Zarins, C.K. (1998b). Finite element modeling of three-dimensional pulsatile flow in the abdominal aorta: relevance to atherosclerosis. *Ann. Biomed. Engrg.*, 158:975–987.

Tezduyar, T. (2001a). Finite element interface-tracking and interface-capturing techniques for flows with moving boundaries and interfaces. In *Proceedings of the ASME Symposium on Fluid-Physics and Heat Transfer for Macro- and Micro-Scale Gas-Liquid and Phase-Change Flows (CD-ROM)*, ASME Paper IMECE2001/HTD-24206, New York, New York. ASME.

Tezduyar, T., Aliabadi, S., Behr, M., Johnson, A., Kalro, V., and Litke, M. (1996). Flow simulation and high performance computing. *Computational Mechanics*, 18:397–412.

Tezduyar, T., Aliabadi, S., Behr, M., Johnson, A., and Mittal, S. (1993). Parallel finite-element computation of 3D flows. *Computer*, 26(10):27–36.

Tezduyar, T. and Osawa, Y. (2001). Fluid–structure interactions of a parachute crossing the far wake of an aircraft. *Computer Methods in Applied Mechanics and Engineering*, 191: 717–726.

Tezduyar, T.E. (1992). Stabilized finite element formulations for incompressible flow computations. *Advances in Applied Mechanics*, 28:1–44.

Tezduyar, T.E. (2001b). Finite element methods for flow problems with moving boundaries and interfaces. *Archives of Computational Methods in Engineering*, 8:83–130.

Tezduyar, T.E. (2003a). Computation of moving boundaries and interfaces and stabilization parameters. *International Journal for Numerical Methods in Fluids*, 43:555–575.

Tezduyar, T.E. (2003b). Stabilized finite element formulations and interface-tracking and interface-capturing techniques for incompressible flows. In Hafez, M.M., editor, *Numerical Simulations of Incompressible Flows*, pages 221–239, New Jersey. World Scientific.

Tezduyar, T.E. (2003c). Stabilized finite element methods for computation of flows with moving boundaries and interfaces. In *Lecture Notes on Finite Element Simulation of Flow Problems (Basic - Advanced Course)*, Tokyo, Japan. Japan Society of Computational Engineering and Sciences.

Tezduyar, T.E. (2003d). Stabilized finite element methods for flows with moving boundaries and interfaces. *HERMIS: The International Journal of Computer Mathematics and its Applications*, 4:63–88.

Tezduyar, T.E. (2004a). Finite element methods for fluid dynamics with moving boundaries and interfaces. In Stein, E., Borst, R.D., and Hughes, T.J.R., editors, *Encyclopedia of Computational Mechanics*, Volume 3: Fluids, chapter 17. John Wiley & Sons.

Tezduyar, T.E. (2004b). Moving boundaries and interfaces. In Franca, L.P., Tezduyar, T.E., and Masud, A., editors, *Finite Element Methods: 1970's and Beyond*, pages 205–220. CIMNE, Barcelona, Spain.

Tezduyar, T.E. (2007a). Finite elements in fluids: Special methods and enhanced solution techniques. *Computers & Fluids*, 36:207–223.

Tezduyar, T.E. (2007b). Finite elements in fluids: Stabilized formulations and moving boundaries and interfaces. *Computers & Fluids*, 36:191–206.

Tezduyar, T.E., Behr, M., and Liou, J. (1992a). A new strategy for finite element computations involving moving boundaries and interfaces – the deforming-spatial-domain/space–time procedure: I. The concept and the preliminary numerical tests. *Computer Methods in Applied Mechanics and Engineering*, 94(3):339–351.

Tezduyar, T.E., Behr, M., Mittal, S., and Johnson, A.A. (1992b). Computation of unsteady incompressible flows with the finite element methods – space–time formulations, iterative strategies and massively parallel implementations. In *New Methods in Transient Analysis*, PVP-Vol.246/AMD-Vol.143, pages 7–24, New York. ASME.

Tezduyar, T.E., Behr, M., Mittal, S., and Liou, J. (1992c). A new strategy for finite element computations involving moving boundaries and interfaces – the deforming-spatial-domain/space–time procedure: II. Computation of free-surface flows, two-liquid flows, and flows with drifting cylinders. *Computer Methods in Applied Mechanics and Engineering*, 94(3):353–371.

Tezduyar, T.E., Cragin, T., Sathe, S., and Nanna, B. (2007a). FSI computations in arterial fluid mechanics with estimated zero-pressure arterial geometry. In Onate, E., Garcia, J., Bergan, P., and Kvamsdal, T., editors, *Marine 2007*, Barcelona, Spain. CIMNE.

Tezduyar, T.E., Liou, J., and Ganjoo, D.K. (1990). Incompressible flow computations based on the vorticity-stream function and velocity-pressure formulations. *Computers & Structures*, 35(4): 445–472.

Tezduyar, T.E., Mittal, S., Ray, S.E., and Shih, R. (1992d). Incompressible flow computations with stabilized bilinear and linear equal-order-interpolation velocity-pressure elements. *Computer Methods in Applied Mechanics and Engineering*, 95:221–242.

Tezduyar, T.E., Mittal, S., and Shih, R. (1991). Time-accurate incompressible flow computations with quadrilateral velocity-pressure elements. *Computer Methods in Applied Mechanics and Engineering*, 87:363–384.

Tezduyar, T.E. and Osawa, Y. (2000). Finite element stabilization parameters computed from element matrices and vectors. *Computer Methods in Applied Mechanics and Engineering*, 190:411–430.

Tezduyar, T.E. and Park, Y.J. (1986). Discontinuity capturing finite element formulations for nonlinear convection-diffusion-reaction equations. *Computer Methods in Applied Mechanics and Engineering*, 59:307–325.

Tezduyar, T.E. and Sathe, S. (2006). Enhanced-discretization selective stabilization procedure (EDSSP). *Computational Mechanics*, 38:456–468.

Tezduyar, T.E. and Sathe, S. (2007). Modeling of fluid–structure interactions with the space–time finite elements: Solution techniques. *International Journal for Numerical Methods in Fluids*, 54:855–900.

Tezduyar, T.E., Sathe, S., Cragin, T., Nanna, B., Conklin, B.S., Pausewang, J., and Schwaab, M. (2007b). Modeling of fluid–structure interactions with the space–time finite elements: Arterial fluid mechanics. *International Journal for Numerical Methods in Fluids*, 54:901–922.

Tezduyar, T.E., Sathe, S., Keedy, R., and Stein, K. (2004). Space–time techniques for finite element computation of flows with moving boundaries and interfaces. In Gallegos, S., Herrera, I., Botello, S., Zarate, F., and Ayala, G., editors, *Proceedings of the III International Congress on Numerical Methods in Engineering and Applied Science*. CD-ROM, Monterrey, Mexico.

Tezduyar, T.E., Sathe, S., Keedy, R., and Stein, K. (2006a). Space–time finite element techniques for computation of fluid–structure interactions. *Computer Methods in Applied Mechanics and Engineering*, 195:2002–2027.

Tezduyar, T.E., Sathe, S., Pausewang, J., Schwaab, M., Christopher, J., and Crabtree, J. (2008a). Interface projection techniques for fluid–structure interaction modeling with moving-mesh methods. *Computational Mechanics*, 43:39–49.

Tezduyar, T.E., Sathe, S., Schwaab, M., and Conklin, B.S. (2008b). Arterial fluid mechanics modeling with the stabilized space–time fluid–structure interaction technique. *International Journal for Numerical Methods in Fluids*, 57:601–629.

Tezduyar, T.E., Sathe, S., Schwaab, M., Pausewang, J., Christopher, J., and Crabtree, J. (2008c). Fluid–structure interaction modeling of ringsail parachutes. *Computational Mechanics*, 43:133–142.

Tezduyar, T.E., Sathe, S., Senga, M., Aureli, L., Stein, K., and Griffin, B. (2005). Finite element modeling of fluid–structure interactions with space–time and advanced mesh update techniques. In *Proceedings of the 10th International Conference on Numerical Methods in Continuum Mechanics (CD-ROM)*, Zilina, Slovakia.

Tezduyar, T.E., Sathe, S., and Stein, K. (2006b). Solution techniques for the fully-discretized equations in computation of fluid–structure interactions with the space–time formulations. *Computer Methods in Applied Mechanics and Engineering*, 195:5743–5753.

Tezduyar, T.E., Sathe, S., Stein, K., and Aureli, L. (2006c). Modeling of fluid–structure interactions with the space–time techniques. In Bungartz, H.-J. and Schafer, M., editors, *Fluid–Structure Interaction – Modelling, Simulation, Optimization*, volume 53 of *Lecture Notes in Computational Science and Engineering*, pages 50–81. Springer.

Tezduyar, T.E., Schwaab, M., and Sathe, S. (2007c). Arterial fluid mechanics with the sequentially-coupled arterial FSI technique. In Onate, E., Papadrakakis, M., and Schrefler, B., editors, *Coupled Problems 2007*, Barcelona, Spain. CIMNE.

Tezduyar, T.E., Schwaab, M., and Sathe, S. (2009a). Sequentially-Coupled Arterial Fluid–Structure Interaction (SCAFSI) technique. *Computer Methods in Applied Mechanics and Engineering*, 198:3524–3533.

Tezduyar, T.E., Senga, M., and Vicker, D. (2006d). Computation of inviscid supersonic flows around cylinders and spheres with the SUPG formulation and YZβ shock-capturing. *Computational Mechanics*, 38:469–481.

Tezduyar, T.E., Takizawa, K., Brummer, T., and Chen, P.R. (2011). Space–time fluid–structure interaction modeling of patient-specific cerebral aneurysms. *International Journal for Numerical Methods in Biomedical Engineering*, 27:1665–1710.

Tezduyar, T.E., Takizawa, K., and Christopher, J. (2009b). Multiscale Sequentially-Coupled Arterial Fluid–Structure Interaction (SCAFSI) technique. In Hartmann, S., Meister, A., Schaefer, M., and Turek, S., editors, *International Workshop on Fluid–Structure Interaction — Theory, Numerics and Applications*, pages 231–252. Kassel University Press.

Tezduyar, T.E., Takizawa, K., and Christopher, J. (2009c). Sequentially-coupled FSI technique. In Kvamsdal, T., Pettersen, B., Bergan, P., Onate, E., and Garcia, J., editors, *Marine 2009*, Barcelona, Spain. CIMNE.

Tezduyar, T.E., Takizawa, K., Christopher, J., Moorman, C., and Wright, S. (2009d). Interface projection techniques for complex FSI problems. In Kvamsdal, T., Pettersen, B., Bergan, P., Onate, E., and Garcia, J., editors, *Marine 2009*, Barcelona, Spain. CIMNE.

Tezduyar, T.E., Takizawa, K., Moorman, C., Wright, S., and Christopher, J. (2010a). Multiscale sequentially-coupled arterial FSI technique. *Computational Mechanics*, 46:17–29.

Tezduyar, T.E., Takizawa, K., Moorman, C., Wright, S., and Christopher, J. (2010b). Space–time finite element computation of complex fluid–structure interactions. *International Journal for Numerical Methods in Fluids*, 64:1201–1218.

Tongchitpakdee, C., Benjanirat, S., and Sankar, L.N. (2005). Numerical simulation of the aerodynamics of horizontal axis wind turbines under yawed flow conditions. *Journal of Solar Energy Engineering*, 127:464–474.

Torii, R., Oshima, M., Kobayashi, T., Takagi, K., and Tezduyar, T.E. (2004). Influence of wall elasticity on image-based blood flow simulation. *Japan Society of Mechanical Engineers Journal Series A*, 70:1224–1231. in Japanese.

Torii, R., Oshima, M., Kobayashi, T., Takagi, K., and Tezduyar, T.E. (2006a). Computer modeling of cardiovascular fluid–structure interactions with the Deforming-Spatial-Domain/Stabilized Space–Time formulation. *Computer Methods in Applied Mechanics and Engineering*, 195:1885–1895.

Torii, R., Oshima, M., Kobayashi, T., Takagi, K., and Tezduyar, T.E. (2006b). Fluid–structure interaction modeling of aneurysmal conditions with high and normal blood pressures. *Computational Mechanics*, 38:482–490.

Torii, R., Oshima, M., Kobayashi, T., Takagi, K., and Tezduyar, T.E. (2007a). Influence of wall elasticity in patient-specific hemodynamic simulations. *Computers & Fluids*, 36:160–168.

Torii, R., Oshima, M., Kobayashi, T., Takagi, K., and Tezduyar, T.E. (2007b). Numerical investigation of the effect of hypertensive blood pressure on cerebral aneurysm — Dependence of the effect on the aneurysm shape. *International Journal for Numerical Methods in Fluids*, 54:995–1009.

Torii, R., Oshima, M., Kobayashi, T., Takagi, K., and Tezduyar, T.E. (2008). Fluid–structure interaction modeling of a patient-specific cerebral aneurysm: Influence of structural modeling. *Computational Mechanics*, 43:151–159.

Torii, R., Oshima, M., Kobayashi, T., Takagi, K., and Tezduyar, T.E. (2009). Fluid–structure interaction modeling of blood flow and cerebral aneurysm: Significance of artery and aneurysm shapes. *Computer Methods in Applied Mechanics and Engineering*, 198:3613–3621.

Torii, R., Oshima, M., Kobayashi, T., Takagi, K., and Tezduyar, T.E. (2010a). Influence of wall thickness on fluid–structure interaction computations of cerebral aneurysms. *International Journal for Numerical Methods in Biomedical Engineering*, 26:336–347.

Torii, R., Oshima, M., Kobayashi, T., Takagi, K., and Tezduyar, T.E. (2010b). Role of 0D peripheral vasculature model in fluid–structure interaction modeling of aneurysms. *Computational Mechanics*, 46:43–52.

Torii, R., Oshima, M., Kobayashi, T., Takagi, K., and Tezduyar, T.E. (2011). Influencing factors in image-based fluid–structure interaction computation of cerebral aneurysms. *International Journal for Numerical Methods in Fluids*, 65:324–340.

van Brummelen, E.H. and de Borst, R. (2005). On the nonnormality of subiteration for a fluid-structure interaction problem. *SIAM Journal on Scientific Computing*, 27:599–621.

van Brummelen, E.H., van der Zee, K.G., Garg, V.V., and Prudhomme, S. (2012). Flux evaluation in primal and dual boundary-coupled problems. *Journal of Applied Mechanics*. Published online. DOI: 10.1115/1.4005187.

Vignon-Clementel, I., Figueroa, C., Jansen, K., and Taylor, C. (2006). Outflow boundary conditions for three-dimensional finite element modeling of blood flow and pressure in arteries. *Computer Methods in Applied Mechanics and Engineering*, 195:3776–3796.

Wall, W. (1999). *Fluid–Structure Interaction with Stabilized Finite Elements*. PhD thesis, University of Stuttgart.

Wang, D. and Xuan, J. (2010). An improved NURBS-based isogeometric analysis with enhanced treatment of essential boundary conditions. *Computer Methods in Applied Mechanics and Engineering*, 199:2425–2436.

Wang, W., Zhang, Y., Xu, G., and Hughes, T. (2012). Converting an unstructured quadrilateral/hexahedral mesh to a rational T-spline. *Computational Mechanics*. Published online. DOI: 10.1007/s00466-011-0674-6.

Warren, J. and Weimer, H. (2002). *Subdivision Methods for Geometric Design*. Morgan Kaufmann Publishers.

Wilcox, D.C. (1998). *Turbulence Modeling for CFD*. DCW Industries, La Canada, CA.

Womersley, J.R. (1955). Method for the calculation of velocity, rate of flow and viscous drag in arteries when the pressure gradient is known. *Journal of Physiology*, 127:553–563.

Wootton, D.M. and Ku, D.N. (1999). Fluid mechanics of vascular systems, diseases, and thrombosis. *Annual Review of Biomedical Engineering*, 1:299.

Wriggers, P. (2008). *Nonlinear Finite Element Methods*. Springer.

Zahle, F., Sørensen, N.N., and Johansen, J. (2009). Wind turbine rotor-tower interaction using an incompressible overset grid method. *Wind Energy*, 12:594–619.

Zhang, Y., Bazilevs, Y., Goswami, S., Bajaj, C., and Hughes, T.J.R. (2007). Patient-specific vascular nurbs modeling for isogeometric analysis of blood flow. *Computer Methods in Applied Mechanics and Engineering*, 196:2943–2959.

Zhang, Y., Wang, W., Liang, X., Bazilevs, Y., Hsu, M.-C., Kvamsdal, T., Brekken, R., and Isaksen, J. (2009). High-fidelity tetrahedral mesh generation from medical imaging data for fluid-structure interaction analysis of cerebral aneurysms. *Computer Modeling in Engineering and Sciences*, 42:131–150.

Index

advection–diffusion equation, 5
  ALE formulation of, 89
    stabilization parameter, 89
  DSD/SST formulation of, 91
    stabilization parameter, 91
  Galerkin formulation of, 58
  semi-discrete formulation of, 59
  space–time formulation of, 91
  stabilized formulation of, 59
  SUPG formulation of, 59
advection-dominated flows, 6
AEVB computation technique, 142, 143
airfoil attached to a torsion spring (2D flow), 174
ALE formulation of FSI, 114
  ALE-VMS in FSI formulation, 115
  continuous form, 115
  generalized-α time integration of the ALE FSI equations, 118
  predictor–multicorrector algorithm for the ALE FSI equations, 120
  semi-discrete, 115
  vector form of the ALE FSI equations, 117
ALE FSI, 114
ALE methods, 84
  advection–diffusion equation, 89
    stabilization parameter, 89
  ALE form, 31
  mesh velocity ($\hat{\mathbf{u}}$), 89
  Navier–Stokes equations, 92
    ALE-VMS, 93
    conservative form, 34
    convective form, 34
    stabilization parameters (τ_{SUPG} and ν_{LSIC}), 93

    SUPG/PSPG, 93
    variational formulation, 92
    vector form, 94
    weak boundary conditions, 93
  separation of time and space derivatives, 85
ALE-VMS, 93
automatic mesh moving, 107
  equations of elasticity, 107
  Jacobian-based stiffening, 108
  stiffening power (χ), 108

B-splines, 74
  basis functions, 74
  continuity, 75
  Cox-de-Boor recursion for, 75
  knot, 74
  knot vector, 74
block-iterative coupling, xiii, 140
  more robust version, 141
blood flow, 191
  just CFD, 191
    overestimation of WSS, 192
  turbulence, 191
boundary conditions in fluid mechanics, 4
  Dirichlet, 4
  essential, 4
  external boundaries, 8
    free-stream conditions, 8
    inflow, 8
    lateral (side), 8
    outflow, 8
    side (lateral), 8
  free surface, 8
  kinematic, 4

Computational Fluid–Structure Interaction: Methods and Applications, First Edition.
Yuri Bazilevs, Kenji Takizawa and Tayfun E. Tezduyar.
© 2013 John Wiley & Sons, Ltd. Published 2013 by John Wiley & Sons, Ltd.

boundary conditions in fluid mechanics, (*continued*)
   natural, 4
   Neumann, 4
   no-slip boundary, 7
   solid surface, 7
   traction, 4
boundary conditions in structural mechanics, 17
   Dirichlet, 17
   elastic foundation, 18
   essential, 17
   follower pressure load, 18
   natural, 17
   Neumann, 17
boundary layer, 6

cable model, 30
cardiovascular FSI, 191
   ALE-VMS, 192
   aneurysm tissue prestress, 236
   constitutive models for the arterial wall, 192
   DSD/SST, 192
   flexible-wall simulation, 192
   importance of including the wall deformations, 191, 192
   left ventricular assist device (LVAD), 234
   mass balance, 221
   OSI with refined meshes, 224, 226
   patient-specific geometry, 191
   rigid-wall simulation, 192
   SSTFSI, 193
   structural mechanics material model – Fung material, 22, 205
   structural mechanics material model – hyperelastic material, 19, 205
   structural mechanics mesh performance, 215
      hexahedral – one layer across the arterial wall, 215
      hexahedral – three layers across the arterial wall, 215
      hexahedral – two layers across the arterial wall, 215
      tetrahedral – two layers across the arterial wall, 215
   total cavopulmonary connection (TCPC), 233
   virtual surgery, 227
   WSS with refined meshes, 224, 226
cardiovascular FSI – special techniques, 194
   arterial wall thickness, 199, 200, 226, 230
      Laplace's equation over the surface mesh, 200
   arterial-surface extraction from medical images, 198, 225
   estimated zero-pressure (EZP) arterial geometry, 193, 199, 226
      10% wall-thickness ratio, 199, 226
      Laplace's equation over the surface mesh, 200
      patches, 199, 226
   EZP arterial geometry, 193, 199, 226
   inclined inflow and outflow planes, 197
      mesh-moving equations, 197
      structural mechanics equations, 197, 205
   inflow boundary conditions, 205
      scaling the flow rate for a target WSS at the inflow, 206, 230
      Womersley parameter, 206
   inflow boundary mapping technique, 194
   Laplace's equation over the surface mesh covering the lumen, 200
      EZP process, 200, 230
      layers of refined mesh near the arterial wall, 200, 230
   layers of refined mesh near the arterial wall, 199, 226
      Laplace's equation over the surface mesh, 200, 230
      OSI with refined meshes, 222, 226
      WSS with refined meshes, 222, 226
   lumen extraction from medical images, 198, 225
   mesh generation, 199
   Multiscale Sequentially-Coupled Arterial FSI (SCAFSI) technique, 213
      spatially multiscale, 213
      temporally multiscale, 213
   OSI calculation, 196

Index

excluding rigid-body rotation, 196, 230, 232
outflow boundary conditions – explicit, 207
  Windkessel model, 207
outflow boundary conditions – implicit, 207
preconditioning technique, 195
Sequentially-Coupled Arterial FSI (SCAFSI) technique, 210
  multiscale SCAFSI techniques, 213
simulation sequence, 209
  S→F→S→FSI, 209
structural mechanics boundary conditions, 197, 205
tissue prestress, 203, 236
wall thickness reconstruction, 201
WSS calculation, 195
Cauchy stress in fluid mechanics, 3
Cauchy stress in structural mechanics, 16
Cauchy–Green deformation tensor, 13
CFD, 191
CFM (porosity unit), 126
classical laminated-plate theory, 28
composite materials, 28, 339
computational fluid dynamics, 191
Computer-Aided Design, xiv, 73
conservation of mass, 15
constant angular velocity on a circular arc, 156
constitutive modeling, 19, 192
  bulk modulus, 20
  elastic moduli, 20
  elastic-energy density, 19
  Fung model, 22
  hyperelasticity, 19, 191
  Lamé constants, 20
  Mooney–Rivlin model, 21
  Neo-Hookean model, 20
  Poisson's ratio, 20
  second Piola–Kirchhoff stress tensor, 19
  shear modulus, 20
  St. Venant–Kirchhoff model, 20, 27
  Young's modulus, 20
contact algorithm for FSI (SENCT technique), 163
  SENCT-D, 164
  SENCT-F, 164
  SENCT-FC, 163
    contact detection and node sets, 164
    contact force and reaction force, 165
    solving for contact force, 167
  SENCT-FC-M1, 164
  SENCT-M1, 164
current configuration, 12

deformation gradient, 13
  determinant of, 13
Deforming-Spatial-Domain/Stabilized Space–Time method, 86
  advection–diffusion equation, 91
    stabilization parameter, 91
  DSD/SST-SUPS, 102
    DSD/SST-DP, 105
    DSD/SST-SP, 105
    DSD/SST-SV, 105
    DSD/SST-TIP1, 105
    stabilization parameters, 103
    SUPG test function option, 104
    TIP1 integration of incompressibility constraint, 106
  DSD/SST-VMST formulation, 101
    conservative form, 101
    convective form, 101
    stabilization parameters, 103
  integration points in time, 88
  Jacobian matrix, 87
  mesh velocity ($\hat{\mathbf{u}}$), 88
  Navier–Stokes equations, 98
  shape function derivatives, 88
  space–time element, 87
  space–time shape functions, 86
  space–time slab, 86
  Space–Time VMS (alternate name for DSD/SST-VMST), 101
  ST-SUPS (alternate name for DSD/SST-SUPS), 102
  ST-VMS, 101
  temporal element coordinate, 86
diffusion-dominated flows, 6
diffusivity, 5
direct coupling, xiii, 142
DSD/SST method, 86
  DSD/SST-DP, 105
  DSD/SST-SP, 105
  DSD/SST-SUPS, 102

DSD/SST method, (*continued*)
  DSD/SST-SV, 105
  DSD/SST-TIP1, 105
  DSD/SST-VMST, 101
  SUPG test function option, 104
    WTSA, 104
    WTSE, 104
DSD/SST with temporal NURBS basis
    functions, 150
  constant angular velocity on a circular
    arc, 156
  data representation in space–time
    computations, 151
  design of temporal NURBS basis
    functions, 153
  mesh computation, 158
  mesh representation, 149
    no-slip condition on a prescribed
      boundary, 159
  path on a circular arc, 154
  path representation with temporal
    NURBS basis functions, 154
    sampling points, 154
  remeshing, 158
  secondary mapping, 149, 150
  Simple-Shape Deformation Model
    (SSDM), 157
    simple shape (SS), 157
  specified speed along
    a path, 150
  starting-mesh representation with
    temporal NURBS basis functions,
    160, 187
    cubic NURBS, 161
    quadratic NURBS, 161
  time representation, 150

element-vector-based (EVB) computation
    technique, 142
  analytical EVB (AEVB) computation
    technique, 142, 143
  mixed AEVB/NEVB computation
    technique, 143
  numerical EVB (NEVB) computation
    technique, 142
Euler equations, 5
EVB computation technique, 142
EZP geometry, 193, 199, 226

FEM, 37
  basis functions, 39
    continuity of, 50
    element shape functions, 43
    global, 43, 49
    interpolation property of, 43
    partition of unity property of, 43
    support of, 50
  bilinear form, 41
  Bubnov–Galerkin method, 42
  data structures, 51
    element assembly, 52
    element matrices, 52
    element vectors, 52
    ID array, 52
    IEN array, 51
  elements, 39
    boundary, 39
    interior, 39
    parametric, 43
    physical, 43
  essential boundary conditions, 37
  Galerkin method, 37
  isoparametric construction, 44
  matrix problem, 41
  mesh, 39
  nodal unknowns, 40
  nodes, 39
  Petrov–Galerkin method, 42
  semilinear form, 38, 41
  spatial domain, 37
finite element method, 37
first Piola–Kirchhoff stress tensor, 16
flapping wings (aerodynamics of), 181
  flapping-motion representation with
    temporal NURBS basis functions,
    185
  fluid mechanics computation, 187
    streamlines, 188
    surface pressures, 190
    vorticity, 189
  locust in a wind tunnel, 181
  mesh motion, 186
  remeshing, 186
  sampling points, 185
  starting condition, 187
  surface and volume meshes, 181
  tracking points, 185

Index

fluid subdomain, 112
fluid–structure coupling, xii
  block-iterative, xiii, 140
  direct coupling, xiii, 142
  loosely-coupled, xiii, 140
  monolithic, xiii, 140
  quasi-direct, xiii, 141
  staggered, xiii, 140
  strongly-coupled, xiii, 140
fluid–structure interaction – general definition, xi
fluid–structure interface, 112
  current configuration, 112
  initial configuration, 112
fluid–structure interface meshes, xii
  matching, xii, 116
  nonmatching, xii, 116
FSI, xi
FSI Directional Geometric Smoothing Technique, 137
FSI formulation at the continuous level, 111
  continuity of the traction at the interface, 114
  fluid–structure interface, 112
  interface conditions for structures made of a porous material, 114
  kinematic constraint at the interface, 113
FSI Geometric Smoothing Technique, 136
FSI-DGST, 137
  parachute-specific, 261
FSI-GST, 136, 260
Fung material, 22, 205

Galerkin formulation, 58, 62
  advection–diffusion equation, 58
  linear elastodynamics, 62
Gaussian quadrature, 55
  accuracy of, 56
  quadrature points, 56
  quadrature weights, 56
  rules, 56
generalized-α time integration, 95, 118
  ALE FSI, 118
  high-frequency dissipation, 119
  predictor–multicorrector algorithm, 120
  second-order accuracy, 119
  tangent matrices, 122
  unconditional stability, 119
ALE-VMS, 95
  high-frequency dissipation, 96
  predictor–multicorrector algorithm, 96
  second-order accuracy, 95
  tangent matrices, 97
  unconditional stability, 95
GMRES method, 96
Green–Lagrange strain tensor, 13

hexahedral element, 47
  trilinear hexahedron, 47
HMGP, 260, 262
  HMGP in its original form, 265
  HMGP-FG, 260, 266
Homogenized Modeling of Geometric Porosity (HMGP), 260, 262
  four-gore slice, 260, 263
  HMGP in its original form, 265
  HMGP-FG, 260, 266
  homogenized porosity, 260, 263
  n-gore slice, 260, 264
  periodic n-gore model, 267
    16-gore model, 268
    8-gore model, 268
    four-gore model, 268
    mesh refinement increase, 268
  rotational-periodicity techniques, 260, 264
hyperelastic material, 19, 205

IGA, xiii, 73
incompressibility constraint, 2
index notation, 3, 13
infinitesimal strain, 24, 62
inflation of a balloon, 175
  mass balance, 177
  quasi-direct coupling technique, 175
interface-capturing technique, 83
interface-tracking technique, 83
interpolation by finite elements, 53
  error estimates, 55
  H^1-projection, 54
  L^2-projection, 54
  nodal interpolation, 54
Isogeometric Analysis, xiii, 73

Jacobian-based stiffening, 108

kinematic viscosity, 5
kinematics (continuity of), 113

Lagrange polynomials, 46
least-squares on incompressibility
    constraint, 68
    LSIC stabilization parameter (ν_{LSIC}), 66, 68
linear-elastodynamics equations, 22
    Galerkin formulation of, 62
    matrix form of, 63
    weak form of, 23, 62
linear-elastostatics equations, 24, 107
locust in a wind tunnel, 181
LSIC stabilization, 68

mass balance in arterial FSI, 221
mass balance in inflation of a balloon, 177
    quasi-direct coupling technique, 175
membrane model, 29
mesh computation with temporal NURBS
    basis functions, 158
mesh generation, 106
    accurate representation of the boundary layers, 106
    automatic mesh generator, 106
    special-purpose mesh generator, 106
    structed layers of elements around solid objects, 106
mesh quality measures, 130
    element area change, 130
    element shape change, 130
mesh representation with temporal NURBS
    basis functions, 150
    no-slip condition on a prescribed boundary, 159
mesh update, 106
    mesh moving, xii, 106
        automatic mesh moving, 107
        boundary layer mesh resolution, 129
        mesh quality measures, 130
        Solid-Extension Mesh Moving Technique, 129
        special-purpose mesh moving, 107
        structed layers of elements near solid surfaces, 129

Move-Reconnect-Renode Mesh Update Method, 132
    reconnect, 133
    renode, 133
remeshing, xii, 106
    cost of automatic mesh generation, 107
    cost of search, 107
    frequency of remeshing in practical computations, 109
    full, 107
    node regeneration, 107
    parallel efficiency, 109
    partial, 107
    pressure clipping, 134
    projection errors, 107
    reducing the frequency of remeshing, 107
    regeneration of just the element connectivities, 107
MITICT, 84
Mixed Interface-Tracking/Interface-Capturing Technique, 84
moving-mesh technique, 83
MRRMUM, 132
multiscale SCAFSI techniques, 213
    spatially multiscale, 213
        SCAFSI M1C, 214
        SCAFSI M1SC, 213
Multiscale SCFSI techniques, 214
    SCFSI M1C, 214
    SCFSI M2C, 214, 275
Multiscale Sequentially-Coupled FSI (SCFSI) techniques, 214
    SCFSI M1C, 214
    SCFSI M2C, 214, 275

Navier–Stokes equations of incompressible flows, 1
    ALE formulation of, 92
        ALE-VMS, 93
        stabilization parameters (τ_{SUPG} and ν_{LSIC}), 93
        SUPG/PSPG, 93
    conservative form of, 2
    convective form of, 3
    DSD/SST-SUPS, 102

Index

DSD/SST-DP, 105
DSD/SST-SP, 105
DSD/SST-SV, 105
DSD/SST-TIP1, 105
  stabilization parameters, 103
  SUPG test function option, 104
  TIP1 integration of incompressibility constraint, 106, 151
DSD/SST-VMST formulation, 101
  conservative form, 101
  convective form, 101
  stabilization parameters, 103
finite element formulation of, 65
space–time formulation of, 98
Space–Time VMS (alternate name for DSD/SST-VMST), 101
ST-SUPS (alternate name for DSD/SST-SUPS), 102
ST-VMS, 101
SUPG/PSPG formulation of, 68
  LSIC stabilization parameter, 68
  PSPG stabilization parameter, 68
  SUPG stabilization parameter, 68
variational multiscale formulation of, 65
NEVB computation technique, 142
Newton–Raphson method, 41, 96
Non-Uniform Rational B-splines, xiv, 73
  basis functions, 75
  control mesh, 76
  control points, 76
  control variables, 80
  curves, 76
  h–refinement, 78
  interpolation error estimate, 81
  k–refinement, 78
  p–refinement, 78
  solids, 77
  surfaces, 77
  weights, 76
nonmoving-mesh technique, 83
numerical integration, 55
NURBS, xiv, 73, 172

oscillatory shear index, 193, 196
OSI, 193, 196

parachute FSI, 259
  breathing, 284

cluster, 259
cluster computations, 293
  asynchronous disreef, 294
  parametric study: initial coning angles, 293
  parametric study: initial parachute diameters, 293
  parametric study: payload-model configurations, 293
  simulated disreef, 294
gap, 259, 263
gore, 259, 263
light structure, 259
PA, 280
PM 11, 280
PM 5, 280
reefed stage, 276
ring, 259, 262
ringsail parachute, 259, 262
sail, 259, 262
single-parachute computations, 280
  canopy configuration: PA, 280
  canopy configuration: PM 11, 280
  canopy configuration: PM 5, 280
  various suspension line length ratios, 288
slit, 259, 263
spacecraft parachutes, 259
parachute FSI – special techniques, 259
  added-mass extraction, 261, 311
    determining added mass, 312
    relevance to parachutes, 311
  cable drag, 270
  contributors to parachute descent speed, 299
    cluster of parachutes, 304
    results, 308
    single parachute, 299
  dynamical analysis, 260, 299
    cluster of parachutes, 304
    results, 308
    single parachute, 299
  four-gore slice, 260, 263
  FSI Geometric Smoothing Technique, 136, 260
  FSI-DGST – parachute-specific, 261
  FSI-GST, 136, 260
  HMGP, 260, 262

parachute FSI – special techniques, (*continued*)
   HMGP in its original form, 265
   HMGP-FG, 260, 266
   Homogenized Modeling of Geometric Porosity (HMGP), 260, 262
   homogenized porosity, 260, 263
   line drag, 270
   model-parameter extraction, 261, 299, 311
   multiscale SCFSI M2C computations, 276
     cable symmetrization, 277
     fabric stress computations, 278
     structural mechanics solution for the reefed stage, 275
   n-gore slice, 260, 264
   nonmatching fluid and structure meshes at the interface, 260
   periodic n-gore model, 267
     16-gore model, 268
     8-gore model, 268
     four-gore model, 268
     mesh refinement increase, 268
   rotational-periodicity techniques, 260, 264
   SENCT-FC-M1, 164, 264
   Separated Stress Projection, 128, 260
   shape determination, 272
   SSP, 128, 260
   starting point for FSI computation, 261, 272
     shape determination, 272
   symmetric FSI, 260, 274
Peclet number, 6
   mesh Peclet number, 59
   numercial oscillations, 59
pressure clipping, 134
Pressure-Stabilizing/Petrov–Galerkin formulation, 68
   PSPG stabilization parameter (τ_{PSPG}), 68
PSPG formulation, 68

quadrilateral element, 46
   bilinear quadrilateral, 46
quasi-direct coupling, xiii, 141
   Selective Scaling, 142

reference configuration, 12
remeshing, xii, 106
remeshing with temporal NURBS basis functions, 158
Reynolds number, 6
rotational-periodicity techniques, 260, 264, 320

sampling points, 154, 185
SCAFSI technique, 210
   multiscale SCAFSI techniques, 213
     spatially multiscale, 213
     temporally multiscale, 213
SCFSI technique, 214
segregated equation solvers and preconditioners, 144
   Segregated Equation Solver for Fluid–Structure Interaction (SESFSI), 146
   Segregated Equation Solver for Linear Systems (SESLS), 145
   Segregated Equation Solver for Nonlinear Systems (SESNS), 144
Selective Scaling, 142
   between momentum conservation and incompressibility constraint equations, 142
SEMMT, 129
   SEMMT–MD, 129
   SEMMT–SD, 129
SENCT, 163
   SENCT-D, 164
   SENCT-F, 164
   SENCT-FC, 163
     contact detection and node sets, 164
     contact force and reaction force, 165
     solving for contact force, 167
   SENCT-FC-M1, 164
   SENCT-M1, 164
Sequentially-Coupled FSI (SCFSI) technique, 214
SESFSI, 146
SESLS, 145
SESNS, 144
shape derivatives, 123
Shear-Slip Mesh Update Method (SSMUM), 317

shell model, 25
    bending stiffness, 28
    coupling stiffness, 28
    curvature tensor, 25
    extensional stiffness, 28
    Kirchhoff–Love, 25, 334
    local Cartesian basis, 26
    membrane strain tensor, 25
    midsurface, 25
Solid-Extension Mesh Moving Technique, 129
    mesh deformation tests, 131
        prescribed bending, 131
        rigid-body rotation, 131
        rigid-body translation, 131
    SEMMT – Multiple Domain, 129
    SEMMT – Single Domain, 129
solution of coupled FSI equations, 139
    block-iterative coupling, 140
        more robust version, 141
    direct coupling, 142
        mixed AEVB/NEVB computation technique, 144
    quasi-direct coupling, 141
        Selective Scaling, 142
space–time domain, 32
    deformation gradient, 32
    Piola transformation, 31
space–time formulation of FSI, 123
space–time method with temporal NURBS basis functions, 150
    constant angular velocity on a circular arc, 156
    data representation in space–time computations, 151
    design of temporal NURBS basis functions, 153
    mesh computation, 158
    mesh representation, 150
        no-slip condition on a prescribed boundary, 159
    path on a circular arc, 154
    path representation with temporal NURBS basis functions, 154
        sampling points, 154
    remeshing, 158
    secondary mapping, 149, 150

Simple-Shape Deformation Model (SSDM), 157
    simple shape (SS), 157
specified speed along a path, 150
starting-mesh representation with temporal NURBS basis functions, 160, 187
    cubic NURBS, 161
    quadratic NURBS, 161
time representation, 151
space–time methods, 86
SS (simple shape), 157
SSDM (Simple-Shape Deformation Model), 157
SSMUM, 317
SSP, 128, 260
SSTFSI technique, 123
    SSTFSI-DP, 126
    SSTFSI-SP, 126
    SSTFSI-SUPS, 126
    SSTFSI-SV, 126
    SSTFSI-TIP1, 126
    SSTFSI-VMST, 123
        conservative form, 126
        convective form, 123
ST-SUPS (alternate name for DSD/SST-SUPS), 102
ST-VMS, 101
stabilization parameters, 60
    ALE formulation of advection–diffusion equation, 89
    ALE formulation of Navier–Stokes equations, 93
    DSD/SST formulation of advection–diffusion equation, 91
    DSD/SST formulation of Navier–Stokes equations, 103
        LHC ($\nu_{\text{LSIC-LHC}}$), 104
        LSIC (ν_{LSIC}), 104
        SUPS (τ_{SUPS}), 103
        TC2 ($\nu_{\text{LSIC-TC2}}$), 104
        TGI ($\nu_{\text{LSIC-TGI}}$), 104
    DSD/SST-VMST formulation, 101
        LHC ($\nu_{\text{LSIC-LHC}}$), 104
        LSIC (ν_{LSIC}), 104
        SUPS (τ_{SUPS}), 103
        TC2 ($\nu_{\text{LSIC-TC2}}$), 104
        TGI ($\nu_{\text{LSIC-TGI}}$), 104

stabilization parameters, (*continued*)
    LSIC stabilization parameter (ν_{LSIC}), 66, 68
    PSPG stabilization parameter (τ_{PSPG}), 68
    space–time formulation of advection–diffusion equation, 91
    space–time formulation of Navier–Stokes equations, 103
    SUPG stabilization parameter (τ_{SUPG}), 60, 68
    SUPS stabilization parameter (τ_{SUPS}), 67, 68
stabilized formulation, 59
    advection–diffusion equation, 59
    discontinuity capturing, 62
    Navier–Stokes equations, 65
    SUPG stabilization parameter, 60
        element length, 61
        mesh size, 60
    SUPG/PSPG formulation, 68
        LSIC stabilization parameter (ν_{LSIC}), 68
        PSPG stabilization parameter (τ_{PSPG}), 68
        SUPG stabilization parameter (τ_{SUPG}), 68
Stabilized Space–Time FSI technique, 123
    CFM (porosity unit), 126
    DSD/SST-VMST in FSI formulation, 123
    interface conditions, 125
    interface projection, 125
        direct substitution, 127
        equivalence to monolithic method when the interface meshes are matching, 127, 128
        FSI Directional Geometric Smoothing Technique, 137
        FSI Geometric Smoothing Technique, 136, 260
        mass lumping, 128
        numerical substitution, 127
        projection over a reference configuration, 127
        Separated Stress Projection, 128, 260
    membrane with fabric porosity, 126

    membrane-edge stabilization with split nodal values for pressure, 127
    porosity coefficient, 126
    SSTFSI-DP, 126
    SSTFSI-SP, 126
    SSTFSI-SUPS, 126
    SSTFSI-SV, 126
    SSTFSI-TIP1, 126
    SSTFSI-VMST, 123
        conservative form, 126
        convective form, 123
    ST-VMS in FSI formulation, 123
starting-mesh representation with temporal NURBS basis functions, 160
    cubic NURBS, 161
    quadratic NURBS, 161
Stokes equations, 5
strain-displacement matrix, 64
strain-rate tensor, 2
Streamline-Upwind/Petrov–Galerkin formulation, 59
    advection–diffusion equation, 59
    Navier–Stokes equations, 68
        SUPG stabilization parameter, 68
    stabilization parameter, 60
        element length, 61
        mesh size, 60
    SUPG test function option, 104
        WTSA, 104
        WTSE, 104
stress tensor in fluid mechanics, 2
structural damping, 64
structural mechanics equations, 12
    linearization of, 22
    strong form of, 16, 18
    total Lagrangian formulation of, 18
    updated Lagrangian formulation of, 18
    variational formulation of, 14, 16, 17, 29
structure subdomain, 112
subdivision surfaces, 74
SUPG formulation, 59, 68
    SUPG test function option, 104
        WTSA, 104
        WTSE, 104
SUPG test function option, 104
    WTSA, 104
    WTSE, 104
SUPG/PSPG, 68, 93

Surface-Edge-Node Contact Tracking (SENCT) technique, 163
  SENCT-D, 164
  SENCT-F, 164
  SENCT-FC, 163, 264
    contact detection and node sets, 164
    contact force and reaction force, 165
    solving for contact force, 167
  SENCT-FC-M1, 164, 264
  SENCT-M1, 164, 263
symmetric FSI, 260, 274

T-splines, 73
tangent stiffness, 23
  geometric stiffness, 23
  material stiffness, 23
temporal NURBS basis functions, 149
  constant angular velocity on a circular arc, 156
  data representation in space–time computations, 151
  design of temporal NURBS basis functions, 153
  mesh computation, 158
  mesh representation, 149
    no-slip condition on a prescribed boundary, 159
  path on a circular arc, 154
  path representation with temporal NURBS basis functions, 154
    sampling points, 154
  remeshing, 158
  secondary mapping, 149, 150
  Simple-Shape Deformation Model (SSDM), 157
    simple shape (SS), 157
  specified speed along a path, 150
  starting-mesh representation with temporal NURBS basis functions, 160, 187
    cubic NURBS, 161
    quadratic NURBS, 161
  time representation, 151
test (weighting) functions, 11
tetrahedral element, 49
  linear tetrahedron, 49
total time derivative, 13
traction vector (continuity of), 114

trial functions, 10
triangular element, 48
  linear triangle, 49
turbulence modeling, 69

variation, 14
variational multiscale formulation, 65
  ALE-VMS formulation of Navier–Stokes equations, 93
    stabilization parameters (τ_{SUPG} and ν_{LSIC}), 93
  DSD/SST-VMST formulation, 101
    conservative form, 101
    convective form, 101
    stabilization parameters, 103
  fine scales, 66
  LSIC stabilization parameter (ν_{LSIC}), 66
  Navier–Stokes equations, 65
  RBVMS, 66
  residual-based, 66
  scale separation, 65
  space–time formulation, 98
  Space–Time VMS (alternate name for DSD/SST-VMST), 101
  ST-VMS, 101
  stabilization parameters, 67
  subgrid scales, 66
  SUPS stabilization parameter (τ_{SUPS}), 67, 68
virtual work (principle of), 14, 27
VMS formulation, 65
Voight notation, 62

wall shear stress, 192, 195
weak boundary conditions, 70
  penalty parameter, 72
weak form of the Navier–Stokes equations, 10
  traction boundary conditions, 11
  velocity boundary conditions, 11
weighting (test) functions, 11
wind-turbine aerodynamics and FSI, 315
  aerodynamics, 315
    Betz law, 324
    DSD/SST computations, 325
    DSD/SST-SUPS, 327
    DSD/SST-VMST, 327

wind-turbine aerodynamics and FSI, (*continued*)
    layers of refined mesh near the rotor surface, 327
    Mesh-2, 325
    Mesh-3, 325
    Mesh-4, 325
    FSI, 315
        blades, 315
        blades: composite lay-up, 339
        blades: pre-bending, 344
        blades: structural modeling, 334
    rotor, 315
        aerodynamic torque, 322
        geometry definition, 317
        geometry definition: NREL Phase VI, 329
        geometry definition: offshore 5MW baseline, 318
wind-turbine aerodynamics and FSI – special techniques, 317
    bending-strip method, 334
    bending-strip patches, 337
    directional bending stiffness, 337
    rotational-periodicity techniques, 260, 264, 320
Windkessel model, 207
windsock computation, 177
    FSI Directional Geometric Smoothing Technique, 179
    FSI-DGST, 179
Womersley parameter, 206
WSS, 192, 195